Transportation of Hazardous Materials

A Practical Guide to Compliance

Frank R. Spellman
Joanne Drinan
Nancy E. Whiting

Government Institutes
Rockville, Maryland

Government Institutes, a Division of ABS Group Inc.
4 Research Place, Rockville, Maryland 20850, USA
Phone: (301) 921-2300
Fax: (301) 921-0373
Email: giinfo@govinst.com
Internet: http://www.govinst.com

Copyright © 2001 by Government Institutes. All rights reserved.

03   02   01   00   5   4   3   2   1

No part of this work may be reproduced or transmitted in any form or by any means, electronic or mechanical, including photocopying, recording, or the use of any information storage and retrieval system, without permission in writing from the publisher. All requests for permission to reproduce material from this work should be directed to Government Institutes, 4 Research Place, Suite 200, Rockville, Maryland 20850, USA.

The reader should not rely on this publication to address specific questions that apply to a particular set of facts. The author and publisher make no representation or warranty, express or implied, as to the completeness, correctness, or utility of the information in this publication. In addition, the author and publisher assume no liability of any kind whatsoever resulting from the use of or reliance upon the contents of this book. Government Institutes, a Division of ABS Group Inc.

Library of Congress Cataloging-in-Publication Data

Spellman, Frank R.
    Transportation of hazardous materials : a practical guide to compliance / by Frank R. Spellman, Joanne Drinan, Nancy E. Whiting.
      p. cm.
    ISBN 0-86587-712-2
    1. Hazardous substances–Transportation–United States. I. Drinan, Joanne. II. Whiting, Nancy E. III. Title.

T55.3.H3 S72 2000
604.7—dc21

00-047687

To Hazmat Responders everywhere

To Anita Baker

Contents

List of Figures and Tables xi

Preface xiii

Prelude xv

Chapter 1
Introduction to the DOT and Hazmat Regulations 1

U.S. Department of Transportation (DOT) 1
A Brief History of DOT 2
    Cabinet-Level Department Proposed 3
    Department of Transportation Established 4
    Nixon Administration 5
    Ford and Carter Administrations 6
    Reagan Administration 7
    Bush Administration 9
    Clinton Administration 10
Motor Carrier Safety Regulations: A Brief History 11
    Ensuring Highway Safety 14
Hazardous Materials 16
Summary 18
Notes 19

Chapter 2
Understanding Title 49 21

Introduction 21
Using the Code 22
Summary 24

Chapter 3
Classification of Hazardous Materials 25

Introduction 25
Hazmat Definitions 26
Classification of Hazardous Materials 30
    DOT Hazard Classes (Listing) 31
Hazardous Materials Table 36
    Column 1: Symbols 37

  Column 2: Hazardous Materials Descriptions and Proper Shipping Names 37
  Column 3: Hazard Class or Division 38
  Column 4: Identification Numbers 39
  Column 5: Packing Group 39
  Column 6: Labels 39
  Column 7: Special Provisions 40
  Column 8: Packaging Authorizations 40
  Column 9: Quantity Limitations 41
  Column 10: Vessel Stowage Requirements 41
Summary 42
Notes 42

Chapter 4
General Awareness Training 43

Introduction 43
  Hazmat Employer 44
  Hazmat Employee 44
OSHA's Hazard Communication Standard 45
  Purpose of the Hazcom Standard 46
  Requirements of the Hazcom Standard 47
Hazard Determination 48
  Chemical Inventory List 50
Hazard Communication and Training 51
Material Safety Data Sheets (MSDS) 55
  MSDS: An Answer Sheet 56
  The MSDS Format 56
  MSDS Contents: An Example 57
  Availability of MSDSs 65
Written Hazard Communication Program 66
Summary 67
Notes 67

Chapter 5
Labeling, Placarding, and Emergency Response 69

Introduction 69
Labeling/Marking Requirements 69
Labeling of Hazardous Chemicals 72
  Non-Specific Chemical Labels 73
  Labeling Systems 74
Labeling of Hazardous Materials (§172.400) 75
  Where Labeling Is Required 84

    Where Labeling Is Not Required 85
    Additional Labeling Requirements (§172.402) 86
    Labels for Mixed and Consolidated Packaging (§172.404) 86
    Authorized Label Modifications (§172.405) 86
    Placement of Labels (§172.406) 87
    Label Specifications (§172.407) 88
    Exceptions to the Labeling Requirements 89
Placarding 90
    General Placarding Requirements 98
    Exceptions to the Placarding Requirements 100
    Permissive Placarding 102
    Prohibited Placarding (§172.502) 102
    Placarding for Subsidiary Hazards (§172.505) 103
    Providing and Affixing Placards: Highway (§172.506) 103
    Special Placarding Provisions: Highway (§172.507) 104
    Providing and Affixing Placards: Rail (§172.508) 104
    Special Placarding Provisions: Rail (§172.510) 104
    Freight Containers and Aircraft Unit Load Devices 105
    Bulk Packagings (§172.514) 106
    Visibility and Display of Placards (§172.516) 106
    Types of Placards 107
DOT Emergency Response (Subpart G—§172.600) 108
    Using the North American Emergency Response Guidebook 110
Summary 112
Notes 112

Chapter 6
Packaging and Shipping 113

Introduction 113
Packagings and Packages: General Requirements (§173.24) 114
    Authorized Packagings 115
    Specification Packagings 115
    Compatibility 115
Standard Packaging 116
Performance-Oriented Packaging (POP) 117
    POP Design Tests 117
    POP Marking Codes 118
Shipping Papers (§172.200) 120
Summary 122
Notes 122

Chapter 7
Function-Specific Training 123

Introduction 123
Loading/Unloading (§177.834) and Storage 125
    General Loading and Unloading Requirements 125
Separation of Hazardous Materials (177.842) 128
Segregation of Hazardous Materials (§177.848) 129
    Using the Segregation Table 131
Summary 131

Chapter 8
Driver Training 133

Introduction 133
DOT-Required Driver Training 134
    Pre-Trip Safety Inspection (§392.7) 135
    Use of Vehicle Controls and Equipment, Including Operation of
        Emergency Equipment (§392.7, §192.8, and §393.95) 135
    Operation of the Vehicle 137
    Procedures for Maneuvering about Tunnels, Bridges, and Railroad
        Crossings 138
    Requirements for Vehicle Attendance, Parking, Smoking, Routing, and
        Incident Reporting 138
    Loading/Unloading of Materials 140
    Cargo Tank and Portable Tank Operators 141
Summary 141
Notes 141

Chapter 9
Controlled Substances and Alcohol Use and Testing 143

Introduction 143
Controlled Substances and Alcohol Use and Testing (Part 382) 148
    Subpart A—General 148
    Subpart B—Prohibitions 151
    Subpart C—Tests Required 153
    Subpart D—Handling of Test Results, Record Retention and
        Confidentiality 164
    Subpart E—Consequences for Drivers Engaging in Substance
        Use-Related Conduct 175
    Subpart F—Alcohol Misuse and Controlled Substances Use Information,
        Training, and Referral 177

Procedures for Transportation Workplace Drug and Alcohol Testing
    Programs (Part 40) 183
    Applicability (§40.1) 183
    Subpart B—Drug Testing 183
    Subpart C—Alcohol Testing 186
    Subpart D—Non-Evidential Alcohol Screening Devices 187
Sample Substance Abuse Policy 187
Summary 188
Notes 188

Afterword: Taming the Suicide Strip 189

Appendices 191

Appendix A
Hazardous Materials Table (§172.101) 191

Appendix B
Sample Written Hazard Communication Program 265

Hazard Communication Program 265
    I. Introduction 265
    II. Responsibilities 266
    III. Definition of Terms 267
    IV. "Right to Know" Work Stations 270
    V. Hazardous Chemical Inventory List 270
    VI. Material Safety Data Sheet (MSDS) 271
    VII. Hazard Warnings and Labeling 272
    VIII. Training 273
    XI. On-site Contractors/Visitors 275

Appendix C
Sample Substance Abuse Policy 277

    I. Purpose and Overview 277
    II. Employee Assistance 278
    III. Medically Prescribed Medications 278
    IV. Pre-Employment Testing 278
    V. Requirements and Substance Abuse Testing for Employees Required
        to Hold a Commercial Drivers License (CDL) 279
    VI. Post Accident and Reasonable Suspicion Substance Abuse Testing ... 280
    VII. Substance Abuse Testing Procedures 280
    VIII. Return to Work 283
    IX. Training 284

X. Disciplinary Action 284

Appendix D
Hazmat Definitions 285

List of Figures and Tables

Figure 1.1. DOT and individual operating administrations 2
Table 1.1. Chronology of Significant U.S. DOT Dates 12
Table 1.2. Ten Federal Statutes Controlling Hazardous Materials 17
Table 2.1. Content of the Code of Federal Regulations (CFRs) 22
Table 2.2. Contents of the Parts of Title 49 22
Table 2.3. Contents of Hazardous Materials Regulations 23
Table 3.1. DOT Hazard Classes 32
Figure 4.1. Material Safety Data Sheet (MSDS) 58
Figure 4.2. Employee "right to know" station 66
Figure 5.1. Typical danger sign 70
Figure 5.2. Typical caution sign 71
Figure 5.3. Typical warning sign 71
Figure 5.4. Methane gas warning sign 74
Figure 5.5. Fire diamond label 75
Figure 5.6. Explosive 1.1, 1.2, and 1.3 labels (§172.411) 75
Figure 5.7. Explosive 1.4, 1.5, and 1.6 labels and subsidiary labels (§172.411) 76
Figure 5.8. Non-flammable gas label (§172.415) 77
Figure 5.9. Poison gas label (§172.416) 77
Figure 5.10. Flammable gas label (§172.417) 77
Figure 5.11. Flammable liquid label (§172.419) 78
Figure 5.12. Flammable solid label (§172.420) 78
Figure 5.13. Spontaneously combustible label (§172.422) 78
Figure 5.14. Dangerous when wet label (§172.423) 79
Figure 5.15. Oxidizer label (§172.426) 79
Figure 5.16. Organic peroxide label (§172.427) 79
Figure 5.17. Poison inhalation hazard label (§172.416) 80
Figure 5.18. Poison label (§172.430) 80
Figure 5.19. Infectious substance label (§172.432) 80
Figure 5.20. Radioactive white-I label (§172.436) 81
Figure 5.21. Radioactive yellow-II label (§172.438) 81
Figure 5.22. Radioactive yellow-III label (§172.440) 81
Figure 5.23. Corrosive label (§172.442) 82
Figure 5.24. Class 9 (miscellaneous hazardous materials) label (§172.446) 82
Figure 5.25. Cargo aircraft only label (§172.448) 82
Figure 5.26. Empty label (§172.450) 83
Table 5.1. Label Names and Designs for Each Hazard Class 84
Table 5.2. Subsidiary Hazard Labels 86
Figure 5.27. Dangerous placard (§172.521) 91
Figure 5.28. Explosives 1.1, 1.2, and 1.3 placards (§172.522) 91
Figure 5.29. Explosives 1.4 placard (§172.523) 92

Figure 5.30. Explosives 1.5 placard (§172.524) 92
Figure 5.31. Explosives 1.6 placard (§172.525) 92
Figure 5.32. Square background placard (§172.527) 93
Figure 5.33. Non-flammable gas placard (§172.528) 93
Figure 5.34. Oxygen placard (§172.530) 93
Figure 5.35. Flammable gas placard (§172.532) 94
Figure 5.36. Poison gas placard (§172.540) 94
Figure 5.37. Flammable placard (§172.542) 94
Figure 5.38. Combustible placard (§172.544) 95
Figure 5.39. Flammable solid placard (§172.546) 95
Figure 5.40. Spontaneously combustible placard (§172.547) 95
Figure 5.41. Dangerous when wet placard (§172.548) 96
Figure 5.42. Oxidizer placard (§172.550) 96
Figure 5.43. Organic peroxide placard (§172.552) 96
Figure 5.44. Poison placard (§172.554) 97
Figure 5.45. Poison inhalation hazard placard (§172.555) 97
Figure 5.46. Radioactive placard (§172.556) 97
Figure 5.47. Corrosive placard (§172.558) 98
Figure 5.48. Class 9 (miscellaneous hazardous materials) placard (§172.560) 98
Table 5.3. 49 CFR 172.504 Table 1 99
Table 5.4. 49 CFR 172.504 Table 2 100
Figure 6.1. Steps to packaging compliance 114
Table 6.1 Most Commonly Used Packaging Codes 118
Table 7.1. Separation Distance Table 129
Table 7.2. Segregation Table for Hazardous Materials 130
Figure 8.1. Driver's vehicle inspection report 136
Table 9.1. Requirements for a Post-Accident Test 155
Figure B.1 Chemical inventory list form 271
Figure B.2 Sample training record 275
Figure C.1 Post incident and reasonable suspicion report 281
Figure C.2 Sample substance abuse policy acknowledgment form 284
Table D.1. Chemical Data/Information Sources 286

Preface

Safety regulation and our reliance on regulation is an ever-growing, constantly demanding, and integral part of modern society. Regulation has become an industry in and of itself, with "compliance" and "enforcement" common buzzwords. Once the process of regulation begins, the rules begin to snowball and knowing where to stop is difficult. Sooner or later, regulation passes the point where the average person says, "This is just common sense!" However, in order to protect everyone, regulations must cover all aspects of a problem.

What does this mean to the person in charge of ensuring compliance? Between OSHA (Occupational Safety and Health Administration), the U.S. EPA (United States Environmental Protection Agency), the FDA (Federal Drug Administration), the CPSC (Consumer Products Safety Commission), and the USDA (United States Department of Agriculture), to name a few, it means wading through the flood of convoluted prose, the miasmic swamp of regulatory legalese, and the effluent of guidelines and regulations that becomes more and more difficult each day. The same can be said for Department of Transportation (DOT) regulations.

DOT regulations are published in Title 49 of the Code of Federal Regulations (CFR). The volumes of the 49 CFR can become a barricade to swift compliance; the sheer bulk of the material is overwhelming. While some people find the CFR requirements difficult to swallow, finding and understanding the written regulations and how to comply with them is even more difficult. The regulations themselves often make compliance difficult because they are not user-friendly or easy to decipher or implement. What is worse, they are open to the caprice of human interpretation, both in terms of implementation and enforcement. Further compounding the complexity and scope of the regulations is their phenomenal rate of growth.

Administrators require a great amount of diligence, persistence, forbearance, and stamina to decipher and implement the many regulations that are their responsibility. This is no less the case for Department of Transportation (DOT) regulations.

Any text explaining complex regulations and requirements should designed to be accessible, concise, and current. The main purpose of *Transportation of Hazardous Materials: A Practical Guide to Compliance* is to ease the pain. This book has been developed and designed for the express purpose of providing a comprehensive and easy-to-use reference to the many federal regulations pertaining to the transportation of hazardous materials (hazmats) on U.S. highways. We provide a general reference that can be used by industry managers to understand the regulations, implement proper programs, and provide hazard awareness training for employees involved in regulated hazardous material transportation. For best results, users of this book should incorporate the generic material provided in the text with company and job-specific safety requirements.

DOT's authority is wide-ranging, but we focus primarily on DOT hazmat safety issues related to overland driving for the trucking industry. The text begins with a brief overview of DOT, including its genesis and development over time to the present. The following chapters present concise information important to understanding the safety requirements of 49 CFR Parts 171 through 180 and the training requirements covered under HM-126F. We also include a comprehensive discussion of Controlled Substances and Alcohol Use and Testing (49 CFR 382/40), including a sample written substance abuse program that has a large advantage over other such programs: it has been tested in the real world.

The purpose of this text is to place needed information in the hands of the person(s) responsible for knowing, understanding, and implementing 49 CFR regulatory requirements pertaining to transportation of hazardous materials. This text is intended to be a handheld working guide for use by those who have the responsibility to ensure hazmat compliance. The "Important Notes" provided throughout this manual are taken from U.S. DOT Interpretations—May 4, 1997, or are based on the authors' experience. While no text could cover every aspect of 49 CFR HMR, this book brings regulatory requirements into proper perspective and allows easy access to hazmat solutions.

Frank R. Spellman
Joanne Drinan
Nancy E. Whiting

Prelude: Suicide Strip

It's an early morning in April 1964 on the Midwestern prairie, and from horizon to horizon stretch vast fields of grain. The only discernable disruption is a two-lane highway that runs east to west and intersects Owl's Creek at a perfect 90 degree angle. If we could see beyond the immediate southern horizon, we would see that approximately eight miles downstream from the bridge, the creek feeds several irrigation ditches that serve agricultural fields on both sides. About four miles further downstream, a small town uses water from Owl's Creek as its principle water supply. Still further downstream, below the town's wastewater outfall, thousands of beef cattle graze on large ranches on both sides of the creek.

But at this creek crossing, there is only a narrow bridge that is in very poor condition and is barely wide enough for two cars to pass at once. The yielding, sponge-like surface of the bridge appears to be better suited to kids hauling little red wagons than to automobiles or trucks. Yet most drivers never worry about the bridge's condition and simply trust that the yielding surface will rebound to its former shape when they have passed.

The bridge separates two very different stretches of highway. The road leading to it from the west is full of potholes. But to the east, drivers find a well-maintained cement surface with only a few minor cracks. The reason for this difference in condition is that Owl's Creek forms the border between two states.

This particular stretch of highway is used both as a local and state-to-state connector and, increasingly, by tractor-trailer truck traffic. The increasing use of this particular highway by 18-wheelers concerns the locals because many local, private vehicles use this highway as well. Rachel's Heights School Complex is about ten miles from the bridge on the east side, and teachers, parents, students, buses, and suppliers use the highway to go to and from the school.

This stretch of road has a bad reputation similar to that found in most towns with a blind curve that catches people unaware and makes them lose control. The white crosses on the shoulders on both sides of the bridge tell of the misery that has been caused by what the locals refer to as "Suicide Strip."

In the distance, a truck approached from the west carrying C.C. "Tex" Cobb and Mary Lou ("Lou") McLemore and their cargo, which was a heterogeneous mixture of dangerous chemicals. Tex and Lou had been trucking together for more than 20 years, and had been companions for longer than that. They were unmarried, but living together and hauling loads from coast to coast for so many years had made them a common-law couple. Neither had ever seriously thought about tying the knot in formal marriage, thinking that marriage was just part of those unnecessary rules and regulations that "busybodies think we should do." And if there was anything Tex and Lou hated more than road hogs, it was regulations.

In the more than twenty years Tex and Lou had been hauling cargo, they had demonstrated a similar disregard for rules concerning what type of materials they hauled or for their own safety. Tex always did the driving, and his driving record was well known to various state and county highway police authorities. He had been cited many times in his twenty years on the road for various local safety regulations. On more than one occasion he had been cited for driving while intoxicated or for driving under the influence of illegal substances. Both Tex and Lou had been arrested several times for smoking marijuana, but they had never been convicted. They traveled too much to be tied down to any jurisdiction, and they had learned whose palms to grease. However, Tex's trucker's license had been suspended and revoked several times over the years. What did Tex have to say about all this? "Who needs a license, anyway? Just another one of them damned regulations. Hell, someone has to haul this stuff, don't they?" Their reputation for hauling materials that no one else would was well known, especially by those unscrupulous shippers who needed to transport sensitive, sometimes dangerous and deadly, hazardous materials across the country.

Only a few days before reaching Owl's Creek, Tex and Lou made their first stop at a chemical shipping outfit 200 miles to the east. They loaded 15 unlabeled 5-gallon glass carboys of concentrated sulfuric acid (H_2SO_4). Pure, concentrated sulfuric acid is a colorless, oily liquid with a density about twice that of water—a highly dangerous corrosive material. Their next stop was about 50 miles to the northeast, where they picked up a shipment of industrial-grade calcium carbide. Calcium carbide is used industrially as a major source of acetylene and for the production of calcium cyanamide, a fertilizer. When calcium carbide reacts with water, acetylene is formed. The principal recommendation for handling calcium carbide is to keep it dry. Calcium carbide that has been exposed to moisture poses a flammable and explosive hazard. After storing the calcium carbide containers snugly alongside the glass carboys of sulfuric acid, Tex and Lou got back on the road and made their next stop, about 100 miles to the east. At this third stop, they took on a load of poisonous liquid elemental mercury and several containers of sodium chlorate. Sodium chlorate is chemically incompatible with sulfuric acid. It reacts violently with it, exploding on contact. During loading, one of the handlers noticed the glass carboys on the left side of the trailer, and asked Tex if the carboys held sulfuric acid. Tex scratched his head and said he thought so. The loader recommended that the sodium chlorate be placed on the right side of the trailer with the containers of calcium carbide and mercury in between, so that the sulfuric acid and sodium chlorate would not be loaded right next to each other. "It'll be safer this way," the loader stated. Tex shrugged and nodded his head in what might have been agreement. With the trailer fully loaded, Tex and Lou got into the tractor cab and pulled the rig out onto the highway.

Tex and Lou were familiar with Owl's Creek bridge; they had traveled this route many times. They weren't too concerned with the condition of the bridge. To the contrary, as they approached it, Tex reached for the bottle and Lou took another hit off the joint. Tex had driven all night, helped with a substantial dose of caffeine pills. "Sleep? Who needs it? Man, we got a valuable load here to deliver. Time is money and money is time." As the truck approached the bridge, Tex could see a gray school bus on the other side. Although he knew the two vehicles could never cross the narrow bridge at the same time, he wasn't worried. He knew that no other vehicle would dare attempt to cross the bridge at the same time his rig did.

He was correct, of course. The conscientious school bus driver, a local who knew the limits of the bridge, had no intention of playing chicken with the tractor-trailer bearing down on her. She simply pulled over about 15 feet from the bridge on what served as a shoulder of the road to wait until the truck passed her and the group of 32 young singers from the Rachel's Heights School Complex who were on their way to a community presentation in celebration of Founders' Day.

The truck entered the bridge crossing at 85 mph. The bridge responded by heaving and bucking with the load, causing the truck to swerv erratically, striking and wiping out the flimsy guardrail on the right, but maintaining tire contact with the bridge surface. By the time the truck was at the end of the bridge passage, Tex eased up on the throttle and fought the wheel to try to stabilize the out-of-control tractor-trailer. Unfortunately, Tex lost his battle to control the truck. As it left the bridge, it veered sharply to his left and struck the idling bus at over 70 mph.

Thirty-five people died almost instantly.

Even if rescuers had been present, they would have had no choice but to stand back and helplessly watch the inferno. After the impact, the immediate fatalities, and the initial horrendous explosions, the liquid chemicals that were not incinerated poured freely onto the ground, down the slope, and into the creek itself. The elements in this toxic mixture reacted with the water in different ways. Sulfuric acid and water produce heat—in this case, massive amounts of heat and explosions. The calcium carbide had a similar reaction, producing even more explosions and fire. When the trickle of sodium chlorate mixed with the flow of sulfuric acid, a violent reaction ensued, completely leveling what was left of Owl's Creek bridge.

It was the mercury, however, that had the most devastating and long-term effect. As a liquid, it flowed like the other chemicals into the creek. The mercury that was not destroyed or changed to other forms by the fire and explosions either drifted to the bottom or was carried on the current downstream. The heavy mercury gradually settled to the creek bed, creating a hydraulically formed plume of poison that stretched for miles downstream from Owl's Creek bridge. With the passage of time and seasons of irrigation, the crops took up some of this mercury. Some of the contaminated water entered the water supply of the small town below the irrigation ditches, lacing the town's potable water with poisonous mercury. The waters continued to flow, eventually reaching the fields below the town where the beef cattle drank the mercury-contaminated water. Beef spiced with poisonous mercury became the dinner of several thousand people who had never heard of Owl's Creek.

Does the Owl's Creek tragedy end here with these events? For those affected by the chemical contamination to crops, livestock, and water, the long-term affects of chemical exposure are hard to trace and even more difficult to prove. Does an incident that occurred 36 years ago still affect the immediate environment? To a degree, probably. The effects of environmental poisoning can be both subtle and far-reaching. Nature has a way of correcting our errors. However, poisons like mercury are persistent. Any major disturbance of a mercury-contaminated streambed might release the mercury.

Though many often say that our society is over-regulated, the Owl's Creek tragedy shows us that regulations are at times necessary. The load (the combination of chemicals) that Tex and Lou were transporting would be strictly prohibited today because of the incompatibility of these chemicals and their inherently dangerous properties. Tex and Lou would also be strictly prohibited from the trucking business. The checks and balances set up in the system would control or halt most of their abuses. If the motor carrier rules and regulations and highway safety rules in effect today had been on the books and properly enforced at the time, the Owl's Creek tragedy never would have occurred.

Chapter 1

Introduction to the DOT and Hazmat Regulations

[T]he Act which I sign today is the most important transportation legislation of our lifetime . . . It is one of the essential building blocks in our preparation for the future . . . . Transportation has truly emerged as a significant part of our national life. As a basic force in our society, its progress must be accelerated so that the quality of our life can be improved.

—President Lyndon B. Johnson, October 15, 1966

U.S. Department of Transportation (DOT)

The mission of the United States Department of Transportation (DOT)—a cabinet-level executive department of the U.S. government established by an Act of Congress and signed into law by President Lyndon B. Johnson on October 15, 1966—is to develop and coordinate policies that will provide an efficient and economical national transportation system, with due regard for need, the environment, and the national defense. This primary agency of the federal government has the responsibility for shaping and administering policies and programs to protect and enhance the safety, adequacy, and efficiency of the transportation system and services.

The Department of Transportation contains the Office of the Secretary and twelve individual operating administrations (see Figure 1.1), each headed by a presidential appointee: (1) the United States Coast Guard, (2) the Federal Aviation Administration, (3) the Federal Highway Administration, (4) the Federal Railroad Administration, (5) the National Highway Traffic Safety Administration, (6) the Federal Transit Administration, (7) the Maritime Administration, (8) the Saint Lawrence Seaway Development Corporation, (9) the Research and Special Programs Administration, (10) the Bureau of Transportation Statistics, (11) the Transportation Administrative Services Center (TASC), and (12) the Surface Transportation Board, an independent adjudicatory body administratively housed within the Department.

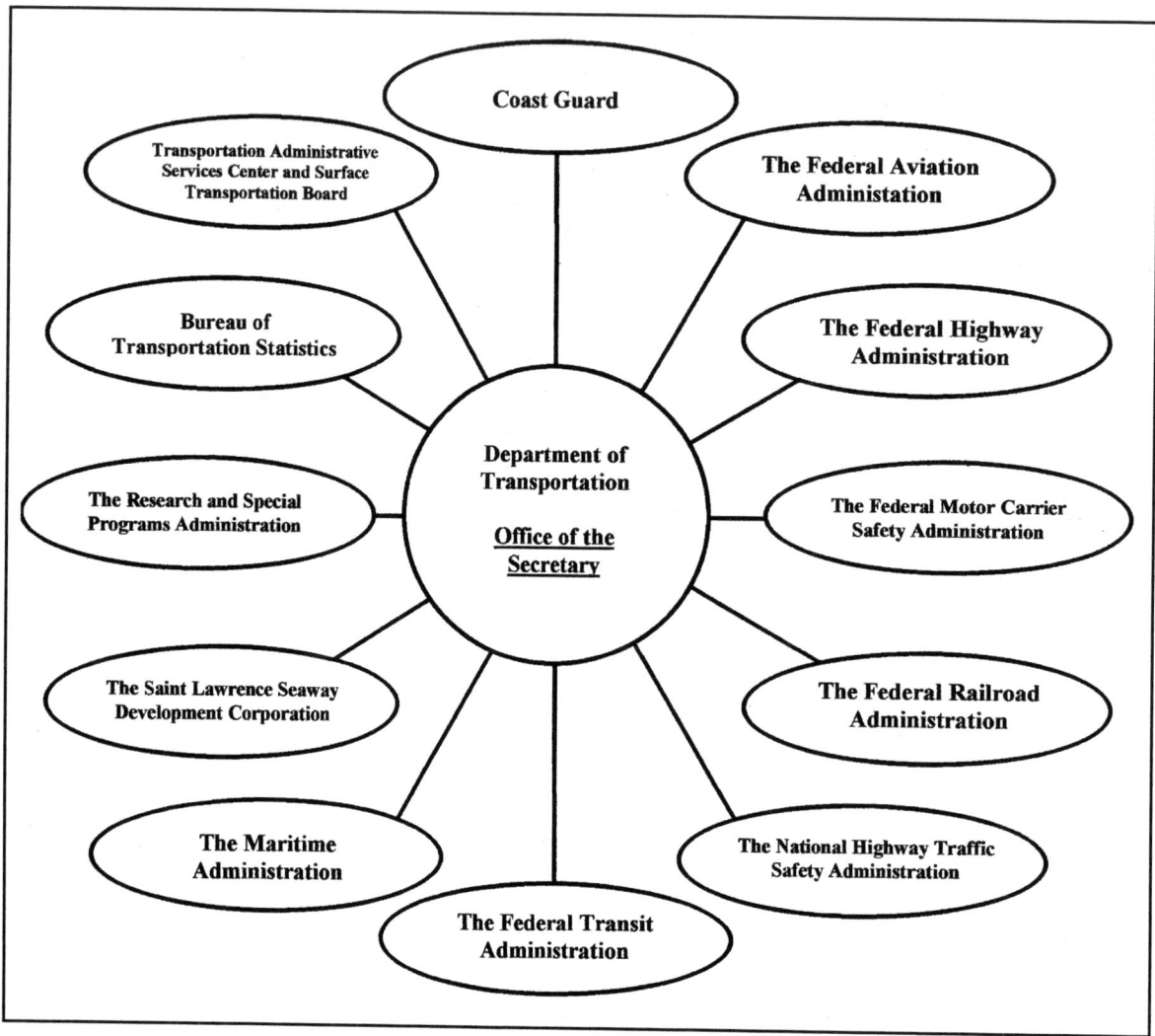

Figure 1.1. DOT and individual operating administrations

A Brief History of DOT[1]

In a modern society such as today's in the United States, we tend to take the favorable traveling conditions and smooth transportation of goods for granted. The United States' massive transportation systems—super interstate highways that stretch from coast to coast; a railway system that parallels many of our superhighways in routes, distance, and efficiency; and an airway transportation system second to none—make this transportation of goods and favorable traveling conditions possible.

From its very inception, the U.S. government wrestled with its role in developing transportation infrastructure and policy. Early efforts to develop transportation infrastructure and policy often resulted in confusion and needless complexity, which led to an overabundance of aid for some types of transportation and inadequate support for others. Transportation policy in the United States was and is very complicated, and the law establishing a cabinet-level Department of Transportation (DOT) did not pass Congress until 92 years after such a law was first introduced.

In reality, the idea of forming DOT or a similar agency is at least as old as Albert Gallatin, Thomas Jefferson's Treasury Secretary, and even before that the Coast Guard and the Army Corps of Engineers had helped to foster trade and transportation. In 1808, Gallatin recommended that the federal government subsidize the National Road to enhance the prosperity of struggling new states and to fulfill the need for rapid, simple, and accessible transportation. From these beginnings, the U.S. DOT eventually emerged.

Cabinet-Level Department Proposed

Much later, in June 1965, Najeeb Halaby, administrator of the independent Federal Aviation Agency (as it was then called), proposed the idea of a cabinet-level Department of Transportation to President Johnson's administrative planners. Halaby argued that the department should assume functions then under the authority of the Secretary of Commerce for Transportation. He also recommended that the Federal Aviation Agency become part of that department. "I guess I was a rarity—an independent head proposing to become less independent," he later wrote.[2]

Halaby's motivation for making such an unprecedented suggestion was driven by frustration. His frustration stemmed from his thought that the Defense Department had locked the Federal Aviation Agency out of the administration's supersonic transport decision-making. Halaby decided that a Department of Transportation was essential to secure decisive transportation policy development. After four and a half years as administrator, he concluded that the agency could do a better job as part of an executive department that incorporated other government transportation programs. He wrote to Johnson: "One looks in vain for a point of responsibility below the president capable of taking an evenhanded, comprehensive, authoritarian approach to the development of transportation policies or even able to assure reasonable coordination and balance among the various transportation programs of the government."[3]

Creation of the new Department of Transportation was pushed by Charles Schultze, director of the Bureau of the Budget, and Joseph A. Califano, Jr., special assistant to the president. They urged Alan S. Body, then Undersecretary of Commerce for transportation, to explore the prospects of having a transportation department initiative prepared as part of Johnson's 1966 legislative program. On October 22, 1965, the Body Task Force submitted recommendations that advocated establishing a Department of Transportation that would include the Federal Aviation Agency, the Bureau of Public Roads, the Coast Guard, the Saint Lawrence Seaway Development Corporation, the

Great Lakes Pilotage Association, the Car Service Division of the Interstate Commerce Commission, the subsidy function of the Civil Aeronautics Board, and the Panama Canal.

After modifications were made, Johnson agreed, and on March 6, 1966, he sent Congress a bill to establish a department. The new agency's mission would be to coordinate and effectively manage transportation programs, provide leadership in the resolution of transportation problems, and develop national transportation policies and programs. The department would accomplish this mission under the leadership of a Secretary, an Undersecretary, and four staff assistant secretaries whose functions, though unspecified, expedited the line authority between the Secretary and Undersecretary and the heads of the operating administrations.

President Johnson understood the predicament the American transportation system faced. While it was the best-developed system in the world, it wasted lives and resources and had proved incapable of meeting the needs of the time. "America today lacks a coordinated transportation system that permits travelers and goods to move conveniently and efficiently from one means of transportation to another, using the best characteristics of each."[4] Johnson maintained that an up-to-date transportation system was essential to the national economic health and well-being, including employment, standard of living, accessibility, and the national defense. Thus, along with the proposed legislation to create the Department of Transportation, Johnson sent a carefully worded message recommending that Congress enact the bill as part of his attempt to improve public safety and accessibility.

Department of Transportation Established

After considerable compromise with a Congress that was protective of its constitutional power of the purse and its relationship with the older bureaucracies, Johnson signed into law the Department of Transportation Enabling Act of October 15, 1966. Compromise made the final version of the bill less than what the White House wanted. Nevertheless, it was a significant move forward, producing the most sweeping reorganization of the federal government since the National Security Act of 1947.

On April 1, 1967, five and a half months after President Johnson had signed the enabling legislation, DOT opened for business by celebrating the "Pageant of Transportation." Dignitaries from the department, the Smithsonian Institution, the transportation industry, and the public gathered for ceremonies on the Mall celebrating the new department. Alan S. Boyd, named by Johnson as DOT's first Secretary, guaranteed that the new department would "make transportation more efficient, more economical, more expeditious and more socially responsible."[5]

DOT was suddenly the fourth largest governmental agency, with a blueprint of organization, an order providing for essential authorizations, and several leading officials on the job. (Formation of DOT brought under one roof more than thirty transportation agencies and functions scattered throughout the government, and about ninety-five thousand employees, most of whom had been with the Federal Aviation Agency, the Coast Guard, and the Bureau of Public Roads.)

To Alan S. Boyd, the former Civil Aeronautics Board Chairman and Undersecretary of Commerce for Transportation, fell the challenge of setting up the new department by: (1) structuring it around Congress's recommendations in the enabling act, (2) organizing it, and (3) setting it in motion. The new Secretary faced a host of problems: (1) creating his own immediate office; (2) providing appropriate missions for his assistant secretaries; (3) building the new Federal Highway Administration and the Federal Railroad Administration; (4) helping to start the National Transportation Safety Board, and (5) setting up an organization and management plan for the entire department.

Meanwhile, in the White House, the president understood the connection between transportation systems and the needs of urban areas. The White House drafted a plan to transfer urban mass transit functions to the DOT that formerly resided in the Department of Housing and Urban Development (HUD). As mandated by the Department of Transportation Act, Johnson directed the Secretaries of Housing and Urban Development and Transportation to inform Congress where the most "logical and efficient organization and location of urban mass transportation functions within the Executive Branch" would be. When this failed to resolve the issue, Johnson transferred most of HUD's mass transit capacity to the DOT, effective July 1, 1968. Responsibility for these programs resided in the newly established Urban Mass Transportation Administration (now known as the Federal Transit Administration).

At the end of Boyd's administration, DOT embraced the Coast Guard, the renamed Federal Aviation Administration, the Federal Highway Administration, the Federal Railroad Administration, the Saint Lawrence Seaway Development Corporation, the Urban Mass Transportation Administration, and, tangentially, the National Transportation Safety Board. Boyd's most significant accomplishment was establishing and organizing the department and getting it operating as an effective governmental entity.

Nixon Administration

During his administration, Richard M. Nixon presided over several transportation-related matters, including the bailout of the Penn Central Railroad, the launching of Amtrak, and the attempted extension of federal support for supersonic transport. He nominated as his Secretary of Transportation the moderate thrice-elected governor of Massachusetts, John A. Volpe (a modern Horatio Alger). Volpe headed a construction firm that built hospitals, schools, shopping centers, public buildings, and military installations along the Eastern Seaboard and in other parts of the country. In 1968, the former federal highway administrator had been a rumored vice presidential nominee—until Spiro Agnew received the nod.

In 1970, the Highway Safety Act authorized the establishment of the National Highway Traffic Safety Administration. Although the law added somewhat to the department's safety mission, the Federal Highway Administration originally had handled most of the functions that the new agency assumed. Besides establishing another operating administration and adding to the Secretary's span of control and coordination workload, the Highway Safety Act separated highway administration into two parts: (1) design, construction, and maintenance; and (2) highway and automobile safety. Such organization ran counter to the original departmental organizing concept for the various modes of transportation: unlike the Coast Guard and the Federal Aviation Administration, for example, the Federal Highway Administration no longer bore responsibility both for facilities and infrastructure or for safety programs.

Volpe's highest priorities were to coordinate the missions of the diverse agencies and to develop a "balanced" transportation policy. Symbolic of this effort was the establishment of the Transportation Systems Center in Cambridge, Massachusetts. Volpe thought he had effectively begun to coordinate separate agencies, each of which had its own constituencies on Capital Hill, in industry, and among the public. For years, these agencies had acted autonomously, with little coordination or teamwork. He believed he had begun to forge them into a united transportation agency.

Volpe's tenure at DOT was marked by resolving high profile national transportation problems. These included airline hijackings, the sick-out of the fledgling Professional Air Traffic Controllers Organization, the decision to end federal support for production of the supersonic transport and to handle applications for Concorde landing slots, the financial insolvency of the Penn Central Railroad and the creation of Amtrak, and the Coast Guard's handling of the case of the defection of the Lithuanian seaman Simas Kudirka.

On December 6, 1972, Nixon named Dr. Claude S. Brinegar to succeed Volpe. Brinegar, a senior vice president of the Los Angeles-based Union Oil Company, had a Ph.D. in econometrics and was a self-styled professional manager. Brinegar was reserved in management style and conservative in political philosophy and successfully steered DOT through Watergate and the energy crisis of 1973-1974.

Ford and Carter Administrations

When he became president, Gerald R. Ford named William T. Coleman, Jr., to succeed Brinegar. Coleman had served on several airline and transit boards, including the Southeastern Pennsylvania Transportation Authority, Philadelphia's transit system. Coleman was a distinguished lawyer who with Thurgood Marshall had played a major role in landmark civil rights cases, including Brown v. the Board of Education of Topeka, which ended de jure (by right) school segregation in 1954. During Coleman's tenure, on April 1, 1975, Congress granted the National Transportation Safety Board, which had been established within the Department, its independence from the department.

When Jimmy Carter became president in 1976, he chose Brock Adams, a six-term member of the House of Representatives from Washington State, for Secretary of Transportation. Adams, a leading authority on transportation matters in the House, had been

Brinegar's nemesis and the primary author of the legislation that reorganized the bankrupt northeastern rail lines into the government-backed Conrail system.

Adam's establishment of the Research and Special Programs Administration (RSPA) on September 23, 1977, was a significant institutional development. When Adams created RSPA, he combined the Transportation Systems Center, the hazardous materials transportation and pipeline safety programs, as well as diverse program activities from the Office of the Secretary that did not readily fit in any of the existing operating administrations. The establishment of the RSPA set a precedent in that it was a creation of the Secretary, not Congress. [Note: Passage of the Pipeline Safety Act of 1992 gave RSPA equal statutory standing with the other operating administrations.] RSPA simultaneously moved crosscutting research and development pursuits from the Office of the Secretary to an autonomous operating administration.

During Adam's administration, the Inspectors General Act of 1978 imposed on the department and most other executive agencies an inspector general, appointed by the President and confirmed by the Senate. The mission of the Inspector General was to help the Secretary cope with waste, fraud, and abuse. Although housed in the Department and given the rank of Secretary, the Inspector General was generally autonomous.

Before he left office, Adams recommended that the Federal Highway Administration and the Urban Mass Transportation Administration be reorganized into a Surface Transportation Administration, an idea to which James Burney and Federico Pena would later return.

Adams was succeeded by Neil E. Goldschmidt, Mayor of Portland, Oregon, since 1972, and later President of the United States Conference of Mayors. Meanwhile, successful deregulation in transportation included the Railroad Regulatory Act (better known as the Staggers Rail Act), the Truck Regulatory Reform Act, the International Airlines Reform Act, and the Household Goods Regulatory Reform Act.

Goldschmidt was interested in government industrial policy, an early example of which was the Chrysler Corporation Assistance Program, worked out largely by the Treasury Department. When Congress drafted the Chrysler Loan Guarantee Act of 1979, he began a review of the automobile industry's problems. Goldschmidt also established the Office of Small and Disadvantaged Business Utilization in the Office of the Secretary. It was responsible for carrying out policies and procedures consistent with federal statutes to provide policy guidance for minority, women-owned, and disadvantaged businesses taking part in the department's procurement and federal financial assistance activities.

Reagan Administration

Andrew L. ("Drew") Lewis, Jr., a management consultant and political leader from Pennsylvania, became Ronald Reagan's first Secretary of Transportation. During his tenure, Lewis successfully negotiated the transfer of the Maritime Administration from the Commerce Department to DOT and provided the Department with the maritime connection it needed to formulate national transportation policy. The Department assumed greater visibility during the air traffic controllers' strike in August 1981, during which Lewis spoke for the administration. After personally negotiating with the Professional Air Traffic Controllers Organization in the days leading up to the strike, Lewis forcefully explained the government's response to the strike-firings.

Elizabeth Hanford Dole, who had been Reagan's assistant for public liaison, became Lewis's successor at DOT. Dole, a consumer adviser in two administrations and a member of the Federal Trade Commission during the Nixon and Ford administrations, brought to her new position experience in consumer and trade matters. At DOT, she focused on many safety-related issues, including drunk driving and the so-called "Dole brake light." Responding to a Supreme Court ruling, Dole authorized deadlines for the installation of air bags and other passive restraints in motor vehicles, which resulted in major increases in seat belt usage by the public and incentives to manufacturers to equip new cars with air bags.

While Dole was Secretary, the Commercial Space Launch Act of 1984 gave the department a multifaceted new mission to promote and to regulate commercial space launch vehicles. Because no operating administration had a comparable mission and because of its modest funding, Dole located the Office of Commercial Space Transportation in the Office of the Secretary.

The Airline Deregulation Act of 1978 and the Civil Aeronautics Board Sunset Act of 1984 had abolished the Board and transferred to the Department many of its functions relating to the economic regulation of the airline industry. Specifically, these included the aviation economic fitness program, functions related to consumer protection, antitrust oversight, airline data collection, and the review of international route negotiations and route awards to carriers. On January 1, 1985, the Office of the Secretary took over most of these functions, under the jurisdiction of the Office of the Assistant Secretary for Policy and International Affairs.

DOT divested itself of entities that it thought should be in the private sector (a trend begun when the Department had transferred the Alaska Railroad to the state of Alaska). Dole moved to end the Federal Railroad Administration ownership of Conrail, finally realized in April 1987. She also encouraged the establishment of the Metropolitan Washington Airports Authority in June 1987, transferring administration of Washington National Airport and Dulles International Airport from the Federal Aviation Administration to that authority.

Reagan chose James H. Burnley IV (Dole's deputy and former general counsel) to replace Dole. While Deputy Secretary, Burnley had helped to negotiate the sale of Conrail, directed the privatization of Amtrak, enabled the transfer of the Washington airports to the regional authority, and helped to assemble an air traffic control work force in the wake of the 1981 strike. He also helped to produce the Department's policies on aviation safety and security.

Because of the Federal Aviation Administration's apparent foot-dragging on safety regulations, and because he sought to increase the Secretary's management oversight capacity within the Department, Burnley proposed to curtail the autonomy of the operating administrations. A working paper recommended integration of the functions of the Maritime Administration, the Federal Aviation Administration, and the surface transportation administrations under three under secretaries, for water, air, and surface transportation, respectively. Burnley offered his reorganization proposal at the conclusion of Ronald Reagan's second term in the hope that it would provide Congress, his successor, and the public an alternative to proposals according to which one agency or another would leave the Department.

Bush Administration

Burnley's successor, Samuel K. Skinner, a George Bush appointee, chose instead to emphasize the establishment of a National Transportation Policy. Skinner also welcomed expansion of the Department's role in crisis management response. His handling of a succession of disasters, both natural and manufactured, earned Skinner the Washington moniker "the Master of Disaster." For Skinner, it began with additional evidence that a terrorist bomb had destroyed Pan American Airways Flight 103 over Lockerbie, Scotland on December 21, 1988. In rapid sequence followed the machinists' strike at Eastern Airlines (March 1989) and the airline's subsequent bankruptcy, the Exxon Valdez oil spill (March 1989), the Loma Prieta earthquake (October 1989), and Hurricane Hugo (September 1990), all high-profile incidents that took place during Skinner's first twenty-one months in office.

Establishing a national transportation policy became Skinner's highest priority. In Moving America, national transportation policymakers outlined six objectives: (1) to maintain and expand America's national transportation system; (2) to nurture a sturdy financial footing for transportation; (3) to keep the nation's transportation industry vigorous and competitive; (4) to guarantee that the transportation system enhances public safety and national security; (5) to maintain the environment and quality of life; and (6) to ready American transportation technology and expertise for the next century. By March 1990, conditions had persuaded Skinner that to realize these goals, diverse departmental offices would have to work together. Consequently, the Secretary launched the National Transportation Policy-Phase 2 (NTP-Phase 2) under the leadership of Thomas Larson, administrator of the Federal Highway Administration. NTP-Phase 2 activities combined to help the Department inventory its strengths and weaknesses and identify room for improvement.

On December 18, 1991, President Bush signed into law the Intermodal Surface Transportation Efficiency Act (ISTEA), derived in part from the NTP, which provided a six-year reauthorization to restructure the Department's highway, highway safety, and transit programs. One effect of this legislation was that the Urban Mass Transportation Administration became the Federal Transit Administration. The ISTEA legislation also required the Department to establish two new organizational entities: the Bureau of

Transportation Statistics, meant to provide timely transportation-related information through the compilation, analysis, and publishing of comprehensive transportation statistics, and the Office of Intermodalism, in the Office of the Assistant Deputy Secretary, which was charged with coordinating and initiating federal policy on intermodal transportation.

Meanwhile, Skinner became White House chief of staff, and Bush named Andrew H. Card, Jr. (his deputy White House chief of staff) as Secretary of Transportation. Disaster response to Hurricane Andrew, which hit southern Florida in August 1992, highlighted Card's term at the Department.

Clinton Administration

When Bill Clinton became President in 1993, he initially selected Federico Pena to head the "cluster group" that dealt with transportation issues during the transition, and ultimately to manage the Department of Transportation.

In March 1993, Clinton announced a plan for a six-month National Performance Review (NPR) of the federal government. Following a program analysis by Texas governor Anne Richards, Clinton asked Vice President Al Gore to head his administration's effort to improve the quality of the government and to reduce the cost of delivering services to the American taxpayer. The NPR challenged federal agencies to identify what worked and what did not, to propose new ways of doing jobs that would eliminate red tape and improve both operations and customer service, and to think about doing their work in smarter, more cost-effective ways.

Responding to several congressional initiatives, including the Chief Financial Officers Act of 1990, the Federal Managers' Financial Integrity Act, and the Government Performance and Results Act of 1993, the NPR laid the groundwork for "reinventing government." The outcome was the DOT Strategic Plan, which Pena announced in January 1994.

The plan delineated DOT's mission and enumerated seven broad strategic goals to carry out "tying America together" through an effective intermodal transportation system, investing strategically in transportation infrastructure, creating a new alliance between the nation's transportation and technology industries to make them more efficient and economically competitive, promoting safe and secure transportation, actively enhancing the environment, "putting people first" in the transportation system, and transforming the department. Meanwhile, the Department continued to be at the center of the federal government's crisis management response team, as exemplified by its response to flooding in the Mississippi River Basin in the summer of 1993 and the Northridge earthquake of January 1994.

The NPR had promised a government that not only did its job better, but cost less as well. Consequently, a small tax cut was proposed, one that would be funded in part by restructuring several federal departments and agencies, including the DOT. At the same time, Pena outlined a plan to restructure the Department by the end of the decade. After a brief period of workshops and discussions with Congress, the public, and Department employees throughout the country, Pena announced a restructuring plan for the Department. Pending Congressional approval, three operating administrations (a Federal Aviation Administration, a new Intermodal Transportation Administration, and the Coast Guard) would replace the current ten. Where Congressional approval was not necessary, Pena moved ahead, transferring the Office of Commercial Space Transportation from the Office of the Secretary to the Federal Aviation Administration, and launching the Transportation Administrative Services Center (TASC) to provide fee-based administrative services previously financed by the Working Capital Fund, both within DOT and to other government agencies.

At the beginning of his second term, Clinton selected Federal Highway Administrator Rodney E. Slater to succeed Pena at DOT. Slater was instrumental in getting ISTEA reauthorized, with the passage of the Transportation Equity Act for the 21st Century, the largest public works legislation in history. Early in his tenure, airline and railroad mergers again became fashionable. Department negotiators helped to avert a strike against Amtrak, and Congress mandated that Corporation's overhaul; the National Highway Traffic Safety Administration issued regulations allowing consumers to turn off their airbag switches where necessary; and the U.S. finalized a long-sought, liberalized aviation agreement with Japan.

In keeping with Slater's conviction that transportation was about "more than concrete, asphalt, and steel," the Garrett A. Morgan Technology and Transportation Futures program was enacted to encourage students to choose careers in transportation. Also enacted was a "Safe Skies for Africa" initiative to promote sustainable improvements in aviation safety and airport security in Africa. On October 8, 1998, Slater proposed the idea of creating a unified Department, ONE DOT, able to act as an integrated, purposeful leader, and increasing transportation efficiency and effectiveness.

A brief chronology of significant U.S. DOT dates is listed in Table 1.1.

Motor Carrier Safety Regulations: A Brief History

The United States Federal Government's involvement in supervising transportation began in the first half of the 19th century with the inspection of steamboats. In 1893, the U.S. Government became further involved in transportation when the first railroad Safety Appliance Act was enacted. After the turn of the century, railroad safety-related legislation continued to be enacted, and in 1908 the Transportation of Explosives Act was enacted. The act was revised in 1921 and brought into being the first regulations affecting (indirectly) motor carriers. However, not until several years later (1935) did legislation (The Motor Carrier Safety Act—now part of the Intestate Commerce Act) specifically include motor carrier safety regulations.

The 1935 Motor Carrier Safety Act made exercising regulatory authority the duty of the Interstate Commerce Commission. Specifically, the Act made common and con-

Table 1.1. Chronology of Significant U.S. DOT Dates

Date	Event
August 4, 1790	President George Washington signed into law a bill authorizing the construction of ten 50-foot, two-masted boats to guard the coast against smugglers.
January 12, 1874	Representative Laurin D. Woodworth (R-OH) introduces the first post-Civil War legislation to establish a federal bureau of transportation.
March 3, 1893	President Benjamin Harrison signed into law the Agriculture Appropriations Act of 1894, $100,000 of which was to be used to launch the Office of Road Inquiry, predecessor agency to the Bureau of Public Roads and Federal Highway Administration.
July 11, 1916	President Woodrow Wilson signed the Federal-aid Road Act, launching the Federal-aid highway program, with grants to the states for the construction of roads used to deliver the mail.
February 5, 1949	The First Hoover Commission Report called for coordination activities, under the auspices of the Commerce Department. President Harry S. Truman would respond by putting the office of Under Secretary for Transportation in the Department of Commerce.
November 20, 1950	A Commerce Department order established the Office of the Under Secretary for Transportation, with supervisory responsibility over transportation functions exercised by various departmental components.
June 29, 1956	Eisenhower signed into law the Federal-Aid Highway Act of 1956 and the Highway Revenue Act of 1956, authorizing the National System of Interstate and Defense Highways, and creating the Federal Highway Trust Fund.
August 23, 1958	Eisenhower signed into law the Federal Aviation Act of 1958, establishing the Federal Aviation Agency (Administration, after the DOT Act passed), to take effect on January 1, 1959. In addition, the bill freed the Civil Aeronautics Board (CAB) from its administrative connections with the Department of Commerce.
June 26, 1961	The U.S. Senate Committee on Commerce issued a staff report on National Transportation Policy, commonly known as the Doyle Report, calling for, among other things, the establishment of a Department of Transportation.
June 30, 1961	President John F. Kennedy signed into law the Housing Act of 1961, which acknowledged, for the first time, a federal role in mass transportation by establishing the Office of Transportation within the Housing and House Finance Agency.
January 12, 1966	In his State of the Union Address, President Johnson announced his intention to seek the establishment of the Department of Transportation.
March 2, 1966	Proclaiming that "in a nation that spans a continent, transportation is the web of union," President Johnson sent Congress a bill, recommending that the United States reorganize its entire transportation policy-making apparatus and created a Department of Transportation. That same day, Representative Chet Holifield (D-CA) and Senator Warren G. Magnuson (D-WA) introduced that measure in the House and Senate.
April 13, 1966	President Johnson signed into law the Uniform Time Act. The DOT Enabling Act charged the Secretary of Transportation with the administration of this act. In turn, he delegated this authority to the Department's General Counsel.
October 15, 1966	President Johnson signed into law Public Law 89-670, establishing the Department of Transportation.
January 6, 1967	Secretary of Commerce John T. Connor was appointed to the first National Motor Vehicle Advisory Council.
January 16, 1967	Alan S. Boyd was administered the oath of office as the nation's first Secretary of Transportation. Simultaneously, the interagency Department of Transportation Task Force adjourned, and many of its members left to take up tasks in the nascent Department of Transportation.
April 1, 1967	The new cabinet-level Department of Transportation was officially open for business.
December 9, 1999	President Clinton signed into law the Motor Carrier Safety Improvement Act of 1999.
March 8, 2000	Slater formally inaugurated the new Federal Motor Carrier Safety Administration, whose mission is to significantly improve truck and bus safety on the nation's highways.

tract carriers subject to economic and safety regulations in Section 204(a)(1) and (2). Briefly, this section provided for the establishment of "reasonable requirements" for "qualifications and maximum hours of service for employees, and safety of operations and equipment" for common and contract carriers.

The 1935 Act set the stage for its various provisions for future revisions, but at the time did not formulate motor carrier safety regulations.

To administer the provisions of the Motor Carrier Act, a Bureau of Motor Carriers was established within the Interstate Commerce Commission. Included within this organization was a section dealing with motor carrier safety. In its early years, this section was heavily involved in the development of the initial motor carrier safety regulations applicable to common and contract carriers. The Interstate Commerce Commission initiated a series of proceedings to determine the nature and extent of safety regulations to be adopted for various categories of motor carriers engaged in interstate and foreign commerce.

One-sided or partisan (ex parte) proceedings were held in the beginning by the Interstate Commerce Commission with public hearings held at various points throughout the nation. The Commission found that it had not only the authority but also a very real need for prescribing qualifications for drivers, initial requirements related to parts and accessories of vehicles, driving of commercial vehicles, reporting of accidents by common and contract carriers, and maintenance requirements.

Not until 1937 did the first Motor Carrier Safety Regulations, formulated by the Bureau of Motor Carriers, become effective. Until 1939 the Commission was concerned primarily with the determination of applications for certificates of public convenience, necessity, and contract carrier permits and related requirements. The first group of Motor Carrier Safety Inspectors, twenty in all, was also employed and assigned to field offices during this time frame.

Soon after the early hearings, findings pointed to the need for prescribing maximum hours of service for employees and the inclusion of private carriers of property in the qualifications for drivers and maximum hours of service regulations. Authority was limited to those employees whose activities were safety sensitive (i.e., activities that affect the safety of operations).

Private carriers were included for several reasons. In the first place, findings indicated that as many as 3 million private carriers were engaged in interstate or intrastate transportation of property. Approximately 20 percent of these were operating in interstate or foreign commerce, and this exceeded the number of vehicles operated by both common and contract carriers in such transportation. In several states, private carriers of property were not subject to the same safety requirements imposed on common and contract carriers—a dangerous exception to regulations. In 28 states, no limitations were placed on hours of service for drivers of private carrier vehicles. In addition, in some states, persons under the age of 21 (sometimes as young as 16) were allowed to drive trucks.

In 1940, the Commission applied to private carriers of property essentially the same regulations as had previously been prescribed for common and contract carriers. Private carriers were, however, excluded from reporting of accidents and excess hours of service for drivers. Farm trucks and their drivers, and other selected trucks and drivers, were made exceptions to or excluded entirely from certain provisions. The minimum age of drivers of farm vehicles weighing less than 10,000 pounds was placed at 18 (rather than 21) years of age, and the physical examination requirement was waived.

The hours of service requirements were modified for driver-salesmen who spent more than half their time selling and less than half driving, loading, and/or unloading.

The next major piece of safety legislation affecting motor carriers was the Transportation of Explosives Act of 1948. Although the Act reached only common carriers, the dangerous articles provisions were applied to contract and private carriers of property through the "safety of operations" provisions of Part II of the Interstate Commerce Act.

Attention was again focused on adequacy of government regulation in 1953. During this period, explosions of a number of explosive-laden trucks and other serious truck and bus accidents brought to question the adequacy of government regulation of commercial vehicle safety. Departmental examination determined that between 1939 and 1954, the number of Bureau of Motor Carrier Safety Inspectors had decreased from 20 to 18. Obviously, the increase in serious accidents and in explosions along with the decrease in inspectors was cause for concern. This concern resulted in an increase of safety inspectors to 100 by 1957. However, not until after the result of a very serious explosion in 1959 involving a truck owned by a private carrier was Congress prodded to enact an amendment (in 1960) to the Transportation of Explosives Act of 1948. This amendment placed contract and private carriers under the same statutes covering the transportation of explosives and other dangerous materials as originally applied only to the common carriers.

Ensuring Highway Safety

During the early 1960s, some members of Congress as well as some private individuals became concerned with the sharp rise in the number of deaths on streets and highways, and the reversal of the downward trend in the fatality rate. Their contention was that additional means were needed to combat this very alarming trend, and extensive investigation and a series of Congressional hearings resulted in the enactment and approval of two significant statues in 1966: The National Traffic and Motor Vehicle Safety Act and the National Highway Safety Act.

The 1966 National Traffic and Motor Vehicle Safety Act assigned to the Secretary of Commerce the administration of the law, which provided for the promulgation of Federal Motor Vehicle Safety Standards. Such provisions as the labeling of tires to inform purchasers of their safety attributes, requiring manufacturers to meet prescribed standards, and establishing severe penalties for violations were included. The Act also declared that neither the states nor the Interstate Commerce Commission could adopt or continue in effect standards different from any vehicle safety standards prescribed under the terms of this act.

The 1966 National Highway Safety Act provided for assistance to individual states by the Federal government in improving their highway safety programs. Again, the Secretary of Commerce was charged with the administration of the provisions of this act, including the grant-in-aid provisions to assist the states in financing these improvements.

Public Law 89-670, the Department of Transportation Act, which established the Department of Transportation, was also approved in 1966. The Act transferred all safety functions (motor carrier, railroad, and hazardous materials) of the Interstate Commerce Commission to the new department. The Act also provided for a Federal Highway administrator to exercise the functions, powers and duties of motor carrier safety responsibilities of the Secretary of Transportation. A Bureau of Motor Carrier Safety was created within the Federal Highway Administration soon after the establishment of the Department. In addition, the Act also transferred the responsibility for administration of the National Traffic and Motor Vehicle Safety Act and the National Highway Safety Act from the Secretary of Commerce to the National Highway Safety Bureau within the Federal Highway Administration.

The National Highway Safety Bureau was removed from the Federal Highway Administration by the 1970 Highway Safety Act. The Act created a separate operating administration in the DOT as the National Traffic Safety Administration. Included in the Act was the responsibility for carrying out the provisions of the National Traffic and Motor Vehicle Safety Act of 1966 and the Highway Safety Act of 1966. The Act also provided for the apportionment of Federal funds to support state and community highway safety programs on the basis of 75 percent by population and 25 percent by public road mileage, and mandated that two-thirds of all funds authorized and expended for highway safety research and development were to be appropriated from the Highway Trust Fund; new highway safety standards were to be submitted to Congress prior to their effective date.

Also in 1970, the Hazardous Materials Transportation Control Act authorized the Secretary of Transportation to establish facilities and staff to evaluate the problems surrounding hazardous materials shipments. The Secretary of DOT was charged with submitting to the President each year a comprehensive report on the transportation of hazardous materials and recommendations for additional legislation.

In the same year, the Occupational Health and Safety Act (OSH Act) was enacted. Primary responsibility for administration and enforcement of the Act was placed with the Secretary of Labor and a new agency, the Occupational Safety and Review Commission, a quasi-judicial board appointed by the President.

The OSH Act, designed "to assure as far as possible every working man and woman in the nation safe and healthful working conditions and to preserve human resources," is primarily aimed at business and industry in general. Although its provisions are applicable to the off-highway operations only, they must be included in any consideration of motor carrier safety regulations.

The Office of the Secretary DOT was centralized in 1974 as the result of the Transportation Safety Act. The Act authorized the promulgation and enforcement of hazardous material regulations for all modes of transportation. In 1976, the Materials Transportation Bureau (MTB), established by the Secretary of DOT in 1975, issued the final document dealing with the consolidation of hazardous materials regulations governing all modes of transportation (air, rail, highway, water), with the entire set of new regulations codified as 49 CFR, Parts 100-189.

The trucking safety regulatory agencies were reorganized into a new structure in 1986. As a result, the Bureau of Motor Carrier Safety and two related agencies ceased to exist. The Office of Motor Carriers (OMC) took over. The OMC consists of four agencies: Office of Motor Carrier Standards (responsible for developing and issuing the Federal Motor Carrier Safety Regulations); the Office of Motor Carrier Safety Field Operations (responsible for safety audits, fitness reviews, and other field activities); the Office of Program Management Support; and the Office of Motor Carrier Information Management and Analysis.

Hazardous Materials

Estimates of the annual volume of hazardous materials (hazmats) transported within the United States range from more than two billion to more than four billion tons. They are called "hazardous materials" because mishandling poses unreasonable risk to health, safety, property and the environment. Although these materials are essential to our modern lifestyle and our industrial economy, in the past decade alone close to 100 people have been killed and nearly $200 million in damages recorded during more than 50,000 hazardous material incidents.

Presently, more than 32,000 commodities are regulated as hazardous materials, ranging from everyday substances such as kerosene to firecrackers and highly explosive or toxic chemicals. Some are products that we use in our homes in limited (usually diluted) quantities (drain cleaners using sodium hydroxide or caustic, for example). All these materials demand care in handling and transportation.

Today, thanks to massive media exposure, we are routinely informed of hazmat incidents, though they occur nationwide more often than some people might expect. The fact is, the highly publicized incidents involving hazmats have led to increasing public concern about the safe shipment of these materials. Safe transportation of hazmats is not a given; it requires the coordinated efforts of both shipper and carrier. It requires planning and careful management with attention to detail. The transportation community and public authorities recognize the need for contingency plans (emergency response) to deal with the situation when things go wrong. One thing is certain: in transportation incidents involving hazmats, the hazmats always make what goes wrong even nastier to deal with—and the consequences more difficult and dangerous to live with.

We all know that the improper transportation and/or handling of hazmats is something that must be guarded against. However, good intentions are never enough, because in the absence of regulation and enforcement, that small minority that seeks to cut corners and costs will always appear, certain that *their* load won't cause a problem. As a result, the transportation of hazmats is a highly regulated activity—as it should be. At the Federal level, the Department of Transportation, the Occupational Safety

Introduction to the DOT and Hazmat Regulations

and Health Administration, and the Environmental Protection Agency are the principal regulatory and enforcement authorities. Various state and local governments are also active in regulating transportation of hazardous materials and in planning for emergency response in case of incidents.

Hazardous materials are controlled under ten federal statutes. Each statute provides protection from hazardous materials in different situations (see Table 1.2).

Table 1.2. Ten Federal Statutes Controlling Hazardous Materials

Statute	Situation
Clean Water Act (CWA)	Water
Clean Air Act (CAA)	Air
Safe Drinking Water Act (SDWA)	Ground Water
Toxic Substances Control Act (TSCA)	Products
Occupational Safety & Health Act (OSHA)	The Workplace
Resource Conservation & Recovery Act (RCRA)	Solid Waste
Emergency Planning and Community Right to Know Act (EPCRA)	Public Domain
Oil Pollution Act (OPA)	Oil Spills
Comprehensive Environmental Response, Compensation, and Liability Act (CERCLA)	Dump Sites
Hazardous Materials Transportation Act (HMTA)	Transportation

The Hazardous Materials Transportation Act of 1975 (HMTA, PL93-633), as amended by the Hazardous Materials Transportation Uniform Safety Act of 1990 (HMTUSA), is the focus of this text because this Act mandates Hazardous Materials Transportation Regulations. The intent of DOT's hazardous materials regulations is to protect transportation personnel and equipment from materials with dangerous properties. To further promote safety within the transportation industry, these regulations are periodically revised and amended. Until 1988, they were highly complex in both content and organization. Then proposals were made to simplify them to produce consistent modal and regional transportation requirements. To the extent possible, the discussion in this text follows these simplified proposals. The Hazardous Materials Transportation Regulations cover transportation of hazardous materials by rail, car, aircraft, vessel, and motor, and cover in particular hazardous substances, hazardous waste, and flammable cryogenic liquids. The regulation enforcement program is administered by the Research and Special Programs Administration (RSPA) of the DOT.

 Important Note: Because the hazardous materials regulations change from time to time, users of the regulations should consult Title 49 of the *Code of Federal Regulations*, Parts 171 through 180, for the latest amendments.

The dominant purpose of HTMA is to improve overall public safety related to hazardous materials transportation, as explained in Section 102 of the 1975 Act:

> *It is declared to be the policy of Congress in this title to improve the regulatory and enforcement authority of the Secretary of Transportation to protect the Nation adequately against the risks to life and property which are inherent in the transportation of hazardous materials in commerce.*[6]

HMTA applies to all interstate or foreign commerce, insofar as permitted by the commerce clause of the Constitution. Under this statute, "hazardous material" is defined as any substance or material so designated by the Secretary of Transportation.[7] The Secretary is required, by regulation, to determine the quantity and form of material that, when transported in commerce, "may pose an unreasonable risk to health and safety or property."[8] Transportation is defined to include any movement of property by any mode, including any loading, unloading, or incidental storage.[9] HMTA gave to the Secretary of Transportation broad power to adopt regulations for the transportation and handling of hazardous materials. Among the transportation acts, it is unique in applying not only to carriers, but also to shippers and to all persons or companies that manufacture, maintain, recondition, repair, test, or sell any packages or containers for use in the transportation of such material.[10]

 Important Note: HMTA contains a wide variety of criminal and civil enforcement remedies that may be invoked by DOT for violations of either the act or the regulations.

Summary

Since the early days when the Interstate Commerce Commission was tasked with the duty of establishing motor carrier safety regulations, the chore of maintaining effective transportation regulation has continued with continuous refinement, adoption, and extension of regulation coverage. One of the most significant recent changes has been the inclusion of private carriers of property in the requirements for reporting and recording of accidents.

The bottom line is that even though we are a society that abhors regulations, and even though many problems remain unsolved, today's motor carrier safety regulations have gone a long way toward focusing attention on making U.S. highways and operations on the highways safer for all. When we consider events such as those described in the prelude to this text, we are compelled to see the necessity for HAZMAT regulations.

Notes

[1] Adapted from *The United States Department of Transportation: A Brief History.* Washington, DC: Departmental Historian, U.S. DOT, April 3, 2000; http://isweb.tasc.dot.gov/Historian/historian.htm, 1-7.

[2] Ibid, 1-7.

[3] Ibid, 1-7.

[4] Ibid, 1-7.

[5] Ibid, 1-7.

[6] 49 U.S.C. App. Sec. 1801.

[7] 49 U.S.C. App. §1802(4).

[8] 49 U.S.C. App. §1803.

[9] 49 U.S.C. App. §1802(15).

[10] 49 U.S.C App. §1804(a).

Chapter 2
Understanding Title 49

Transportation is about more than concrete, asphalt, and steel—it's about people. It's about giving people the means to get to their jobs safely and efficiently. It's about giving U.S. businesses the competitive edge they need to compete and win in today's global economy. Yes, this is what transportation is all about—providing opportunity.

—Secretary of Transportation Rodney E. Slater

Introduction

When we need to find information on any subject, several different reference sources are available, including dictionaries, abstracts, encyclopedias, and the Internet. If we need to define a basic term such as "transportation," we know to use a standard dictionary. A typical entry for transportation would probably include at least the following definitions: (1) The act of transporting; (2) The state of being transported; (3) A means of transport; conveyance. We all know how and when to use a dictionary because using a dictionary and other standard reference works is rather straightforward.

But just as we need to know how to use a dictionary in order to find the meaning of a term, in order to find information and answer questions about Hazardous Materials Regulations, we must know where to look and how to find the correct material. One thing is certain: We won't find answers to questions about Hazardous Materials Regulations in a standard dictionary or encyclopedia. We need to refer to Title 49 of the Code of Federal Regulations (CFR). Those that have never used a CFR may have difficulty finding information, and even those that have used a CFR before may have difficulty understanding the information they find. In this chapter, we discuss information to help you understand and use Title 49.

Using the Code

The key to compliance with Hazardous Materials Regulations is knowing how the regulations are organized in the Code of Federal Regulations so that they may be referenced, deciphered, and put into effect. The Code of Federal Regulations is basically an encyclopedia of government regulations, a collection of different numbered volumes. Each numbered issue contains the regulations specific to certain agencies and subjects. For example, Title 29 is specific to the Department of Labor and Occupational Safety and Health regulations (OSHA). Title 40 is specific to the United States Environmental Protection Agency (U.S. EPA) and Environmental regulations. A complete listing of CFRs is presented in Table 2.1.

Table 2.1. Content of the Code of Federal Regulations (CFRs)

Title	Specific To	Title	Specific To
1, 2	General Provisions	27	Alcohol, Tobacco Products and Firearms
3	The President	28	Judicial Administration
4	Accounts	29	Labor/OSHA
5, 6	Administrative Personnel	30	Mineral Resources
7	Agriculture	31	Money and Finance: Treasury
8	Aliens and Nationality	32	National Defense
9	Animals and Animal Products	33	Navigation & Navigable Waters
10	Energy	34	Education
11	Federal Elections	35	Panama Canal
12	Banks and Banking	36	Parks, Forests, and Public Property
13	Business Credit and Assistance	37	Patents, Trademarks, and Copyrights
14	Aeronautics and Space	38	Pensions, Bonuses, and Veterans' Benefits
15	Commerce and Foreign Trade	39	Postal Service
16	Commercial Practices	40	Environment
17	Commodity and Securities Exchanges	41	Public Contracts and Property Management
18	Conservation of Power and Water Resources	42	Public Health
19	Customs Duties	43	Public Lands: Interior
20	Employees' Benefits	44	Emergency Management and Assistance
21	Food and Drugs	45	Public Welfare
22	Foreign Relations	46	Shipping
23	Highways	47	Telecommunications
24	Housing and Urban Development	48	Federal Acquisitions Regulations System
25	Indians	49	**Transportation**
26	Internal Revenue	50	Wildlife & Fisheries

Title 49 (more commonly known as 49 CFR), part of which contains the Hazardous Materials Regulations of USDOT, is the subject matter of this text. The various parts that make up 49 CFR are listed in Table 2.2. Table 2.3 lists the parts and sections referring to Hazardous Materials Regulations.

Table 2.2. Contents of the Parts of Title 49

49 CFR 1-99	Secretary of Transportation
49 CFR 100-185	Hazmat Transportation
49 CFR 186-199	Pipeline Safety
49 CFR 200-399	Federal Highway Administration
49 CFR 400-999	National Highway Traffic Safety Administration
49 CFR 1000-1199	Interstate Commerce Commission
49 CFR 1200-END	Interstate Commerce Commission

Understanding Title 49

Table 2.3. Contents of Hazardous Materials Regulations

Part/Section	Content
171.8	Definitions
171.15	Incident Reporting
Part 172	HM Table and Provisions
172.101-102	Table and Special Provisions
172.200-205	Shipping Papers
172.300-338	Marking
172.400-450	Labeling
172.500-560	Placarding and Appendix C Dimensions
172.600-604	Emergency Response
172.700-704	Training Requirements
173.1—173.478	Packaging
174.1—174.840	Shipping by Rail
175.1—175.705	Carriage by Aircraft
176.1—176.906	Carriage by Vessel
177.800—177.861	Carriage by Public Highway
177.816	Driver Training
178	Packaging Specifications: Tanks, Cylinders
180	Packaging Maintenance

All CFRs are organized into sub-titles, chapters, subchapters, parts, subparts, sections, paragraphs, and subparagraphs. For example, DOT's Hazardous Materials Regulations are contained in Subchapter C of Chapter I of Subtitle B of Title 49 CFR. Subchapter C is organized into parts that run numerically. The overall organization is as follows:

> Title 49 - Transportation
> > Subtitle B - Other Regulations Pertaining to Transportation
> > > Chapter I - Research and Special Programs Administration
> > > > Parts 171 - 180

Let's take a closer look at a specific section of Title 49, 49 CFR 172.101 (c)(10), and put it in a manageable, more understandable form.

49	CFR	172	.101	(c)	(10)
Title	Code of Federal Regulations	Part	Section	Paragraph	Subparagraph

Parts	These are organized into Sections, which run numerically. 172.101 refers to the 101st Section within Part 172.
Sections	The Section starts with the first series of numbers to the right of the period after the Part. A Section reference must include the Part because each Part could have a Section "1" in it. Example: 171.1 = refers to the first Section within Part 171
Paragraphs and Subparagraphs	Sections are organized into Paragraphs, Subparagraphs, etc. The first lower-case letter in parentheses to the right of the Section denotes the start of the Paragraph. The first number in parentheses to the right of the Paragraph is the Subparagraph. Example: 172.101 (c)(10) The (c) refers to a Paragraph reference and the (10) is a Subparagraph reference.
Sub-Subparagraphs	Subparagraphs are organized into Sub-Subparagraphs. These are first in lower case Roman Numeral in parentheses after the Subparagraph and Sub-Sub-Subparagraphs are denoted by a capital letter in parentheses. Example: 172.101 (c)(10)(i)(A) (c)(10)(i)(A) = refers to Sub-Sub-Subparagraph (A) of Sub-Subparagraph (i) of Subparagraph (10) of Paragraph (c).

Subparts   Parts are divided into Subparts containing Sections. These are lettered with a capital letter alphabetically within the Part.
Examples: Subpart A, Subpart B, etc

Summary

Locating specific regulatory information in the Code of Federal Regulations can be very difficult for readers unfamiliar with the organization of CFRs. To understand a CFR citation readers must be able to recognize the part, section, paragraph, and sub-paragraph being referenced.

Chapter 3
Classification of Hazardous Materials

Introduction

One of the most menacing health and physical injury threats that concerns those who work in industry—or sometimes even in the home—is the potential exposure to hazardous materials (hazmats). As with the hazardous material asbestos, exposure to many hazmats may be undetectable; they can easily go unnoticed. Like cancer, the devastating effects of exposure to many hazmats are insidious. But not all hazardous substances take that long to show their effects. If we are exposed to a highly corrosive substance such as concentrated sulfuric acid, the results (pain and tissue destruction) are immediately obvious to us.

The fact is that many hazmats are used in the traditional workplace. Some of these are obviously dangerous, but the dangers of many more are not so obvious. These hazardous substances may be biological, chemical, or physical. Examples of include explosives, flammables, toxics, etiological agents, corrosive substances, irritants, and oxidizers.

Unfortunately, exposure to hazardous materials is also possible outside the traditional workplace. This is especially true whenever hazardous materials are transported from manufacturer to shipper to user. Much of this transportation is accomplished by way of large overland trucking operations. Transporting potentially hazardous materials over public highways takes the hazard potential from the traditional workplace setting out into far-reaching areas.

Several concerns are important in transporting hazmats on public highways: the potential exposure to those involved with the transportation effort, the potential exposure to those in the vicinity of a hazmat incident on the highway, and potential exposure to the environment (soil, water, and air). Transporting hazmats in and of itself is dangerous enough, but when we factor in transport conditions such as substandard roads, adverse weather conditions, and incompetent or unfit drivers, the dangers are compounded.

Because of concern for the health, safety, and well-being of drivers, the public, and the environment, DOT has instituted Hazmat Transportation Regulations in a holistic way. Basically, hazmat regulations stipulate that the materials must be evaluated throughout their life cycle—cradle-to-grave. Today, the challenge is to determine the risks of potentially hazardous materials from the point of origins, through usage to final destination, whether in a discarded material or as a trace contaminant in air, water, or food supplies.

To be up to the challenge, we must have a basic understanding of hazmats and their potential to harm. The intent of this chapter is to provide the basic background information on hazmats that will make it possible to reduce the risks associated with these materials.

Hazmat Definitions[1]

Hazardous materials can be defined in a number of ways. However, the hazmat definitions we are concerned with in this text are defined related to their transport. A hazardous material is any substance or material in a quantity or form that poses an unreasonable risk to safety, health, or property when transported in commerce, or any substance that must be placarded when moving in interstate commerce.

The starting point in understanding DOT hazmat regulations is knowing the key concepts. We define the most important of these here:

Aerosol	Any non-refillable metal receptacle containing a gas compressed, liquefied or dissolved under pressure, the sole purpose of which is to expel a nonpoisonous (other than a Division 6.1 Packing Group III material) liquid, paste or powder, and fitted with a self-closing release device allowing the contents to be ejected by the gas.
Agricultural product	A hazardous material, other than a hazardous water, whose end use directly supports the production of an agricultural commodity, including but not limited to a fertilizer, pesticide, soil amendment or fuel. An *agricultural product* is limited to a material in Class 3, 8 or 9, Division 2.1, 2.2, 5.1, or 6.1, or an ORM - D Material.
Asphyxiant gas	A gas that dilutes or replaces oxygen normally in the atmosphere.
Atmospheric gases	Air, nitrogen, oxygen, argon, krypton, neon and xenon.
Bag	A flexible packaging made of paper, plastic film, textiles, woven material or other similar materials.
Bottle	An inner packaging having a neck of relatively smaller cross section than the body and an opening capable of holding a closure for retention of the contents.
Box	A packaging with complete rectangular or polygonal faces, made of metal, wood, plywood, reconstituted wood, fiberboard, plastic, or other suitable material. Holes appropriate to the size and use of the packaging, for purposes such as ease of handling or opening, or to meet classification requirements, are permitted as long as they do not compromise the integrity of the packaging during transportation.

Classification of Hazardous Material

Break-bulk	Packages of hazardous materials that are handled individually, palletized, or unitized for purposes of transportation as opposed to bulk and containerized freight.
Bulk packaging	A packaging, other than a vessel or a barge, including a transport vehicle or freight container, in which hazardous materials are loaded with no intermediate form of containment and which has: 1. A maximum capacity greater than 450 L (119 gallons) as a receptacle for a liquid; 2. A maximum net mass greater than 400 kg (882 pounds) and a maximum capacity greater than 450 L (119 gallons) as a receptacle for a solid; or 3. A water capacity greater than 454 kg (1000 pounds) as a receptacle for a gas.
Cargo tank motor vehicle	A motor vehicle with one or more cargo tanks permanently attached to or forming an integral part of the motor vehicle.
Carrier	A person engaged in the transportation of passengers or property by: 1. Land or water, as a common, contract, or private carrier, or 2. Civil aircraft.
Class	Hazard class (see *hazard class* in the glossary) *Class 1* *Class 2* *Class 3* *Class 4* *Class 5* *Class 6* *Class 7* *Class 8* Class 9 (Each of the above classes is discussed later in this chapter.)
Combination packaging	A combination of packaging for transport purposes, consisting of one or more inner packagings secured in a non-bulk outer packaging. It does not include a composite packaging.
Combustible liquid	A liquid with a flash point above 140°F according to National Fire Protection Association (NFPA) definition.
Compatibility group letter	A designated alphabetical letter used to categorize different types of explosive substances and articles for purposes of stowage and segregation.
Composite packaging	A packaging consisting of an outer packaging and an inner receptacle, so constructed that the inner receptacle and the outer packaging form an integral packaging. Once assembled it remains thereafter an integrated single unit; it is filled, stored, shipped and emptied as such.
Compressed gas	Any material or mixture having in the container an absolute pressure exceeding 40 psia at 70°F or regardless of the pressure at 70°F, having an absolute pressure exceeding 104 psia at 130°F.
Corrosive material	Any liquid or solid that can destroy human skin tissue, or a liquid that has a severe corrosion rate on steel.
Cryogenic liquid	A liquified gas stored at temperatures approaching absolute zero. Normally, they have a boiling point of about -100°C.
Cylinder	A pressure vessel designed for pressure higher than 40 psia and having a circular cross section. It does not include a portable tank, multi-unit tank car tank, cargo tank, or tank car.
Dangerous when wet material	See Class 4.3 definition.
Division	A subdivision of a hazard class.
Domestic Transportation	Transportation between places within the U.S. other than through a foreign country.
Drum	A flat-ended or convex-ended cylindrical packaging made of metal, fiberboard, plastic, plywood, or other suitable materials. This definition also includes packagings of other shapes made of metal or plastic (e.g., round taper-necked packagings or pail-shaped packagings) but does not include cylinders, jerricans, wooden barrels or bulk packagings.

Elevated temperature material	A material that, when offered for transportation or transported in a bulk packaging: 1. Is in a liquid phase and at a temperature at or above 100°C (212°F); 2. Is in a liquid phase with a flash point at or above 37.8°C (100°F) that is intentionally heated and offered for transportation or transported at or above its flash point; or 3. Is in a solid phase and at a temperature at or above 240°C (464°F).
Etiologic agent	A living microorganism that may cause human disease; germs.
Explosive	A solid or liquid substance possessing the faculty under certain circumstances of undergoing instantaneous decomposition, extending throughout the entire mass. The process is accompanied by a considerable disengagement of heat, the substance being wholly or partially converted into gaseous products.
Flammable gas	Any gas that will burn.
Flammable liquid	Any liquid with flash point below 100°F (37.7°C).
Flammable solid	Any material, other than an explosive, liable to cause fires through friction, retained heat from manufacturing or processing, or that can be ignited readily and when ignited burns so vigorously and persistently as to create a serious transportation hazard.
Flash Point	The temperature when a liquid gives off flammable vapors sufficient to form an ignitable mixture near the surface of the liquid; combustion is not continuous at the flash point.
Gas	A material that has a vapor pressure greater than 300 Kpa (43.5 psi) at 50°C (122°F) or is completely gaseous at 20°C (68°F) at a standard pressure of 101.3 Kpa (14.7 psi).
Hazard class	The category of hazard assigned to a hazardous material under DOT definitional criteria. A material may meet the defining criteria for more than one hazard class, but is assigned to only one hazard class.
Hazardous substance	Any material, including its mixtures and solutions, that is listed in Appendix B to 49 CFR 172.101. These materials are in a concentration greater than or equal to 1. 10% by weight for most of the materials listed; and 2. 1% by weight for materials identified as severe marine pollutants.

> ✓ **Important Note:** OSHA uses the term hazardous substances differently than U.S. EPA and DOT. Hazardous substances, as used by OSHA, cover every chemical regulated by both DOT and U.S. EPA.

Hazardous waste	Any material subject to the Hazardous Waste Manifest Requirements of the U.S. Environmental Protection Agency specified in 40 CFR Part 262.
Hazmat employee	A person employed by a hazmat employer, who in the course of employment directly affects hazardous materials transportation safety. This term includes an owner-operator of a motor vehicle that transports hazardous materials in commerce. This item includes an individual, including a self-employed individual, employed by a hazmat employer who, during the course of employment: 1. Loads, unloads, or handles hazardous materials; 2. Manufactures, tests, reconditions, repairs, modifies, marks, or otherwise represents containers, drums, or packagings as qualified for use in the transportation of hazardous materials; 3. Prepares hazardous materials for transportation; 4. Is responsible for safety of transporting hazardous materials; or 5. Operates a vehicle used to transport hazardous materials.

Classification of Hazardous Material

Hazmat employer	An employer that uses one or more of its employees in connection with: transporting hazardous materials in commerce; causing hazardous materials to be transported or shipped in commerce; or representing, marking, certifying, selling, offering, manufacturing, reconditioning, testing, repairing, or modifying containers, drums, or packagings as qualified for use in the transportation of hazardous materials. This term includes an owner-operator of a motor vehicle that transports hazardous materials in commerce. This term also includes any department, agency, or instrumentality of the U.S., State, political subdivision of a State, or Indian tribe engaged in an activity described in the first sentence of this definition.
Irritating material	Liquids or solid substances that give off dangerous or intensely irritating fumes on contact with fire or when exposed to air.
Liquid	A material, other than an elevated temperature material, with a melting point or initial melting point of 208°C (68°F) or lower at a standard pressure of 101.3 Kpa (14.7 psi). A viscous material for which a specific melting point cannot be determined must be subjected to the procedures specified in ASTM D 4359 "Standard Test Method for Determining Whether a Material is Liquid or Solid."
Marine pollutant	A material that, when in a solution or mixture of one or more marine pollutants, is packaged in a concentration which equals or exceeds: 1. 10% by weight of the solution or mixture for listed materials; or 2. 1% by weight of the solution or mixture for materials that are identified as severe marine pollutants.
Marking	Descriptive names, identification numbers, instructions, cautions, weights, specifications, UN marks, or combinations thereof, required by this subchapter on outer packagings of hazardous materials.
Mixture	Any material consisting of more than one chemical compound or ingredient. Mixtures can be liquids, solids, or gases.
N.O.S.	Not otherwise specified.
ORM	Other regulated material.
Overpack	With few exceptions, means an enclosure that is used by a single consignor to provide protection or convenience in handling of a package, or to consolidate two or more packages. Overpack does not include a transport vehicle, freight container, or aircraft unit load device. Examples of overpacks are one or more packages: 1. Placed or stacked onto a load board such as a pallet and secured by strapping, shrink wrapping, stretch wrapping, or other suitable means; or 2. Placed in a protective outer packaging such as a box or crate.
Oxidizer	A substance that yields oxygen readily to stimulate the combustion of organic matter and inorganic matter.
Oxidizing gas	A gas that may, generally by providing oxygen, cause or contribute to the combustion of other material more than air does.
Packaging	A receptacle and any other components or materials necessary for the receptacle to perform its containment function in conformance with the minimum packing requirements.
Packing group	A grouping according to the degree of danger presented by hazardous materials. Packing Group I indicates greater danger; Packing Group II, medium danger; Packing Group III, minor danger.
Primary Hazard	The hazard class of a material as assigned in Table 3.1.
Pyrophoric liquid	A characteristic of those materials that, if ground into fine particles, will spontaneously ignite when exposed to air
Reportable quantity (RQ)	The minimum quantity of hazardous waste generated as a result of a discharge or spill that must be reported (for the purpose of this text, the quantity specified in column 2 of the appendix to 49 CFR §172.101, for any material identified in column 1 of the appendix).
Solid	A material that is not a gas or liquid.
Solution	Any homogeneous liquid mixture of two or more chemical compounds or elements that will not undergo any segregation under conditions normal to transportation.
Subsidiary hazard	A hazard of a material other than the primary hazard (see *primary hazard* in the glossary).

Viscous liquid	A liquid material that has a measured viscosity in excess of 2500 centistoke at 25°C. (77°F.) when determined in accordance with the procedures specified in ASTM Method D 445-72 "Kinematic Viscosity of Transparent and Opaque Liquids (and the Calculation and Dynamic Viscosity)" or ASTM Method D 1200-70 "Viscosity of Paints, Varnishes, and Lacquers by Ford Viscosity Cup."
Volatility	The relative rate of evaporation of materials to assume the vapor state.
Water reactive material	Substances that react in varying degrees when mixed with water or when they come in contact with humid air; generally flammable solids.

Classification of Hazardous Materials

For classification purposes, we regard hazardous materials in several different ways. From the perspective of fire science, a hazardous material is often regarded as any substance or mixture that, if improperly handled, may be damaging to our health and well-being or to the environment. While our concern in this text is with the way the U.S. Department of Transportation classifies hazardous materials (which is much more inclusive than the approach usually followed in fire science), we set the stage with the traditional approach, which is a useful introduction to the characteristics likely to cause a substance to be regarded as a hazardous material.

The following are the seven classes of hazardous materials conventionally noted in fire science:[2]

1. *Flammable materials.* Any solid, liquid, vapor, or gaseous materials that ignite easily and burn rapidly when exposed to an ignition source. Examples of flammable materials within this broad definition include certain solvents like benzene and ethanol, dusts like flour and certain finely dispersed powders like aluminum, and gases like hydrogen and methane.

2. *Spontaneously ignitable materials.* Solid or liquid materials that ignite spontaneously without an ignition source, usually (but not necessarily) due to the dangerous buildup of heat during storage caused by oxidation or microbiological action. An example of a substance that ignites spontaneously without an ignition source is white phosphorus; examples of substances that spontaneously ignite due to the buildup of heat are fishmeal and grass.

3. *Explosives.* These chemical substances detonate, usually as the result of shock, heat, or some other initiating mechanism. Examples are dynamite and trinitrotoluene (TNT).

4. *Oxidizers.* Substances that evolve or generate oxygen, either at ambient conditions or when exposed to heat. Examples are ammonium nitrate and benzoyl peroxide.

5. *Corrosive materials.* Solids or liquid materials, such as battery acid, that burn or otherwise damage skin tissue at the site of contact.

Classification of Hazardous Material

6. *Toxic materials.* Broadly, poisons that in small doses either kill or cause adverse health effects. Examples of toxic materials of primary concern to firefighters are carbon monoxide and hydrogen cyanide.

7. *Radioactive materials.* These materials are characterized by transformations occurring in their atomic nuclei. Uranium hexafluoride is an example of a radioactive material.

> ✓ **Important Note:** Not only are these seven classes of materials individually hazardous, they pose an even more severe hazard when mixed.

However, Title 49 requires shippers to properly classify materials according to the criteria established by DOT. This means that we must understand the DOT classification process. The properties of the material are compared to the hazardous material definitions. After assessing the specific hazard class, packing group, and subsidiary risk, the appropriate proper shipping name is chosen from the Hazardous Materials Table (49 CFR 172.101) discussed later in this chapter. We must then determine whether the material is a hazardous substance or a marine pollutant.

> ✓ **Important Note:** A material may meet the definition of more than one hazard class. In this case, the material will be described in accordance with the precedence table established by DOT in 49 CFR 173.2a.

DOT Hazard Classes (Listing)

The hazard class of a hazardous material is indicated either by its class (or division) number, its class name, or by the letters "ORM-D." Nine hazard classes are used to identify hazardous materials. Table 3.1 lists class numbers, division numbers, and class or division names. The specific definitions of the nine hazard classes follows.

Table 3.1. DOT Hazard Classes

Class #	Division #	Name of Class or Division
None	—	Forbidden Materials
None	—	Forbidden Explosives
1	1.1	Explosives (with a mass explosion hazard)
1	1.2	Explosives (with a projection hazard)
1	1.3	Explosives (with predominantly a fire hazard)
1	1.4	Explosives (with no significant blast hazard)
1	1.5	Very insensitive explosives; blasting agents
1	1.6	Extremely insensitive detonating substances
2	2.1	Flammable gas
2	2.2	Non-flammable compressed gas
2	2.3	Poisonous gas
3	—	Flammable and combustible liquid
4	4.1	Flammable solid
4	4.2	Spontaneously combustible material
4	4.3	Dangerous when wet material
5	5.1	Oxidizer
5	5.2	Organic peroxide
6	6.1	Poisonous materials
6	6.2	Infectious substance (Etiologic agent)
7	—	Radioactive material
8	—	Corrosive material
9	—	Miscellaneous hazardous material
None	—	Other regulated material: ORM-D

Class 1 (§173.50)

This class includes explosive substances—materials that can detonate or are subject to very rapid combustion. Specifically, an *explosive* is defined as any substance or article, including a device, which is designed to function by explosion (i.e., an extremely rapid release of gas and heat) or which, by chemical reaction within itself, is able to function in a similar manner even if not designed to function by explosion, unless the substance or article is otherwise classed.

Class 1 is divided into six divisions: 1.1, 1.2, 1.3, 1.4, 1.5, and 1.6.

> **Division 1.1** consists of explosives that have a mass explosion hazard. A mass explosion is one that affects almost the entire load instantaneously. Division 1.1 includes dynamite, TNT, black powder, and some types of military ammunition.
>
> **Division 1.2** consists of explosives that have a projection hazard but not a mass explosion hazard.
>
> **Division 1.3** consists of explosives that have a fire hazard and either a minor blast hazard or a minor projection hazard or both, but not a mass explosion hazard.
>
> **Division 1.4** consists of explosives that present a minor explosion hazard. The explosive effects are largely confined to the package and no projection of fragments of appreciable size or range is to be expected. An external fire must not

cause virtually instantaneous explosion of almost the entire contents of the package.

Division 1.5 consists of very insensitive explosives. This division is comprised of substances that have a mass explosion hazard but are so insensitive that there is very little probability of initiation or of transition from burning to detonation under normal conditions of transport. [Note: the probability of transition from burning to detonation is greater when large quantities are transported in a vessel.]

Division 1.6 consists of extremely insensitive articles that do not have a mass explosive hazard. This division is comprised of articles that contain only extremely insensitive detonating substances, and which demonstrate a negligible probability of accidental initiation. [Note: the risk from articles of Division 1.6 is limited to the explosion of a single article.]

> ✓ **Important Note:** Unless otherwise provided, no person may offer for transportation or transport an explosive, unless it has been tested, classed, and approved by the Associate Administrator for Hazardous Materials Safety.

Class 2 (§173.115)

Class 2 materials include compressed and liquid gases, in three divisions.

Division 2.1 *Flammable Gases* (examples are propane and hydrogen). A Division 2.1 flammable gas means any material that is a gas at 20°C (68°F) or less and 101.3 Kpa (14.7 psi) of pressure (a material that has a boiling point of 20°C (68°F) or less at 101.3 Kpa (14.7 psi).

Division 2.2 *Nonflammable, Nonpoisonous Compressed Gases.* These include compressed gas, liquefied gas, pressurized cryogenic gas, compressed gas in solution, asphyxiant gas, and oxidizing gas (an example is nitrogen). A Division 2.2 nonflammable, nonpoisonous compressed gas means any material (or mixture) that exerts in the packaging an absolute pressure of 280 Kpa (40.6 psia) or greater at 20°C (68°F), and does not meet the definition of Division 2.1 or 2.3.

Division 2.3 *Poisonous Gases by Inhalation* (an example is chlorine). A material that is a gas at 20°C (68°F) or less and a pressure of 101.3 Kpa (14.7 psi) (a material with a boiling point of 20°C (68°F) or less at 101.3 Kpa (14.7 psi)) and which is known to be so toxic to humans as to pose a hazard to health during transportation, or in the absence of adequate data on human toxicity, is presumed to be toxic to humans because when tested on laboratory animals it has an LC_{50} (Lethal Concentration) value of not more than 5000 ml/m^3.

When the hazard zone is to be determined based on the grouping criteria for Division 2.3, the hazard zones are determined by apply the following criteria:

Hazard Zone	Inhalation Toxicity
A	LC_{50} less than or equal to 200 ppm.
B	LC_{50} greater than 200 ppm and less than or equal to 1000 ppm.
C	LC_{50} greater than 1000 ppm and less than or equal to 3000 ppm.
D	LC_{50} greater than 3000 ppm or less than or equal to 5000 ppm.

Class 3 (§173.120)

This class includes both flammable and combustible liquids.

Flammable liquid. A liquid having a flash point of not more than 141°F, or any material in a liquid phase with a flash point at or above 100°F that is intentionally heated and offered for transportation or transported at or above its flash point in a bulk packaging (an example is gasoline).

Combustible liquid. Any liquid that does not meet the definition of any other hazard class specified, and has a flash point above 141°F and below 200°F (an example is fuel oil).

> ✓ **Important Note:** Certain flammable liquids with a flashpoint of 100°F or more may be reclassified as combustible liquids, and certain combustible liquids in non-bulk packaging are not subject to the Hazardous Materials Regulations.

Class 4 (§173.124)

This class is comprised of three divisions:

Division 4.1 *Flammable Solids* (such as nitrocellulose)—any of three types of materials:

1. Desensitized explosives that when dry are explosives of Class 1 (other than those of compatibility group A), which are wetted with sufficient water, alcohol, or plasticizer to suppress explosive properties; and are specifically authorized.

2. Self-reactive materials are thermally unstable materials that can undergo a strongly exothermic decomposition even without participation of oxygen (air).

3. Readily combustible solids are solids that may cause a fire through friction (such as matches); show a burning rate faster than 2.2 mm per second when tested in accordance with UN Manual of Tests and Criteria; or any metal powders that can be ignited and react over the whole length of a sample in 10 minutes or less, when tested in accordance with UN Manual of Tests and Criteria.

Division 4.2 *Spontaneously Combustible* (such as phosphorus):

1. A *pyrophoric material*: a liquid or solid that, even in small quantities and without an external ignition source, can ignite within five (5) minutes after coming in contact with air when tested according to the UN Manual of Tests and Criteria.

2. A *self-heating material*: a material that, when in contact with air and without an energy supply, is liable to self-heat. A material of this type that exhibits spontaneous ignition or if the temperature of the sample exceeds 200°C during the 24-hour test period when tested in accordance with UN Manual of Tests and Criteria.

Division 4.3 *Dangerous When Wet* (such as sodium and calcium carbide or magnesium)—a material that, by contact with water, is liable to become spontaneously flammable or to give off flammable or toxic gas at a rate greater than 1 liter per kilogram of the material per hour, when tested in accordance with UN Manual of Tests and Criteria.

Class 5 (§173.127 and §§173.128)

This class includes materials that cause or enhance the combustion of other materials. This class has two divisions.

Division 5.1 *Oxidizer* (example: hydrogen peroxide)—a material that may, generally by yielding oxygen, cause or enhance the combustion of other materials.

Division 5.2 *Organic peroxide* (example: benzoyl peroxide)—any organic compound containing oxygen (O) in the bivalent -O-O- structure, and which may be considered a derivative of hydrogen peroxide, where one or more of the hydrogen atoms have been replaced by organic radicals.

Class 6 (§173.132 and §173.134)

Class 6 is comprised of two divisions:

Division 6.1 *Poisonous Materials* are liquids or solids known to be so toxic to humans as to afford a hazard to health during transportation. The skull and crossbones *poison* label is required for materials in Packing Groups I and II (see *Packing Groups* in the glossary). A *Harmful, Keep Away From Foodstuffs* label is used for Packing Group III materials.

Division 6.2 *Infectious Substances* are materials that may cause disease in humans or animals (for example, medical waste).

Class 7 (§173.403)

This class includes radioactive materials that can cause burns and other injuries. Examples are radioisotopes such as uranium and cobalt.

Class 8 (§173.136)

This class consists of corrosive liquids and solids that cause visible full thickness destruction or irreversible alterations in human skin tissue at the site of contact, and liquids that have a severe corrosion rate on steel or aluminum. Examples are sulfuric acid, nitric acid, and sodium hydroxide.

Class 9 (§173.140)

This class includes *miscellaneous hazardous materials*. Specifically, miscellaneous hazardous material means a material that presents a hazard during transportation but that does not meet the definition of any other hazard class. This class includes:

- Any material that has an anesthetic, noxious or other similar property that could cause extreme annoyance or discomfort to a flight crew member so as to prevent the correct performance of assigned duties; or

- Any material that meets the definition for an elevated temperature material, a hazardous substance, a hazardous waste, or marine pollutant.

Other Regulated Materials (ORM)—(§173.144)

ORM-D materials are materials such as any consumer commodity, which, although otherwise subject to the regulations of this subchapter, present a limited hazard during transportation due to their form, quantity, or packaging.

Hazardous Materials Table

The transportation of almost 3000 hazardous materials is regulated by U.S. DOT. These materials are tabulated in the *Hazardous Materials Table* (§172.101), which can be found in Appendix A. For each listed material, the table identifies the hazard class or specifies that the material is forbidden to transport. The Hazardous Materials Table designates the materials as hazardous materials for the purpose of transportation. For each listed material, the table identifies the hazard class or specifies that the material is forbidden in transportation, and gives the proper shipping name or directs the user to the preferred proper shipping name. In addition, the table specifies or references requirements pertaining to labeling, packaging, quantity limits aboard aircraft, and stowage of hazardous materials aboard vessels. The table is organized into 10 columns: symbols, proper shipping names, hazard class or division, identification numbers, packing groups, label codes, special provisions, packaging, quantity limitations, and vessel stowage.

Classification of Hazardous Material

Column 1: Symbols

Column 1 establishes the Proper Shipping Name, Hazard Class, and Packing Group. Column 1 of the table contains six symbols ("+", "A", "D", "I", and "W"), as follows:

- **+** Fixes the proper shipping name, hazard class, and packing group for that entry without regard to whether the material meets the definition of that class or packing group or meets any other hazard class definition. The Associate Administrator for Hazardous Materials Safety may authorize an appropriate alternate proper shipping name and hazard class.

- **A** Indicates the material is regulated as a Hazardous Material only when being transported by air unless it is regulated as a Hazardous Substance or Hazardous Waste, in which case it is regulated by all modes, including air. For example:

Symbol	Hazardous Materials Shipping Name	Hazard Class or Division	Identification Number	Packing Group
A	Calcium Oxide	8	UN1910	III

- **D** Identifies proper shipping names appropriate for describing materials for domestic transportation, but may be inappropriate for international transportation under the provisions of international regulations (e.g. IMO, ICAO). An alternate proper shipping name may be selected when either domestic or international transportation is involved.

- **G** Identifies proper shipping names for which one or more technical names of the hazardous material must be entered in parentheses, in association with the basic description.

- **I** Identifies proper shipping names appropriate for describing materials in international transportation. An alternate proper shipping name may be selected when only domestic transportation is involved.

- **W** Restricts the application of requirements of this subchapter to materials offered or intended for transportation by vessel, unless the material is a hazardous substance or a hazardous waste.

Column 2: Hazardous Materials Descriptions and Proper Shipping Names

Column 2 lists the hazardous materials descriptions and *proper shipping names* of materials designated as hazardous materials. Modification of a proper shipping name may otherwise be required or authorized by this section. Proper shipping names are limited to those shown in Roman type (not italics).

> ✓ **Important Note:** Every hazardous material contained in column 2 of the table has a proper shipping name.

Generic or n.o.s. descriptions. If an appropriate technical name is not shown in the table, selection of a proper shipping name shall be made from the generic or n.o.s. descriptions corresponding to the specific hazard class, packing group, hazard zone, or subsidiary hazard (if any) for the material. The name that most appropriately describes the material shall be used; e.g., an alcohol not listed by its technical name in the table shall be described as "Alcohol, n.o.s." rather than "Flammable liquid, n.o.s." Some mixtures may be more appropriately described according to their application, such as "Coating solution" or "Extracts, flavoring, liquid," rather than by an n.o.s. entry, such as "Flammable liquid, n.o.s." Note, however, that an n.o.s. description as a proper shipping name may not provide sufficient information for shipping papers and package markings. Under the provisions of subparts C and D of this part, the technical name of one or more constituents that make the product a hazardous material may be required in association with the proper shipping name.

> ✓ **Important Note:** The Hazardous Materials Table is arranged in alphabetical order of proper shipping names without regard to prefixes and numbers. For example, for the substance 2.3,4,6-Tetranitrophenyl methyl nitramine, we look under "Tetranitrophenyl methyl nitramine," not "2, 3, 4, 6-" in the Hazardous Materials Table.

The alphabetization of proper shipping names also does not take into account prefixes of chemical names. For example, for n-Propanol, we would look under the P entries for propanol and ignore the prefix n-. Some common single letters you may see on chemical names and their meanings are as follows:

    a - alpha   N - omega
    b - beta   n - sec
    m - meta   o - tert

Column 3: Hazard Class or Division

Column 3 contains a designation of the hazard class or division corresponding to each proper shipping name, or the word "Forbidden."

1. A material for which the entry in this column is "Forbidden" may not be offered for transportation or transported. This prohibition does not apply if the material is diluted, stabilized, or incorporated in a device, and classed in accordance with the definitions of hazardous materials contained in part 173 of this subchapter.

2. When a reevaluation of test data or new data indicates a need to modify the "Forbidden" designation or the hazard class or packing group specified for a

material specifically identified in the table, this data should be submitted to the Associate Administrator for Hazardous Materials Safety.

3. A complete list of the different classes, divisions, and definition sections is contained in §173.2. Table 3.1 lists these classes.

Column 4: Identification Numbers

Column 4 lists the identification number assigned to each proper shipping name. Those proceeded by the letters "UN" are associated with proper shipping names considered appropriate for international transportation as well as domestic transportation. The prefix "NA" indicates proper shipping names not recognized for international transportation, except in certain cases for shipments to and from Canada. Identification numbers in the "NA9000" series are associated with proper shipping names not appropriately covered by international hazardous materials (dangerous goods) transportation standards, or not appropriately addressed by the international transportation standards for emergency response information purposes, except for transportation between the United States and Canada.

Column 5: Packing Group

Column 5 contains the packing group assigned to the hazardous material. Class 2, Class 7, Division 6.2 (other than regulated medical wastes), and ORM-D materials, do not have packing groups. Packing groups I, II, and III indicate that the degree of danger presented by the material is either great (most regulated), medium (moderately regulated), or minor (least regulated), respectively. If more than one packing group is indicated for an entry, the packing group for the hazardous material is determined using the criteria for assignment of packing groups specified in subpart D of part 173.

Column 6: Labels

Column 6 contains the primary and subsidiary hazard (risk) labels that must be affixed to each package or overpack. For example, according to the Hazardous Materials Table column six entries for barium nitrate, both labels "OXIDIZER" and "POISON" must be affixed to each barium nitrate package or overpack.

✓ **Important Note:** When more than one label is affixed to a package containing hazardous materials, the first label represents the *Primary Hazard*, and the second and subsequent label(s) represent the *Subsidiary Hazard(s)*. In addition, section 172.402 may require that labels be affixed to the package or overpack **in addition** to those contained in column 6.

Column 7: Special Provisions

Column 7 contains the codes for the "Special Provisions," if any, which could apply to the hazardous materials covered by the entry. When column 7 refers to a special provision for a hazardous material, the meaning and requirements of that special provision are as set forth in §172.102. The codes generally apply as follows:

Alpha Numeric Codes

> **A** - Air transportation only
> **N** - Non Bulk Packaging only
> **W** - Water transportation only
> **B** - Bulk Packaging
> **R** - Rail only
> **H** - Highway only
> **T** - IM portable tanks only

Numerical Codes

Apply to all modes, bulk and non-bulk packaging, and IM Portable Tanks, if applicable.

Column 8: Packaging Authorizations

Columns 8A, 8B and 8C specify the applicable sections for exceptions, non-bulk packaging requirements, and bulk packaging requirements, respectively. Columns 8A, 8B, and 8C are completed in a manner that indicates that "§173." precedes the designated numerical entry. For example:

Column 2 Hazardous Materials Descriptions and Proper Shipping Names	Column 8 Packaging Authorizations		
	8A Exceptions	8B Non-Bulk Packaging	8C Bulk Packaging
Chlorine	None	304	314, 315
Calcium Nitrate	152	213	240

The word "None" as illustrated for "Chlorine" means no exceptions are permitted except as may be authorized under the special provisions in column 7 of the section 172.101 table. The entry "152" in column 8A for the calcium nitrate entry in the 172.101 table indicates that exceptions authorized for this chemical will be found in section 173.152.

Classification of Hazardous Material

Column 8B contains the section within part 173 containing the non-bulk packagings authorized to be used. The entry "304" in column 8B for the chlorine entry indicates that the non-bulk packaging authorized for this chemical will be found in section 173.202.

Column 8C contains the section within part 173 containing bulk packagings authorized to be used. The entries "314, 315" in column 8C for the chlorine entry indicates that the bulk packaging authorized for this chemical will be found in sections 173.314 and 173.315.

Column 9: Quantity Limitations

Column 9 is organized into two columns, containing the maximum quantity that may be offered **per package** for transportation aboard "Passenger aircraft or rail cars" (column 9A) or "Cargo Aircraft" (column 9B). Note that hazardous materials are not normally transported aboard passenger-carrying rail cars, so column 9A is used mostly with the transportation of hazardous materials aboard passenger aircraft.

Except where otherwise indicated, the quantities are "net," not "gross."

The word "Forbidden" placed in column 9A means that the hazardous material covered by the entry may not be offered for transportation aboard passenger-carrying aircraft or railcars. "Forbidden" in column 9B indicates that the hazardous material may not be offered for transportation aboard passenger aircraft or railcars **or** cargo aircraft.

For example:

Acetone - In column 9A we would find the entry 5 L, indicating that we are permitted up to 5 L of Acetone per package for transportation aboard "Passenger aircraft or rail car." Column 9B for the **Acetone** entry indicates that up to 60 L of acetone may be offered per package for transportation aboard cargo aircraft.

Column 10: Vessel Stowage Requirements

Column 10A [Vessel stowage] specifies the authorized locations on board cargo and passenger vessels. Column 10B [Other provisions] specifies codes for stowage requirements for specific hazardous materials. These codes are defined as follows:

> Stowage category "A" —the material may be stowed "on deck" or "under deck" on a cargo vessel and a passenger vessel.
>
> Stowage category "B" —the material may be stowed "on deck" or "under deck" on a cargo vessel, but must be stowed "on deck" on a passenger vessel.
>
> Stowage category "C" —the material must be stowed "on deck" on a cargo vessel and on a passenger vessel.

Stowage category "D" —the material must be stowed "on deck" on a cargo vessel, but is prohibited on a passenger vessel.

Stowage category "E" —the material may be stowed "on deck" or "under deck" on a cargo vessel, but is prohibited on a passenger vessel.

Summary

Knowing the classifications for hazardous materials and understanding how the paperwork involved in transporting hazardous materials functions to track the materials "cradle-to-grave" is critical to safely handling and carrying these materials. Being familiar with hazmat terminology and how to use the Hazardous Materials Table helps lessen the threat posed by hazardous materials to the general public and employees. Also critical to safe transportation of hazardous materials is mastering the basic skills of safety. These begin with general awareness training—the subject of Chapter 4.

Notes

[1] 49 CFR §171.8

[2] From Meyer, E. 1989. *The Chemistry of Hazardous Materials*. 2nd ed. Englewood Cliffs, NJ: Prentice Hall.

Chapter 4
General Awareness Training

The flagship regulation, Hazardous Materials Transportation Uniform Safety Act (HMTUSA) of 1990 (Public Law 101-615) established standards for training by "hazmat employers" of all "hazmat employees."

Introduction

HMTUSA required the Secretary of Transportation to establish standards for training all hazmat employees by hazmat employers. This key act virtually changed the face (and requirements) of transportation training as it was known before 1990. HM-126F, an amendment to the Hazardous Materials Regulations, sets training requirements for individuals involved in all modes of transportation (over-the-road, rail, aircraft, or vessel) of hazardous materials. This amendment ensures that hazmat employers train their hazmat employees in safe work practices in the following areas:

- loading and unloading
- handling
- storing
- transporting
- emergency preparedness to accidents involving hazardous materials

The implementation of HM-126F added two key definitions to the regulation: hazmat employer and hazmat employee.

Hazmat Employer

- A company/person who utilizes one or more employees to transport or cause to transport hazardous materials in commerce, or

- One who represents, marks, certifies, sells, offers, reconditions, tests, repairs, or modifies containers, drums, or packaging for use in transporting hazardous materials. This includes:

- owners and operators of vehicles that transport hazardous materials

- any department or agency of the U.S.

- a state or political subdivision of a state

- an Indian tribe that deals with hazardous material as a form of business

Hazmat Employee

- One who directly effects the safe transportation of hazardous materials, either as a self-employed person or one who performs duties relating to hazardous materials as part of the job. This includes the owner/operator of a motor vehicle that transports hazardous materials in commerce.

Based on the definitions of hazmat employer and employee, workers who must be trained include those who:

- load or unload hazardous materials

- test, recondition, repair, modify, or mark containers, drums, or packaging used in transporting hazardous materials

- prepare hazardous materials for transporting

- are responsible for the safe transporting of hazardous materials (e.g., supervisors), or

- operate a vehicle transporting hazardous materials.

HM-126F established four categories of training: (1) general awareness, (2) safety, (3) function-specific, and (4) driver. The first three categories are required for all modes of transportation, while the fourth obviously applies to highway transportation. Chapters 5 and 6 cover safety training, Chapter 7 covers function-specific training, and Chapter 8 covers driver training; this chapter covers general training requirements and provides a list of associated terminology by introducing the OSHA requirements under its Hazard Communication Standard (Hazcom) (29 CFR 1910.1200).

As far as emergency response training is concerned, the information provided in this text is not intended to qualify the employee to perform significant emergency response and cleanup. Persons whose primary responsibilities include emergency response are required to receive the Hazmat Emergency Response OSHA training (i.e., Hazmat Levels I, II, or III training). The OSHA DOT required training provided in this text covers elemental emergency response, such as notifying appropriate authorities, or otherwise controlling the situation.

OSHA's Hazard Communication Standard

Though the Hazard Communication Standard (29 CFR 1910.1200), issued in 1985, is long and some parts are very technical, the basic concepts are simple. In fact, the main concept could not be more simple—employees have both a right and a need to know the hazards and identities of the hazardous chemicals to which they are exposed in the workplace. Many employers have been complying with the requirements of Hazcom as a matter of normal operations for several years.

Hazardous materials are common in most workplaces, and to a degree in most homes. In fact, most of us work every day with materials that are potentially hazardous in some way or another, but that are so common that we don't even think of them as posing any danger. The term "hazardous" is of course relative; for this text we define hazardous materials as those substances that (during normal usage or otherwise) have the capability of inflicting harm. These substances are characterized as toxic, corrosive, flammable, reactive, irritants, or strong sensitizers, and these characteristics mean they pose a threat to health and the environment. Typically, safety professionals and others in the profession commonly recognize hazardous materials as materials that warrant special care, caution, and control methods.

Indeed, safety professionals and others in the profession are well aware of the potentially destructive, life-threatening nature of many hazardous materials. Their job is to know about hazardous materials and the repercussions incident to their misuse and mishandling, and to recognize their hazardous and physical properties, incompatibilities, symptoms, emergency treatment, containment, storage and transportation requirements, and control methods. Hazmat employees may or may not know what they should know about hazardous materials and their inherent dangers. The bottom line is, under OSHA/DOT regulations, the hazmat employee has a **right to know** about hazardous materials. In fact, hazmat employees have the right to know about *anything* that might cause them harm in the workplace.

There are several important reasons why workers have a right to know about hazardous materials and the dangers involved with producing, handling, storing, and/or transporting them. We could say that all hazmat employers have a moral or ethical obligation to protect their employees from harm. If we accept this tenet as true, we would expect hazmat employers to ensure that their hazmat employees are properly informed about hazardous materials and their inherent dangers. However, there are still many hazmat employers that believe they don't have any moral or ethical obligation for the protection of workers on the job. Only fear of legal penalties forces these hazmat employers to inform their hazmat employees about hazardous materials and safety-related items associated with their production, handling, storage, and transportation. This is certainly a view held by many hazmat and other employers today—especially in this age of "I'm gonna take you to court for everything you've got." Being found at fault for directly or indirectly causing injury to employees from any job-related casual factor can be quite costly in legal fees and judgements lost.

Other financial repercussions are at work in the issue of not properly protecting workers on the job. Beyond legal fees, medical costs and worker's lost time can be huge financial burdens. The intangible costs of workers injured on the job can affect organizational morale and the overall effectiveness and efficiency of everyone in the workplace. Insurance costs also rise. When an insurance company reviews a company's history of on-the-job injuries and finds that the incidence and severity rates are high, you can count on insurance rates that reflect these high costs, if you can find an insurance company that will take the chance on insuring your organization.

By complying with OSHA's Hazard Communication Standard, employers avoid increased costs by fulfilling an ethical obligation to protect employees. The Hazcom Standard applies to chemical manufacturers and importers, and was issued to ensure that the hazards of all chemicals produced or imported are evaluated for toxicity, and that information concerning their hazards is transmitted to affected customers and employees. By 1986, all employers were required to ensure that their employees handling chemical substances were properly informed of associated hazards. With time, Hazcom became known as the workers' "Right To Know" law.

Hazmat Employers are responsible for their workers. Whether the employer is a chemical manufacturer, a distributor of hazardous materials or products containing hazardous materials, or simply an employer using such products or operating a process that produces such products (hazardous off gases like hydrogen sulfide and/or others, for example), that employer is responsible for both the safety and health of its employees, and for compliance with Hazcom.

> ✓ **Important Note:** In this text, when we use the term "employer," we are referring to DOT's definition of a hazmat employer.

Purpose of the Hazcom Standard

The purpose of the Hazard Communication Standard is to ensure that the hazards of all chemicals produced or imported are evaluated, and that information concerning their hazards is transmitted to employers and employees. This transmittal of information is to be accomplished by means of comprehensive hazard communication programs, which are intended to include container labeling and other forms of warning, material safety data sheets and employee training.[1]

OSHA makes the point that Hazcom is intended to comprehensively address the issue of evaluating the potential hazards of chemicals and communicating information concerning hazards and appropriate protective measures to employees. This comprehensive coverage is a positive. Let's take a closer at what is required.

Requirements of the Hazcom Standard

The standard requires that employees be provided with information and training on hazardous chemicals in the workplace at the time of initial assignment, and whenever new chemicals are introduced. Workers must be informed initially of:

- Requirements for the standard;
- Any operation in their work area containing hazardous chemicals;
- Location and availability of the company's written hazard communication program;
- Methods for detecting the presence or release of hazardous chemicals in the workplace;
- Physical and health hazards of workplace chemicals;
- Appropriate measures to be used by the employees to protect themselves from exposure to chemicals;
- Details of the employer's program developed for labeling and material safety data sheets (MSDSs).

Hazcom requires employers to become aware of the chemical hazards in their workplaces and to relay that information to their employees. Contractors conducting work at the facility of a client must provide chemical information to the client regarding the chemicals brought onto the work site. One of the key requirements under Hazcom is that any information gathered by the employer or his/her representative must be transmitted to employees. To do this, employers obviously must have the information they need. Under Hazcom, the chemical manufacturers must make this information available to the employer for further dissemination to employees.

The standard specifically requires the employer to:

- Develop an inventory or list of hazardous chemicals that employees may be exposed to in the workplace.
- Have material safety data sheets for each hazardous chemical on the inventory and make them available to all employees.
- Develop a written program describing the employers' compliance plan wherein the person in responsible charge of its implementation is clearly identified.
- Provide training to all employees on chemical hazards, safety, and employer and employee rights and responsibilities.
- Ensure that all hazardous substances are labeled correctly.

The Hazard Communication Standard is a *performance standard*, which leaves the specifics on labeling, training, material safety data sheets, and related programs to the employer to develop. OSHA's requirements are federally mandated. However, keep in mind that many state level programs are also in place, and the essence of OSHA's Hazcom requirements may differ slightly (but in no way significantly or to a lesser degree) from the mandated federal requirements. Check with your state OSHA (if you are from one of the more than 20 states that has one) to ensure that the written program and requirements that you are putting together meet both federal and state requirements. Many OSHA and other regulatory standards, regulations, and rules are full of vagaries and ambiguities. Fortunately, the Hazcom standard is not one of these.

In a nutshell, Hazcom requires chemical manufacturers or importers to assess the hazards of chemicals that they produce or import, and employers are to provide information to their employees about the hazardous chemicals to which they are exposed, by means of a Hazard Communication Program, labels and other forms of warning, material safety data sheets, and information and training.

Hazard Determination

Chemical manufacturers and importers shall evaluate chemicals produced in their workplaces or imported by them to determine if they are hazardous. Employers are not required to evaluate chemicals unless they choose not to rely on the evaluation performed by the chemical manufacturer or importer for the chemical to satisfy this requirement.[2]

How do employers determine if their places of employment use or produce hazardous materials? OSHA requires the employer to accomplish this by performing a *hazard determination*. The employer must determine if and when hazardous materials are used, produced, transported, or received in the workplace. In our experience, the only way to do this that really makes sense is to literally go take a look—everywhere. In many cases, this is an accurate description of the chemical hazard determination process in many workplaces. You might be surprised at the number and amount of hazardous chemical products that can accumulate in the average workplace over time. In closets, on shelves, in storage rooms—you may find containers of all types of chemical products that present one or more hazardous properties, from cans of pesticides to cleaning agents.

If the employer orders hazardous materials for use in the workplace, he or she can naturally assume (because of Hazcom requirements) that the manufacturer, distributor, or supplier will include a material safety data sheet or sheets with the chemical purchase. This is a given, to a point. Remember, although OSHA requires that the MSDS be forwarded to whomever the shipment is sent to, this does not mean that it will actually arrive there. Routinely, hazardous materials are shipped to the user (the employer) without an MSDS for a simple reason: mistakes happen. What is important is that you **must** be aware that the MSDS for the hazardous material shipped to your place of employment must be included. If not, you should not accept shipment. This takes care of new shipments. However, it does not cover materials already on site.

The best way to approach this aspect of hazardous materials in the workplace is to completely survey the facility. Depending on the size of the facility, this can be accomplished by one knowledgeable person or by a team or teams of knowledgeable employees. "Knowledgeable employees" means that the person or persons conducting the hazardous materials survey of the workplace must know what they are looking at and for. Most hazardous materials are labeled. Being able to read a label is one requirement, but the expertise needed to understand the label is harder to find. Some labels have many chemical terms, and whoever attempts to decipher whether or not the product is hazardous would therefore need some knowledge of chemistry. Other labels include long lists of active ingredients. Which ones are hazardous and which ones are not? The person conducting the survey must be able to determine this. Some chemical names can seem very confusing—ethylenediamine or crotonaldehyde or aziridine, 2-methyl. Are these hazardous materials? Are they listed hazardous materials? The "knowledgeable employee" needs to be able to answer questions like these. But how? Must we hire a full-time chemist in every workplace to manage the chemicals?

Actually, all that is required is to ensure that someone who is conscientious, thorough, and accurate is assigned to conduct the hazard determination survey. But the survey person can't perform such an important and required function unarmed. He or she needs to have the ability to contact the various chemical manufacturers to obtain the pertinent data needed, and a small library of helpful resource materials for use in making chemical product identifications is a useful tool. For example, CFR 1910, Subpart 2; *Toxic and Hazardous Substances*, OSHA is an excellent reference. The ACGIH's *Threshold Limit Value for Chemical and Physical Agents in the Work Environment* is another excellent reference, as are the National Toxicology Program's *Annual Report on Carcinogens* and the International Agency for Research on Cancer's *Monographs*.

However, keep in mind that in most cases you shouldn't have to do any research at all, or have to answer questions related above for new shipments. The shipper will have ensured that these chemicals were shipped with an MSDS and proper shipping papers that take all the guesswork out of the process. Only when an MSDS is not shipped with the product or it becomes lost will research need to be done. The individual making the survey must be able to determine exactly what he or she is working with during the inventory.

Before the Hazard Communication Standard was implemented, many chemical products were labeled in accordance with the routine procedures used by whatever chemical manufacturer was involved. Old chemicals or chemical products within your workplace may be labeled in ways that make determining exactly what the chemical is or what hazards it presents difficult.

If you do find improperly labeled chemicals or chemical products (by today's standard), after you make a determination of what the hazards are, you should properly dispose of such products. The key words here are "properly dispose." Proper disposal cannot occur without identification.

Once the chemical is identified, the next questions are whether it is hazardous and what kind of hazard does it present? Is the chemical or product a fire hazard or is it a health hazard? Will the product react if exposed to air, water, or some other substance? Is the chemical corrosive, an oxidizer, a carcinogen, or a water reactive product? These are some of the characteristics that the survey personnel must be able to determine.

Labels do not always last forever. What happens when the hazard determination survey person finds a product that is not labeled? Under Hazcom and DOT shipping requirements, this is an illegal, unacceptable situation. **All hazardous materials must be labeled.** While hazmats must be labeled, anyone who has worked in the real world knows that labels fall off and disappear, are destroyed, or become illegible. When this occurs, the employer still has a duty and obligation under Hazcom to determine what the material or substance is, and what hazards it presents. How is this accomplished? The MSDS is useless unless the product is labeled.

When a container is not labeled or the label is not legible, the employer must have the material tested to determine what it is, or at least what hazards the mixture presents. To do this, the material will need to be tested by an environmental laboratory or other facility or agency with the capacity and ability to perform such tests. OSHA is quite specific as to procedures when the employer tests, evaluates, or has another qualified party (a laboratory, for example) test or evaluate unknown substances. Such an evaluation (the actual procedures used) must be in writing, so keep **very good** records. These records should be kept in the driver's personnel record.

 Important Note: Materials found unlabeled in the workplace should be handled with care. The substance might be dangerous, and until proven otherwise, always assume that it is indeed a hazardous material.

Chemical Inventory List

Surveying, locating, and identifying hazardous chemicals and materials in the workplace is only one important part of making a hazard determination survey of the workplace. Once you've completed this task, ensure that the journey will never be as arduous and/or confusing again by keeping a written inventory of each hazardous chemical found and identified.

When the site hazard determination is completed and all hazardous materials are identified, labeled properly, or disposed of in accordance with applicable regulations, the written inventory list **must** become a part of the organization's Hazcom written program, and available to personnel 24 hours a day. This inventory list of workplace hazardous chemicals, beyond its primary purpose, is a critical element of the organization's Hazcom for two other reasons: (1) it is required by OSHA; and (2) it provides an excellent cross-reference to ensure that each listed hazardous chemical has an accompanying MSDS.

Hazard Communication and Training

Employers shall provide employees with effective information and training on hazardous chemicals in their work area at the time of their initial assignment, and whenever a new physical or health hazard the employees have not previously been trained about is introduced into their work area. Information and training may be designed to cover categories of hazards (e.g., flammability, carcinogenicity) or specific chemicals. Chemical-specific information must always be available through labels and material safety data sheets.[3]

In putting together your site-specific hazard communication training program, you'll need to take a number of steps to meet the training/information requirements of the standard. Some standard guidelines apply to any type of training where the goal is to increase knowledge. Note that since a core of health and safety knowledge is essential to the safe performance of any job, the importance of safety and health training cannot be overemphasized.

In accomplishing your training effort, remember that knowledge is best gained when:

- Learning objectives are made clear to the employee
- The facts to be learned are broken down into a logical sequence
- Examples are drawn from the worker's own experience
- The worker has the opportunity to participate or respond
- The trainee is exposed to the material in a variety of formats and teaching methods
- The worker receives feedback on his or her progress by means of quizzes or tests, and enough repetition is built in to provide "imprintation" and thus to delay the process of forgetting

Let's look at a list of important Hazcom-related elements that may help you put together the type of training program you will need to satisfy regulatory and legal requirements.

- Employees must be informed initially of:
  - Requirements of the Hazard Communication Standard
  - Any location in the work area containing hazardous chemicals
  - Location and availability of written Hazard Communication Program

- Employee training must include initially at least:
  - Methods for detecting the presence or release of hazardous chemicals in the work area
  - Physical and health hazards of the workplace chemicals
  - Appropriate measures to be used by employees to protect themselves from exposure to chemicals
  - Details of the Hazard Communication Program developed by the employer, especially the labeling system and MSDS.

More specifically, the training program you put together to inform and train your employees should:

- **Explain the Hazardous Chemical Inventory List**
- **Specify location and availability of MSDS**
- **Describe the workplace labeling system**
- **Explain hazard-warning terms**
- **Focus on the types of chemicals used or likely to be used in the workplace, and the hazards they involve**
- **Provide training any time a new hazardous material is introduced into the workplace**

  In addition, any time a new process, work procedure, policy, administrative procedure, or new information source (e.g., new MSDS) is introduced in the company, workers must be trained or re-trained.

- **Provide means of assessing the effectiveness of the existing training program**

  This involves a performance evaluation that is usually accomplished by giving an examination or short quiz on the material presented to employees. While testing the trainees' knowledge level (what they have learned from the presentation) is always a good idea, checking the information presented to them against how they can and will apply it in real-world situations is very important. Having the worker take a workplace walk-around with the instructor is a very effective method for such an evaluation. For example, after participating in Hazcom training, employees should be taken out into the workplace and directed by the trainer to explain the significance of Hazcom requirements in the workplace. During the walk-around, the trainee should be required to point out hazard warning labels, where the MSDSs are kept, and where the company's written Hazard Communication Program is kept. In addition, throughout the walk-around, or when a hazardous material or process using or producing hazardous materials is encountered, the trainee should be required to explain the hazards involved with such materials. This is an important requirement, because if OSHA audits your site, OSHA will interview one or more workers. During this interview process, the auditor will always discuss and ask questions concerning hazard communication.

- **Provide documentation**

    The most important thing about any employee training is that the training must be documented. You can go out and hire the best, most informed, most influential trainer there is, he or she can do a fantastically professional job of training your personnel on OSHA required training, and the trainer can bring your employees to 100 percent knowledgeable about the safe work practices, but this training is absolutely worthless unless you have proof (documentation) that the training was conducted. You can do all the training you want to, but if it is not documented, in the eyes of OSHA and the legal system, it simply was not done. Training that is not documented did not occur. This certainly is the view the regulators and a court of law will take. Use a form similar to the one shown below to document your Hazcom training.

ATTENDANCE ROSTER

HAZARD COMMUNICATION TRAINING

Date:_____   Instructor:_____

In accordance with the recordkeeping and training requirements of OSHA's 29 CFR 1910.1200, I have received training and knowledge on hazard communication, including labeling requirements, MSDS, and the mandatory training requirements. I have agreed to verify my understanding and training under 1910.1200 by signing and dating this form.

Signature   Date   Workcenter
_____   _____   _____

- **Stay current with any and all changes to the Federal or State Hazard Communication Standard; ensure that employees are informed about such changes**

The organizational training person responsible for training hazmat employees according to DOT's guidelines must determine who must be trained at each of the general awareness, function specific, and safety training levels, and must also determine the required scope of training at each level. The minimum training frequency must also be determined, as well as how to document the training.

> ✓ **Important Note:** Training employees is a standard requirement under OSHA, U.S. EPA, DOT, and other regulatory agencies. The organization responsible for such training must be attentive, especially when it comes to proper documentation of the training. If training is not properly documented, in the eyes of the concerned regulators, the training **never** took place.

As obvious and critical as properly documenting training, the training content is also critically important. Each hazmat employee must know, for example, what a hazardous material is. More specifically, they must know:

- How to determine which materials are classified by DOT as hazardous
- How to distinguish between hazardous materials, hazardous wastes, hazardous chemicals, and hazardous substances
- Hazard class definitions
- How to use the Hazardous Materials Table
- Subsidiary hazard classes and the precedence of the Hazard Table
- How to determine which packing group applies to your hazardous materials: Packing Group I, II, or III
- How to safely manage and handle containers
- Performance-oriented containers
- Non-bulk container codes
- Container performance tests
- Requirements for container reuse
- How to select the appropriate primary and subsidiary labels
- Bulk and non-bulk marking requirements
- Requirements for bills of lading (shipping papers) and hazardous waste manifests
- Step-by-step procedures for completing shipping papers
- How to determine the correct proper shipping name
- Special shipping paper entries for toxics, hazardous substances, and limited quantities
- Practical exercises on selecting shipping names, packaging, marking, labeling, and placarding shipments of hazardous materials
- Separation and segregation requirements

- Safe handling procedures for working with hazardous materials
- How to avoid accidents
- How to select the appropriate placard and when to use the dangerous placard
- Emergency response procedures and spill notification requirements
- How to use emergency response information in the North American Emergency Response Guidebook, shipping papers, MSDSs, and other sources

Hazardous materials certification training for DOT 49 CFR transportation employees (hazmat employees) entails several individual elements that work together comprehensively.

For example, along with training on the laws and regulations applicable to hazardous materials, in all recent dockets including HM-181 and HM-215C, hazmat employees must be informed about newly expanded registration requirements.

Material Safety Data Sheets (MSDS)

Chemical manufacturers and importers shall obtain or develop a material safety data sheet for each hazardous chemical they produce or import. Employers shall have a material safety data sheet in the workplace for each hazardous chemical which they use.[4]

Along with employee training and labeling, material safety data sheets are a vital part of hazard communication. The MSDS is the cornerstone of the regulation—the basic tool that organizations and their employees use as a guide to safe practices and emergency response. The MSDS provides a protocol, answer sheet, and reference map that allows you to reliably predict a possible future. The MSDS lets you imagine what might occur if you were to have a chemical spill, for example. Under Hazcom, you may hope a spill will never occur, but you plan for it in case it does.

We all know that emergencies occur. Chemical emergencies probably occur more than anyone other than safety professionals would suspect. The MSDS allows you and your emergency response team to plan and rehearse (via practice dry runs) exactly how to respond, what equipment is required, and what steps to take to protect yourself, your team members, the public, and the environment. The MSDS does all this for you, and much more. Employers are required to maintain copies of all MSDSs that are received with incoming shipments of the sealed containers of hazardous chemicals. Employers must also ensure that if for any reason MSDSs are not received from the shipper, their program has some type of tracking mechanism in place to ensure that MSDSs are obtained. Employers must ensure that employees handling sealed containers of hazardous chemicals are provided with easy access, 24-hours per day, to all MSDSs. Employees must also be trained to the extent necessary to protect themselves in the case of a spill or leak of a hazardous chemical from a sealed container.

MSDS: An Answer Sheet

The MSDS for any hazardous material is really an answer sheet. It answers questions about the chemical you work with. For example, consider the following questions:

- What product/chemical is this MSDS for?
- What will happen to me if this material:

  Gets into my eyes?

  Gets onto my skin?

  Is swallowed?

  Is breathed in?
- What first aid steps do I follow?
- How much of this material can I safely be exposed to?
- Are there conditions/materials this should not come into contact with?
- If it catches fire, what should I use to put it out?
- What type of protective clothing and equipment should I use?
- What is the proper way to safely handle this chemical?
- What should I do if there is a leak?
- What should I do if there is a spill?
- How do I clean up a spill of this chemical?
- If there is a spill, what must I do?
- How do I recognize this material in the workplace?
- What toxicology information is available for this material?

A complete MSDS answers all of these questions. Each question refers to a particular section of the MSDS. In providing training to workers and others, breaking an MSDS down into questions that reflect each section of the MSDS can aid in the training effort and in ensuring employee understanding.

The MSDS Format

All MSDSs must be in English, but anyone who uses them soon learns that different suppliers' MSDSs are different. OSHA requires that each MSDS provide at least the following standardized information for each hazardous material, but does not set a standard form to follow.

- The material's identity, including its chemical name and common name (for example, product name: *Helium, compressed*; chemical name: *Helium*; synonyms: *Helium gas, Gaseous helium, Balloon gas*)

General Awareness Training

- Hazardous ingredients (even in parts as small as 1 percent)
- Cancer-causing ingredients (even in parts as small as 0.1 percent)
- List of physical and chemical hazards and characteristics (flammable, explosive, corrosive, etc.)
- List of health hazards, including:

    Acute effects such as burns or unconsciousness, which occur immediately

    Chronic effects such as allergic sensitization, skin problems, or respiratory disease, which build up over a period of time

- Limits to which a worker can be exposed, the primary entry routes into the body, specific organs likely to sustain damage, and medical problems that exposure can worsen
- Precautions and safety equipment
- Emergency and first aid procedures
- Identity of the organization responsible for creating the MSDS and the date of issue.

All this is obviously valuable information. Indeed, the MSDS is a valuable tool that can do much good for those who use it. And this is the key—"for those who use it." Employees must be informed that the MSDS is available, and that they are *required* to use it, read it, and follow its recommendations.

MSDS Contents: An Example

Figure 4.1 presents a blank sample MSDS form. Note that only eight sections are presented on the sample form. These eight sections are required for every MSDS and must be filled in completely—blank spaces are not permitted on the MSDS. If you have used an MSDS before, you have probably noticed that many suppliers provide MSDS with more than the eight sections required by the Hazcom standard. Some MSDSs may include 16 or more sections. As long as the eight required sections are included and properly filled in, suppliers can add as many sections as they deem necessary to provide the information the user will need to protect his or her employees or themselves.

Let's take an MSDS as an example and go through all the sections listed, explaining the significance of each. Since we have already mentioned helium, we will use a standard MSDS for helium to explain each MSDS section. For illustrative purposes, we chose the sample MSDS for helium because it contains 16 separate sections (8 more than normally required). In this day of heavy duty liability suits and follow-up litigation, manufacturers and suppliers (as well as employers) are performing due diligence by providing as much hazard, health, protection, and mitigation information as possible.

Material Safety Data Sheet
May be used to comply with
OSHA's Hazard Communication Standard,
29 CFR 1910.1200. Consult Standard for
specific requirements.

U. S. Department of Labor
Occupational Safety and Health Administration
Non-Mandatory Form
Form Approved
OMB No. 1218-0072

Identity (As Used on Label and List)	Note: Blank spaces are not permitted. If any item is not applicable, or no information is available, the space must be marked to indicate that.

Section I

Manufacturer's name	Emergency Telephone Number
Address (Number, Street, City, State and Zip Code	Telephone Number for Information
	Date Prepared
	Signature of Preparer

Section II Hazardous Ingredients/Identity Information

Hazardous Components (Specify Chemical Identity: Common Name)	OSHA PEL	ACGIH TLV	Other limits Recommended	%(Optional)

Section III — Physical/Chemical Characteristics

Boiling Point		Specific Gravity ($N_2 0 - 1$)	
Vapor Pressure (mm Hg.)		Melting Point	
Vapor Density (Air - 1)		Evaporation Rate (Butyl Acetate - 1)	

Solubility in Water
Appearance and Odor

Section IV — Fire and Explosion Hazard Data

Flash Point (Method Used)			
Extinguishing Media			
Special Fire Fighting Procedure			
	Flammable Limits	LEL	UEL
Unusual Fire and Explosion Hazards			

Figure 4.1. Material Safety Data Sheet (MSDS)

General Awareness Training

Material Safety Data Sheet
May be used to comply with
OSHA's Hazard Communication Standard,
29 CFR 1910.1200. Consult Standard for
specific requirements.

U. S. Department of Labor
Occupational Safety and Health Administration
Non-Mandatory Form
Form Approved
OMB No. 1218-0072

Section V — Reactivity Data				
Stability	Unstable		Conditions to Avoid	
	Stable			
Incompatibility (Materials to Avoid)				
Hazardous Decomposition or Byproducts				
Hazardous Polymerization	May Occur		Conditions to Avoid	
	Will Not Occur			

Section VI — Health Hazard Data			
Route(s) of Entry:	Inhalation?	Skin?	Ingestion?
Health Hazards (Acute and Chronic)			
Carcinogenicity?	NTP?	IARC Monographs?	OSHA Regulated?
Signs and Symptoms of Exposure			
Medical Conditions Generally Aggravated by Exposure			
Emergency and First Aid Procedures			

Section VII — Precautions for Safe Handling and Use
Steps to Be Taken in Case Material is Released or Spilled
Waste Disposal Method
Precautions to Be Taken in Handling and Storing
Other Precautions

Section VIII — Control Measures			
Respiratory Protection (Specify Type)			
Ventilation	Local Exhaust	Special	
	Mechanical (General)	Other	
Protective Gloves		Eye Protection	
Other Protective Clothing or Equipment			
Work/Hygienic Practices			

Figure 4.1. Material Safety Data Sheet (MSDS) (continued)

Section 1: Product Information

Product Name:	Helium, compressed
Chemical Name:	Helium
Chemical Formula:	He
Synonyms:	Helium gas, Gaseous Helium, Balloon gas
Manufacturer:	(name goes in this space)
Product Information:	(manufacturer's phone number here)
MSDS Number:	(manufacturer's MSDS number here)
Revision Date	(October 1996)
Revision:	4
Revision Date:	(October 1998)

Section 2: Composition/Information on Ingredients

Helium is sold as a pure product > 99 percent.

Cas Number:	7440-59-7
ACGIH:	Simple asphyxiant
OSHA:	Not established
NIOSH:	Not established

Normally, in this section, the chemical's individual hazardous chemical compounds and their relative percentage of concentration are listed. If established, each chemical's exposure limits are shown. For example, the phrase "8-hr TWA: 80 ppm or 260 mg/m^3" is a guideline establishing an exposure that should not be exceeded when averaged over an eight-hour workday.

The phrase "Rat, Oral, LD_{100}: 100 mg/kg" means that 100 milligrams of the chemical per each kilogram of body weight is the lethal dose that killed 100 percent of a group of test rats. These data are used to help establish the degree of hazard to humans.

These data for helium have not been established, so they are not included on this MSDS.

By law, suppliers are only required to identify hazardous ingredients. However, some trade name products have secret ingredients termed *trade secrets* that are treated as proprietary information. Suppliers of such products are still required to provide health hazard information on the MSDS, and additional information to safety professionals if they have a documented need to know.

Section 3: Physical/Chemical Characteristics

Appearance:	Colorless gas
Odor:	Odorless
Molecular Weight:	4.00
Boiling Point:	-452.1°F
Specific Gravity (Air =1):	@ 70°F and 1 Atm: 0.138
Specific Volume:	96.71 ft^3/lb
Freezing Point/ Melting Point:	@ 1 Atm: None
Vapor Pressure (At 70°F):	Not applicable
Gas Density:	@ 70°F and 1 Atm: 0.0103 lbs/ft^3
Solubility in Water:	Vol./Vol. @ 32°F: 0.0094

Section 4: Fire and Explosion Hazard Data

Flashpoint:	N/A
Autoignition Temperature:	Nonflammable
Flammable Limits:	Nonflammable
Extinguishing Media:	Helium is nonflammable and does not support combustion. Use extinguishing media appropriate for the surrounding fire.
Hazardous Combustion Products:	None
Special Fire Fighting Instructions:	Helium is a simple asphyxiant. If possible, remove helium cylinders from fire area or cool with water. Self-contained breathing apparatus may be required for rescue workers.
Unusual Fire and Explosion Hazards:	Upon exposure to intense heat or flame, cylinder will vent rapidly and or rupture violently. Most cylinders are designed to vent contents when exposed to elevated temperatures. Pressure in a container can build up due to heat and it may rupture if pressure relief devices should fail.

Section 5: Reactivity Data

Chemical Stability:	Stable
Conditions to Avoid:	None
Incompatibility:	None
Hazardous Decomposition Products:	None
Hazardous Polymerization:	Will not occur

Section 6: Health Hazard Data

Emergency Overview:	Helium is a nontoxic, odorless, colorless, non-flammable gas stored in cylinders at high pressure. It can cause rapid suffocation when concentrations are sufficient to reduce oxygen levels below 19.5 percent. It is lighter than air and may collect in high points or along ceilings. Self-Contained Breathing Apparatus (SCBA) may be required to protect workers.
Potential Health Effects:	None
Inhalation:	Simple asphyxiant. Helium is nontoxic, but may cause suffocation by displacing the oxygen in air. Lack of sufficient oxygen can cause serious injury or death.
Eye Contact:	No adverse effect.
Skin Contact:	No adverse effect.
Carcinogenic Potential:	Helium is not listed as a carcinogen or potential carcinogen by NTP, IARC, or OSHA subpart Z.
Exposure Information:	None
Route of Entry:	Inhalation
Target Organs:	None
Effect:	Asphyxiation (suffocation)
Medical Conditions Aggravated by Over-Exposure:	None

General Awareness Training 63

Symptoms: Exposure to an oxygen deficient atmosphere (<19.5 percent) may cause dizziness, drowsiness, nausea, vomiting, excess salivation, diminished mental alertness, loss of consciousness, and death. Exposure to atmospheres containing 8-10 percent or less oxygen will bring about unconsciousness without warning and so quickly that the individuals cannot help or protect themselves.

Section 7: Precautions for Safe Handling and Use

Storage: Cylinders should be stored upright in a well-ventilated, secure area, protected from the weather. Storage area temperatures should not exceed 125°F, and the area should be free of combustible materials. Storage should be away from heavily traveled areas and emergency exits. Avoid areas where salt or other corrosive materials are present. Valve protection caps and valve outlet seals should remain on cylinders not connected for use. Separate full cylinders from empty cylinders. Avoid excessive inventory and storage time. Use a first-in first-out system. Keep good inventory records.

Handling: Do not drag, roll, or slide cylinder. Use a suitable handtruck designed for cylinder movement. Never attempt to lift a cylinder by its cap. Secure cylinders at all times while in use. Use a pressure-reducing regulator or separate control valve to safely discharge gas from cylinder. Use a check valve to prevent reverse flow into cylinder. Do not overheat cylinder to increase pressure or discharge rate. If user experiences any difficulty operating cylinder valve, discontinue use and contact supplier. Never insert an object (e.g., wrench, screwdriver, pry bar, etc.) into valve cap openings. Doing so may damage valve, causing a leak to occur. Use an adjustable strap-wrench to remove over-tight or rusted caps.

Special Requirements: Always store and handle compressed gases in accordance with Compressed Gas Association, Inc. (703-412-0900) pamphlet CGA P-1, *Safe Handling of Compressed Gases in Containers*. Local regulations may require specific equipment for storage or use.

Section 8: Control Measures

Engineering Controls: Provide good ventilation and/or local exhaust to prevent accumulation of high concentrations of gas. Oxygen levels in work area should be monitored to ensure that they do not fall below 19.5 percent.

Respiratory
Protection: None

General use: None required.

Emergency: Use SCBA or positive pressure air line with mask and escape pack in areas where oxygen concentration is <19.5 percent. Air purifying respirators will not provide protection.

Other Protective
Equipment: Safety shoes are recommended when handling cylinders.

Section 9: Accidental Release Measures

Evacuate all personnel from affected area. Increase ventilation to release area and monitor oxygen level. Use appropriate protective equipment (SCBA). If leak is from container or its valve, call the Air Products' emergency telephone number. If leak is in user's system, close cylinder valve and vent pressure before attempting repairs.

Section 10: Toxicological Information

Helium is a simple asphyxiant. Helium is not toxic.

Section 11: Ecological Information

Helium is not toxic. No adverse ecological effects are expected. Helium does not contain any Class I or Class II ozone depleting chemicals. Helium is not listed as a marine pollutant by DOT (49 CFR 171).

Section 12: Disposal

Unused Product/
Empty Container: Return container and unused product to supplier. Do not attempt to dispose of residual or unused quantities.

Disposal: For emergency disposal, secure the cylinder and slowly discharge gas to the atmosphere in a well-ventilated area or outdoors.

Section 13: Transportation

DOT Hazard Class: 2.2

DOT Shipping Label: Nonflammable Gas

DOT Shipping Name: Helium, Compressed

Identification Number: UN 1066

Reportable Quantity (RQ): None

Special Shipping Information: Cylinders should be transported in a secure upright position in a well-ventilated truck. Never transport in passenger compartment of a vehicle.

Section 14: Regulatory Information

Applicable regulatory information is provided in this section.

Section 15: First Aid Information

Inhalation: Persons suffering from lack of oxygen should be moved to fresh air. If victim is not breathing, administer artificial respiration. If breathing is difficult, administer oxygen. Obtain prompt medical attention.

Eye/Skin Contact: Not applicable.

Section 16: Supplemental Information

NFPA Ratings:		**HMIS Ratings:**	
Health:	0	Health:	0
Flammability:	0	Flammability:	0
Reactivity:	0	Reactivity:	0
Special:		Simple asphyxiant:	

Availability of MSDSs

OSHA requires employers to make MSDS available to all employees on site upon request 24-hours per day. Employers are also required to ensure that MSDS workplace copies are maintained for each hazardous material in the workplace. MSDS cannot be filed away in a file cabinet or locked in an office unless all employees have access to the location. In practice, the MSDS, the chemical inventory form, and a copy of the

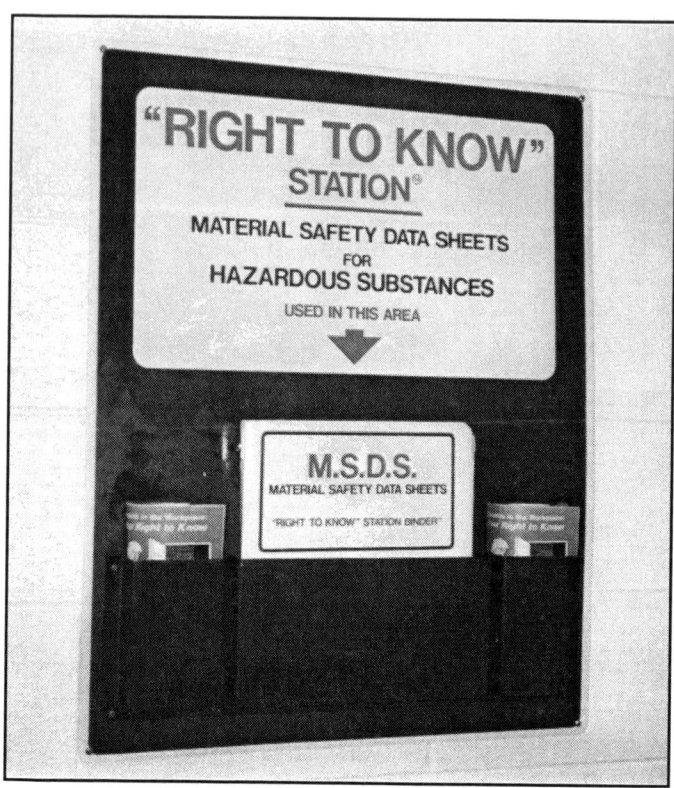

Figure 4.2. Employee "right to know" station

organization's written hazard communication program are kept in an employee "right to know" station like the one shown in Figure 4.2.

Written Hazard Communication Program

Employers shall develop, implement, and maintain at each workplace, a written hazard communication program which at least describes how the criteria specified in [the standard] for labels and other forms of warning, material safety data sheets, and employee information and training will be met... [5]

In early attempts to comply with the requirements of the Hazard Communication Standard (compliance date: May 25, 1985), many organizations felt that all that was needed to comply was to simply make MSDS available to all employees. This routine was also initially followed (in many places) to satisfy the training requirement. Employers would simply hand workers an MSDS, tell them to read it, and consider the required training complete.

This procedure is not what OSHA had in mind with the implementation of its flagship Hazcom standard. In the standard, OSHA specifically states the training requirements, and that an organizational Hazcom program must be in writing. The employer is required to do more than just develop a written hazard communication program. This particular requirement is often cited as the most frequently cited non-compliance violation of OSHA requirements found in industry today. The written Hazcom program must be present and maintained in each workplace (usually placed in the organization's right-to-know station). The written program must contain a section for labels and other warning devices, and material safety data sheets. Employee information must be provided and training conducted. The written program must include a list of hazardous chemicals known to be present using an identity that is referenced on the appropriate material safety data sheet, the methods the employer uses to inform employees of the hazards of non-routine tasks, and the hazards associated with chemicals contained in unlabeled pipes in their work areas.

Included in Appendix B is a sample written hazard communication program that has been used in industry for years and has the advantage of having been tested and amended as required.

Summary

OSHA and the Hazcom standard and the U.S. DOT and hazmat regulations are inextricably entwined. They constantly connect, interfacing as materials are ferried in and out of our workplaces and through our transportation systems. What holds the complicated network of business, government, environment, and community together is a combination of elements that includes stable, intelligently created and applied methods of examination, determination, training, safe work practices, and recordkeeping. Without the background adherence to both OSHA's Hazcom standard and the U.S. DOT's safety training (the subject of Chapter 5), we could never successfully achieve "cradle-to-grave," country-wide handling of the materials that place us at risk.

Notes

[1] OSHA, 29 CFR 1910.1200

[2] OSHA, 29 CFR 1910.1200 (d)(1)

[3] OSHA, 29 CFR 1910.1200 (h)(1)

[4] OSHA, 29 CFR 1910.1200 (g)(1)

[5] OSHA, 29 CFR 1910.1200 (e)(1)

Chapter 5
Labeling, Placarding, and Emergency Response

Introduction[1]

OSHA has had an important impact on the safety of individuals involved in the trucking industry, especially in shipping, handling, producing, and transporting of hazardous materials. However, DOT has also played a significant part in protecting the safety and well-being of those involved in hazmat transportation. In this chapter, we present DOT required hazmat employee training, including labeling of a hazardous material, proper placarding, and dealing with hazmat emergencies.

Labeling/Marking Requirements

As with many other OSHA requirements, marking requirements are frequently non-specific. An example of a non-specific OSHA requirement that is familiar to safety professionals is the requirement for fire extinguishers in the workplace (29 CFR 1910.157). OSHA states that a portable fire extinguisher must be "located, and identified" in the workplace. The kind of identification to be used is not specified, but you must clearly "locate" and "identify" each fire extinguisher in the workplace. Typically, this is accomplished by installing a "fire extinguisher" sign close to the fire extinguisher. While these signs are effective, they have a tendency to disappear with time. To prevent this, fire extinguisher signage is sometimes stenciled in large bold print on the wall or other surface next to or near the fire extinguisher, or an over-sized red rectangle is painted on a flat surface with the fire extinguisher mounted on a bracket within the red rectangle. The red-painted rectangle is quite visible to anyone and makes the fire extinguisher "stand out" in a specific area, making its identification and location much easier.

The red rectangle's purpose is to alert employees to the location of the fire extinguisher and, if it is missing, to prompt them to ask where it is. This is of the greatest importance. Not only should employees be able to identify and locate fire extinguishers, they should also notice when a fire extinguisher is missing. The whole purpose of warning signage is that it must draw attention to itself, and thus to the danger it warns against. Signs increase awareness.

Colors used and warning words (signal words), unless otherwise specified by OSHA or other regulatory authority (DOT, for example) for labels and signs, are required to follow a general pattern. You cannot arbitrarily post or attach any type of label, tag, or sign; you must use markings as specified by OSHA or DOT.

- **DANGER** signs indicate immediate danger and that special precautions are necessary. These signs use the colors red, black, and white. They must be used with a consistent design throughout the facility for a specific danger. A typical danger sign is shown in Figure 5.1.

- **CAUTION** signs alert workers to potential hazards or warn against unsafe practices. These signs must have a yellow background and a black panel with yellow letters (see Figure 5.2).

- **WARNING** signs caution employees of hazards to reduce the risk of serious injury. These signs must have an orange background and a black panel with orange letters (see Figure 5.3).

Figure 5.1. Typical danger sign

Figure 5.2. Typical caution sign

Figure 5.3. Typical warning sign

In addition to the signal word signs just discussed (danger, caution, and warning), other types of signs are used to warn or advise employees. Safety instruction signs convey general instructions and suggestions relative to safety measures. These signs must have a white background and a green panel with white letters. **"Safety first"** signs remind employees to "report all accidents or injuries at once" or "wear safety equipment."

Other warning labels or signs may be quite specific and require specific color arrangements. For example, biological hazard signs are specific and must be orange-red. Radiation symbols must be purple and yellow.

Labeling of Hazardous Chemicals

The chemical manufacturer, importer, or distributor shall ensure that each container of hazardous chemicals leaving the workplace is labeled, tagged or marked with the following information:

(i) Identity of the hazardous chemical(s);

(ii) Appropriate hazard warnings; and

(iii) Name and address of the chemical manufacturer, importer, or other responsible party.[2]

This statement above largely sums up OSHA's requirement for chemical manufacturers, importers, and distributors (shippers) to label, mark, or tag all containers of hazardous chemicals leaving their premises. Note, however, that the labeling/marking requirement applies to *all* hazmat employers, not just to the manufacturer, importer, and/or distributor/shipper. All employers who use, store, handle, produce, or ship hazardous chemicals must label, mark, or tag all containers of hazardous chemicals.

> ✓ **Important Note:** All containers of hazardous chemicals/materials must be labeled in English. Information may be presented in other languages for non-English speaking employees, but English is required.

Under 29 CFR 1910.1200(f)(5), OSHA requires the employer to ensure that every container of hazardous chemicals in the workplace is labeled, tagged, or marked with the chemical name and appropriate hazard warning. Specifically, the appropriate hazard warning must be used (whether words, pictures, symbols, or a combination) that will provide at least general information regarding the hazards of the chemicals. The information on the tag must, in conjunction with the other information immediately available to employees under the hazard communication program, provide employees with the specific information related to the physical and health hazards of the hazardous chemical.

If labels cannot be used (for whatever reason), the employer may use signs, placards, process sheets, batch tickets, operating procedures, or other such written materials in lieu of affixing labels to individual stationary process containers, as long as the alternate method identifies the container to which it is applicable and conveys the required information.

> ✓ **Important Note:** Any written materials used in lieu of warning labels must be readily available to the employees at all times.

> ✓ **Important Note:** Employers are not required to label portable containers into which hazardous chemicals are transferred from labeled containers, and which are intended only for the immediate use of the employee who performs the transfer.

Employees must be informed that labels are not to be removed from chemical containers, and that labels must not be defaced in any way. These labels must be in English and prominently displayed on the container. Employers with employees who speak other languages may also include the information in those languages, but the information must be presented in English. When the place of employment includes workers from different ethnic backgrounds, pictures depicting hazards and the results of hazards may be more understandable than any words. Someone may have difficulty understanding English, but most people have little trouble understanding a picture or symbol of a particular hazard.

Non-Specific Chemical Labels

If the employer requires workers to work around, near, or within some type of process that produces hazardous chemicals or hazardous off-gases, the employer has a duty to inform workers and others of the hazards present. Although this particular stipulation is not specifically addressed or cited within the Hazcom standard, under OSHA's General Duty Clause, the employer is responsible for informing the worker of all workplace hazards.

For example, if the place of employment is a wastewater treatment plant that has vaults, wet wells, and/or other sewage containing vessels or conveyances, the chance exists that organic degradation of sewage materials is actively taking place or there is the potential for such action to take place, and by-products of biodegradation almost always include deadly hydrogen sulfide gas and/or the production of deadly and explosive methane gas. The employer, operator, owner, or manager of such a facility has an obligation and a duty to ensure that all workers are aware of the potential hazards. This information must be readily available to workers.

The best way in which to accomplish this is by labeling the entrance to the space or vessel for the hazards involved. For example, if the danger is potentially explosive methane gas, a warning sign like the one shown in Figure 5.4 is appropriate.

Labeling Systems

Many systems for labeling hazardous chemicals have been developed, including the National Fire Protection Association (NFPA) 704 labeling system (the fire diamond label shown in Figure 5.5) or the Hazardous Material Information System (HMIS). OSHA does not require any one specific type of label, but does require some type of label on all hazardous materials containers.

Figure 5.4. Methane gas warning sign

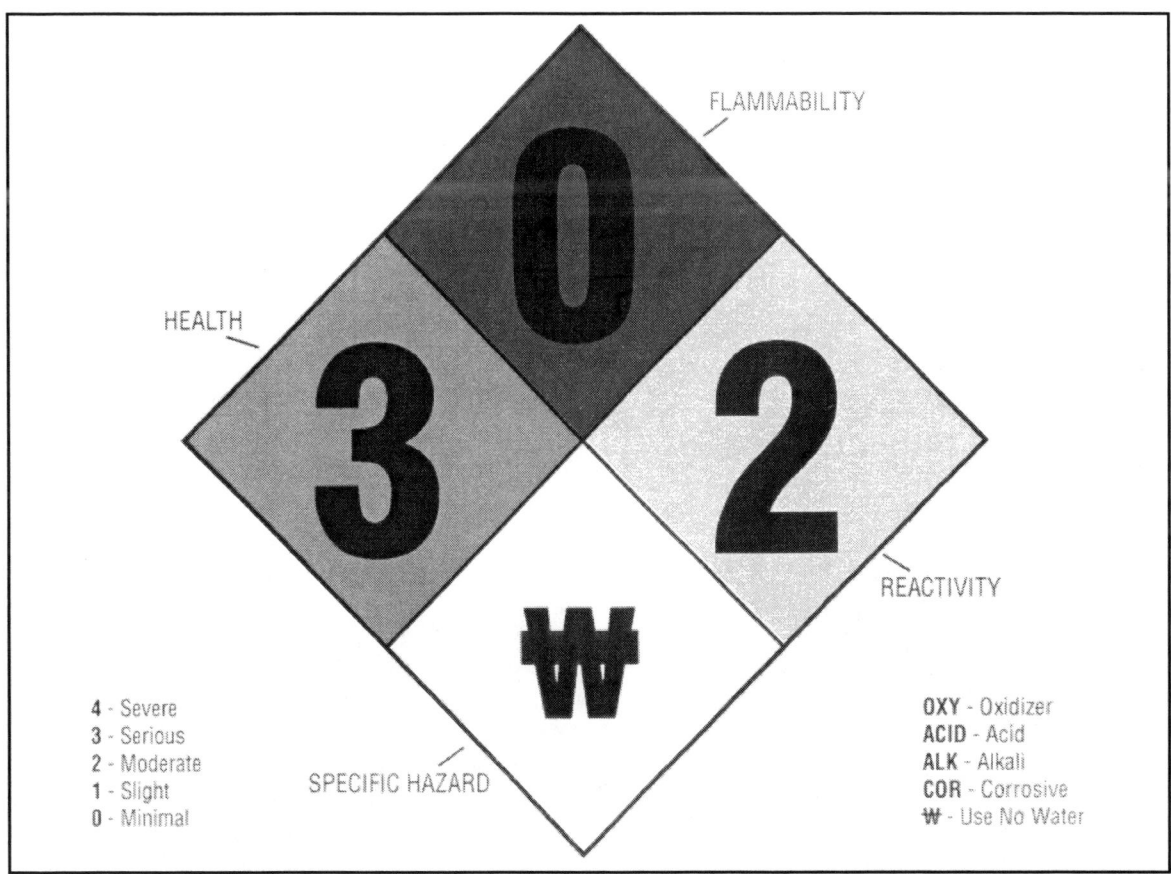

Figure 5.5. Fire diamond label

Labeling of Hazardous Materials (§172.400)

Hazardous material warning labels are designed and color-coded so that the hazards can be quickly recognized. Warning labels correspond to the placards that must appear on each bulk packaging, freight container, unit load device, transport vehicle, or rail car that contains a hazardous material. The labels must include both the hazard class and the division of hazard, if required, according to the Hazardous Material Table. Unless excepted, all hazardous material packages must be labeled. In Figures 5.6 through 5.26, we provide a written and visible description of the primary and subsidiary warning labels required by DOT for use on hazardous materials packages.

Figure 5.6. Explosive 1.1, 1.2, and 1.3 labels (§172.411)
Except for size and color, the explosive 1.1, explosive 1.2, and explosive 1.3 labels must be as shown here. In addition to complying with §172.407, the background color on the explosive 1.1, explosive 1.2, and explosive 1.3 labels must be orange. The "**" must be replaced with the appropriate division number and compatibility group. The compatibility group letter must be the same size as the division number and must be shown as a capitalized Roman letter.

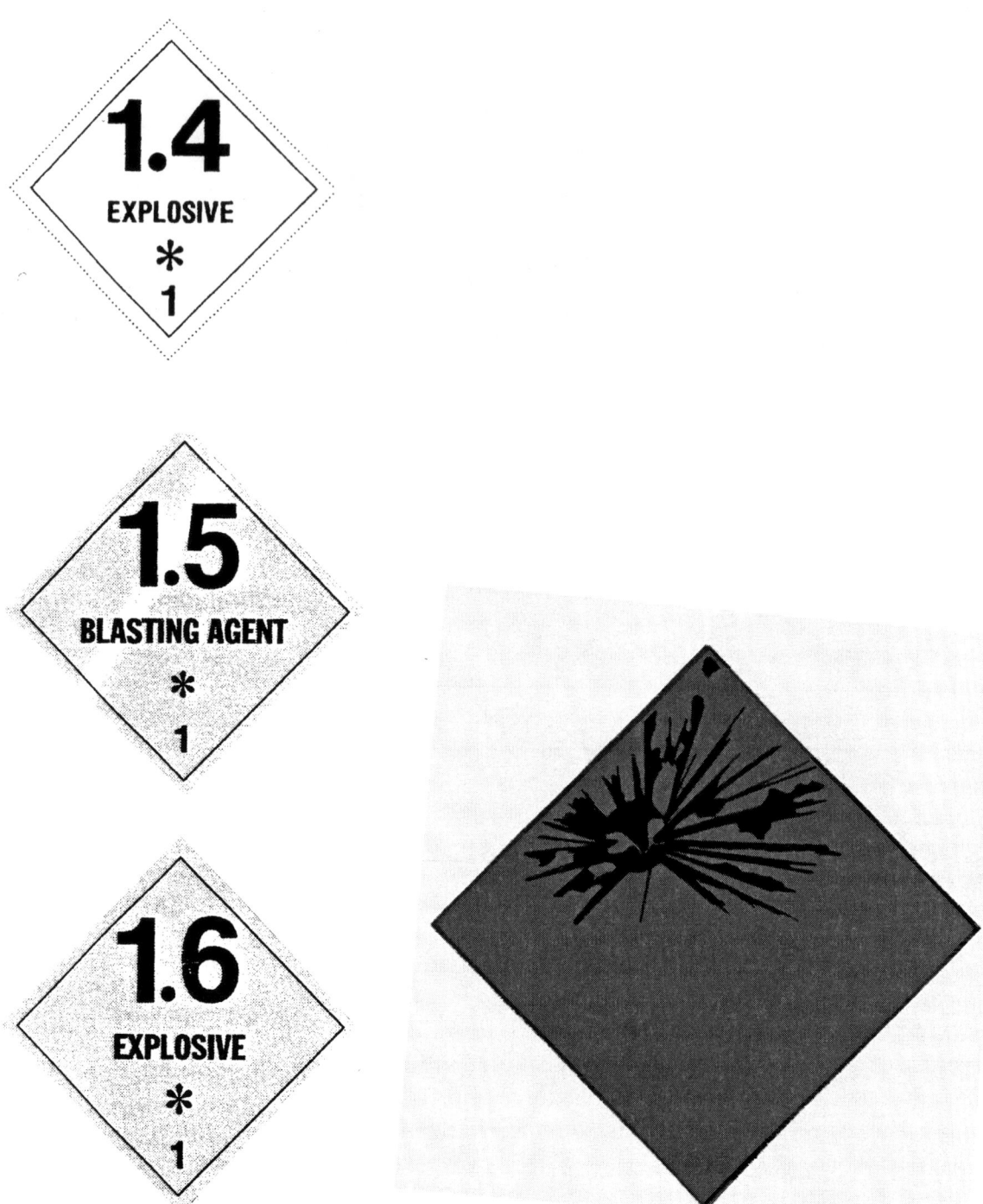

Figure 5.7. Explosive 1.4, 1.5, and 1.6 labels and subsidiary labels (§172.411)
Except for size and color, the explosive 1.4, explosive 1.5, explosive 1.6 labels, and explosive subsidiary labels must be as shown here. No compatibility group letter or Class/Division number may be displayed on this label. In addition to complying with §172.407, the background color on these labels must be orange. Except for the explosive subsidiary label, the "" must be replaced with the appropriate compatibility group. The compatibility group letter must be shown as a capitalized Roman letter. Except for the explosive subsidiary label, division numerals must measure at least 1.2 inches in height and at least 0.2 inches in width.*

Labeling, Placarding, and Emergency Response 77

Figure 5.8. Non-flammable gas label (§172.415)
Except for size and color, the non-flammable gas label must be as shown here. In addition to complying with §172.407, the background color on the non-flammable gas label must be green.

Figure 5.9. Poison gas label (§172.416)
Except for size and color, the poison gas label must be as shown in Figure 5.7. In addition to complying with §172.407, the background on the poison gas label and the symbol must be white. The background of the upper diamond must be black and the lower point of the upper diamond must be 0.54 inches above the horizontal center line.

Figure 5.10. Flammable gas label (§172.417)
Except for size and color, the flammable gas label must be as shown here. In addition to complying with §172.407, the background color on the flammable gas label must be red.

Figure 5.11. Flammable liquid label (§172.419)
Except for size and color, the flammable liquid label must be as shown here. In addition to complying with §172.4.7, the background color on the flammable liquid label must be red.

Figure 5.12. Flammable solid label (§172.420)
Except for size and color, the flammable solid label must be as shown here. In addition to complying with §172.407, the background on the flammable solid label must be white with vertical red stripes equally spaced on each side of a red stripe placed in the center of the label. The red vertical stripes must be spaced so that visually they appear equal in width to the white spaces between them. The symbol (flame) and text (when used) must be overprinted. The text flammable solid may be placed in a white rectangle.

Figure 5.13. Spontaneously combustible label (§172.422)
Except for size and color, the spontaneously combustible label must be as shown here. In addition to complying with §172.407, the background color on the lower half of the spontaneously combustible label must be red and the upper half must be white.

Labeling, Placarding, and Emergency Response

Figure 5.14. Dangerous when wet label (§172.423)
Except for size and color, the dangerous when wet label must be as shown here. In addition to complying with §172.407, the background color on the dangerous when wet label must be blue.

Figure 5.15. Oxidizer label (§172.426)
Except for size and color, the oxidizer label must be as shown here. In addition to complying with §172.407, the background color on the oxidizer label must be yellow.

Figure 5.16. Organic peroxide label (§172.427)
Except for size and color, the organic peroxide label must be as shown here. In addition to complying with §172.407, the background color on the organic peroxide label must be yellow.

Figure 5.17. Poison inhalation hazard label (§172.416)
Except for size and color, the poison inhalation hazard label must be as here. In addition to complying with §172.407, the background on the poison inhalation hazard label and the symbol must be white. The background of the upper diamond must be black and the lower point of the upper diamond must be 0.54 inches above the horizontal center line.

Figure 5.18. Poison label (§172.430)
Except for size and color, the poison label must be as shown here. In addition to complying with §172.407, the background on the poison label must be white. The word "toxic" may be used in lieu of the word "poison."

Figure 5.19. Infectious substance label (§172.432)
Except for size and color, the infectious substance label must be as shown here. In addition to complying with §172.407, the background on the infectious substance label must be white.

Labeling, Placarding, and Emergency Response

Figure 5.20. Radioactive white-I label (§172.436)
Except for size and color, the radioactive white-I label must be as shown here. In addition to complying with §172.407, the background on the radioactive white-I label must be white. The printing and symbol must be black, except for the "I," which must be red.

Figure 5.21. Radioactive yellow-II label (§172.438)
Except for size and color, the radioactive yellow-II label must be as shown here. In addition to complying with §172.407, the background color on the radioactive yellow-II label must be yellow in the top half and white in the lower half. The printing and symbol must be black, except for the "II," which must be red.

Figure 5.22. Radioactive yellow-III label (§172.440)
Except for size and color, the radioactive yellow-III label must be as shown here. In addition to complying with §172.407, the background color on the radioactive yellow-III label must be yellow in the top half and white in the lower half. The printing and symbol must be black, except for the "III," which must be red.

Figure 5.23. Corrosive label (§172.442)
Except for size and color, the corrosive label must be as shown here. In addition to complying with §172.407, the background on the corrosive label must be white in the top half and black in the lower half.

Figure 5.24. Class 9 (miscellaneous hazardous materials) label (§172.446)
Except for size and color, the Class 9 (miscellaneous hazardous materials) label must be as shown here. In addition to complying with §172.407, the background on the Class 9 label must be white with seven black vertical stripes on the top half. The black vertical stripes must be spaced so that visually they appear equal in width to the six white spaces between them. The lower half of the label must be white with the class number "9" underlined and centered at the bottom.

Figure 5.25. Cargo aircraft only label (§172.448)
Except for size and color, the cargo aircraft only label must be as shown here. The cargo aircraft only label must be black on an orange background.

Labeling, Placarding, and Emergency Response

Figure 5.26. Empty label (§172.450)
Each empty label, except for size, must be as shown here. Each side must be at least 6 inches with each letter at least 1 inch in height. The label must be white with black lettering.

Except as specified in exceptions from labeling (§172.400a), DOT requires each person who offers for transportation or transports a hazardous material in any of the following packages or containment devices to label the package or containment device with labels specified for the material in the Hazardous Materials Table (§172.101).

- A non-bulk package
- A bulk packaging, other than a cargo tank, portable tank, or tank car, with a volumetric capacity of less than 640 cubic feet, unless placarded
- A portable tank of less than 1000 gallons capacity, unless placarded
- A DOT Specification 106 or 110 multi-unit tank car tank, unless placarded
- An overpack, freight container or unit load device of less than 640 cubic feet that contains a package for which labels are required, unless placarded or marked

> ✓ **Important Note:** Be careful with the distinction between "labels" and "markings." In hazmat parlance, hazard warning labels are called "labels." Any other information required on the exterior of a package is called a "marking."

Except where otherwise provided, no person may offer for transportation and no carrier may transport a package bearing a label specified in 172.401 unless:

- The package contains a material that is a hazardous material, and
- The label represents a hazard of the hazardous material in the package.

No person may offer for transportation and no carrier may transport a package bearing any marking or label that by its color, design, or shape could be confused with or conflict with a label prescribed by 172.401. The restrictions in Paragraph 1 and 2 of this section do not apply to packages labeled in conformance with:

- Any UN recommendation, including the class number
- The International Maritime Organization (IMO) requirements, including the class number, in the document entitled "International Maritime Dangerous Goods Code"

- The ICAO Technical Instructions
- The TDG Regulations

The provisions of Paragraph (a) of this section do not apply to a packaging bearing a label if that packaging is:

- Unused or cleaned and purged of all residue
- Transported in a transport vehicle or freight container in such a manner that the packaging is not visible during transportation
- Loaded by the shipper and unloaded by the shipper or consignee

Where Labeling Is Required

Labeling is required for a hazardous material that meets one or more hazard class definitions, in accordance with column 6 of the Hazardous Materials Table (§172.101) and the following table:

Table 5.1. Label Names and Designs for Each Hazard Class

Hazard class or division	Label name	Label design or section
1.1	EXPLOSIVES 1.1 172.411	1.1
1.2	EXPLOSIVES 1.2 172.411	1.2
1.3	EXPLOSIVES 1.3 172.411	1.3
1.4	EXPLOSIVES 1.4 172.411	1.4
1.5	EXPLOSIVES 1.5 172.411	1.5
1.6	EXPLOSIVES 1.6	172.411
2.1	FLAMMABLE GAS	172.417
2.2	NONFLAMMABLE GAS	172.415
2.3	POISON GAS	
3 (flammable liquid)	FLAMMABLE LIQUID	172.419
Combustible liquid	(none)	
4.1	FLAMMABLE	172.420
4.2	SPONTANEOUSLY COMBUSTIBLE	172.422
4.3	DANGEROUS WHEN WET	172.423
5.1	OXIDIZER	172.426
5.2	ORGANIC	PEROXIDE
6.1 (inhalation hazard, Zone A or B)	POISON INHALATION HAZARD	
6.1 (other than inhalation hazard, Zone A or B)	POISON	172.430
6.2	INFECTIOUS SUBSTANCE*	
7	RADIOACTIVE	WHITE-I
7	RADIOACTIVE YELLOW-II	172.438
7	RADIOACTIVE YELLOW-III	172.440
7	EMPTY	
8	CORROSIVE	172.442
9	CLASS 9	172.446

* The etiologic agent label specified in regulations of the Department of Health and Human Services at 42 CFR 72.3 may apply to packages of infectious substances.

Where Labeling Is Not Required

Notwithstanding the provisions of §172.400, a label is not required on:

- A cylinder, or a flask conforming to §173.320 of this subchapter containing a Division 2.1 or Division 2.2 gas that is:
  - Not poisonous
  - Carried by a private or contract motor carrier
  - Not overpacked
  - Durably and legibly marked in accordance with CGA Pamphlet C-7, Appendix A.
  - A package or unit of military explosives (including ammunition) shipped by or on behalf of the DOD when in:
  - Freight container load, carload, or truckload shipments, if loaded and unloaded by the shipper or DOD; or
  - Unitized or palletized break-bulk shipments by cargo vessel under charter to DOD if at least one required label is displayed on each unitized or palletized load.
- A package containing a hazardous material other than ammunition that is:
  - Loaded and unloaded under the supervision of DOD personnel, and
  - Escorted by DOD personnel in a separate vehicle.
- A compressed gas cylinder permanently mounted in or on a transport vehicle.
- A freight container, aircraft unit load device or portable tank that:
  - Is placarded in accordance with Subpart F of this part, or
  - Conforms to Paragraph (a)(3) or (b)(3) of §172.512.
- An overpack or unit load device in or on which labels representative of each hazardous material in the overpack or unit load device are visible.
- A package of low specific activity radioactive material, when transported under §173.427(a)(6)(vi) of this subchapter.

> ✓ **Important Note:** Certain exceptions to labeling requirements are provided for small quantities and limited quantities in applicable sections in part 173 of this subchapter. Notwithstanding the provisions of §172.402(a), a subsidiary hazard label is not required on a package containing Class 8 (corrosive) material that has a subsidiary hazard of Division 6.1 (poisonous), if the toxicity of the material is based solely on the corrosive destruction of tissue rather than systemic poisoning.

Additional Labeling Requirements (§172.402)

Subsidiary hazard labels—Each package containing a hazardous material must be labeled with primary and subsidiary hazard labels as specified in column 6 of the §172.101 Hazardous Materials Table. Materials other than Class 1 or Class 2 materials, if not already labeled under other provisions, must be labeled with subsidiary labels in accordance with the following table:

Table 5.2. Subsidiary Hazard Labels

Subsidiary hazard level (packing group)	Subsidiary Hazard (Class or Division)						
	3	4.1	4.2	4.3	5.1	6.1	8
I	X	***	***	X	X	X	X
II	X	X	X	X	X	X	X
III	*	X	X	X	X	X	X

* Required for all modes, except for a material with a flash point at or above 38°C (100°F) transported by rail or highway.
*** Impossible as subsidiary hazard.

Display of hazard class on labels—the appropriate hazard class, or for Division 5.1 or 5.2, the division number, must be displayed in the lower corner of a primary hazard label and may not be displayed on a subsidiary label.

Labels for Mixed and Consolidated Packaging (§172.404)

Mixed packaging—when hazardous materials having different hazard classes are packed within the same packaging, or within the same outside container or overpack, and authorized by §173.21, the packaging, outside container, or overpack must be labeled as required for each class of hazardous material contained therein.

Consolidated packaging—when two or more packages containing compatible hazardous material are placed within the same outside container or overpack, the outside container or overpack must be labeled as required for each class of hazardous material contained therein.

Authorized Label Modifications (§172.405)

For Classes 1, 2, 3, 4, 5, 6, and 8, text indicating a hazard (flammable liquid, for example) is not required on a primary or subsidiary label when:

- The label otherwise conforms to the provisions of this subpart, and

- The hazard class, or for Division 5.1 or 5.2, the division number is displayed in the lower corner of the label, if the label corresponds to the primary hazard class of the hazardous material.

For a package containing oxygen, compressed, or oxygen, refrigerated liquid, the oxidizer label specified in §172.426, modified to display the word "oxygen" instead of "oxidizer," and the class number "2" instead of "5.1," may be used in place of the nonflammable gas and oxidizer labels. The word "oxygen" must appear on the label.

For a package containing a Division 6.1, Packing Group III material, the poison label specified in §172.430 may be modified to display the text "PG III" instead of "poison" or "toxic" below the mid line of the label.

Placement of Labels (§172.406)

Each label required by this subpart must:

- Be printed on or affixed to a surface (other than the bottom) of the package or containment device containing the hazardous material
- Be located on the same surface of the package and near the proper shipping name marking, if the package dimensions are adequate

Except for where otherwise specified, duplicate labeling is not required on a package containment device (such as to satisfy redundant labeling requirements).

Exceptions include:

- A label may be printed on or placed on a securely affixed tag, or may be affixed by other suitable means to:
  - A package that contains no radioactive material and which has dimensions less than those of the required label
  - a cylinder
  - a package that has such an irregular surface that a label cannot be satisfactorily affixed

When primary and subsidiary hazard labels are required, they must be displayed next to each other. Placement conforms to this requirement if labels are within six inches of one another.

Each label must be printed on or affixed to a background of contrasting color, or must have a dotted or solid line outer border.

As stated earlier, generally only one of each different required label must be displayed on a package. However, duplicate labels must be displayed on at least two sides and two ends (other than the bottom of):

- Each package or overpack having a volume of 64 cubic feet or more.

- Each non-bulk package containing a radioactive material.
- Each DOT 106 or 110 multi-unit tank car tank. Labels must be displayed on each end.
- Each portable tank of less than 1000 gallon capacity.
- Each freight container or aircraft unit load device having a volume of 64 cubic feet or more, but less than 640 cubic feet. One of each required label must be displayed on or near the closure.

> ✓ **Important Note:** A label must be clearly visible and not be obscured by markings or attachments.

Label Specifications (§172.407)

Each label, whether printed on or affixed to a package, must be durable and weather resistant. A label on a package must be able to withstand, without deterioration or a substantial change in color, a 30-day exposure to conditions incident to transportation that reasonably could be expected to be encountered by the labeled package.

The design, printing, inner border, and symbol on each label must be as shown in Figures 5.6-5.26 (except for size and color, which are discussed in the figure captions).

The size of each diamond (square-on-point) label prescribed in this subpart must be at least 3.9 inches on each side with each side having a solid line inner border (0.2 to 0.25 inches) from the edge.

The "cargo aircraft only" label must be a rectangle measuring at least 4.3 inches in height by 4.7 inches in width. The word "danger" must be shown in letters measuring at least 0.5 inches in height.

Except as otherwise specified, the hazard class number, or division number, as appropriate, must be at least 0.25 inches and not greater than 0.5 inches.

When text indicating a hazard is displayed on a label, the label name must be shown in letters measuring at least 0.3 inches in height. For "spontaneously combustible" or "dangerous when wet" labels, the words "spontaneously" and "when wet" must be shown in letters measuring at least 0.2 inches in height.

The symbol on each label must be proportionate in size to that shown in the appropriate section of this subpart.

The color and the background color on each label must be as prescribed in Figures 5.6-5.24, as appropriate.

The symbol, text, numbers, and border must be shown on the label in black except that:

- White may be used on a label with a one color background of green, red or blue
- White must be used for the text and class number for the "corrosive" label

Black and any color on the label must be able to withstand, without substantial change, a 72-hour fadeometer test.

A color on a label, upon visual examination, must fall within the color tolerances:

- Displayed on color charts conforming to the technical specifications for charts set forth by DOT
- For labels printed on packaging surfaces, specified by DOT

> ✓ **Important Note:** Color charts conforming to DOT standards are on display in Room 8421, Nassif Building, 400 Seventh Street, SW, Washington, D.C.

The specified label color must extend to the edge of the label in the area designated on each label except for the "corrosive," "radioactive yellow-II," and "radioactive yellow-III" labels on which the color must extend only to the inner border.

For form identification, a label may contain form identification information, including the name of its maker, provided that information is printed outside the solid line inner border in no larger than 10-point type.

Exceptions. A label conforming to specifications in the UN Recommendations may be used in place of a corresponding label that conforms to the requirements of this subpart.

Trefoil symbol. The trefoil symbol on the "radioactive white-I," "radioactive yellow-II," and "radioactive yellow-iii" labels must meet the appropriate DOT specifications.

> ✓ **Important Note:** When compatible materials are packaged together, the labels for each individual material must appear on the exterior of the package.

Exceptions to the Labeling Requirements

Hazcom does provide some exceptions to the requirement for labeling hazardous chemicals. For example, pesticides, fungicides, and rodenticides must be labeled in accordance with the U.S. EPA. Any chemical substance or mixture covered and labeled under the Toxic Substance Control Act is not required to be labeled under Hazcom. Any food, food additive, color additive, drug, cosmetic, or medical or veterinary device or product covered under the Federal Food, Drug, and Cosmetic Act is exempt from Hazcom labeling requirements. Distilled spirits are exempt from Hazcom. Agricultural or vegetable seeds treated with pesticides and labeled in accordance with the Federal Seed Act are Hazcom exempt. Any hazardous waste defined by the Solid Waste Disposal Act is Hazcom exempt. Tobacco products are exempt. Wood and wood products are exempt. Some nuisance products are exempt, along with ionizing and nonionizing radiation and biological hazards.

Placarding

Placarding is covered in Subpart F of the 49 CFR standard. Though similar to labels, placards are larger; they are square-on-point signs used to show the hazard or hazards associated with hazardous materials contained in:

- bulk packages
- transport vehicles, railcars, freight containers, or unit load devices containing bulk or non-bulk packages

For placarding, the following rules apply (§172.500):

> ✓ **Important Note:** The color, design, symbol, and class or division number on a primary placard corresponds to the hazard class or division the placard represents.

Each person who offers for transportation or transports any hazardous material must comply with the following applicable placarding requirements:

- Infectious substances
- Hazardous materials classed as ORM-D
- Hazardous materials authorized by this subchapter to be offered for transportation as limited quantities when identified as such on shipping papers in accordance with §172.203(b)
- Hazardous materials packaged as small quantities under the provisions of §173.4
- Combustible liquids in non-bulk packagings

In Figures 5.27 through 5.45, we briefly describe several types of placards currently used in the transportation of hazardous materials.

Labeling, Placarding, and Emergency Response

Figure 5.27. Dangerous placard (§172.521)
The dangerous placard may be used for combinations of different categories of Table 5.2, materials in non-bulk packages loaded in the same transport vehicle, freight container, unit load device, or rail car. Except for size and color, the dangerous placard must be as shown here. In addition to meeting the requirements of §172.519 and Appendix B, the dangerous placard must have a red upper and lower triangle. The placard center area and 0.5-inch border must be white. The inscription must be black with the 1/8-inch border marker in the white area at each end of the inscription red.

✓ **Important Point:** Note that the dangerous placard cannot be used when an aggregate gross weight of 2,205 pounds of one type of hazardous materials (one category listed in Table 5.2 in this text) is loaded at a single facility. When the 2,205 pound weight limit is reached, a separate placard must be used. This particular requirement may come as a surprise to many experienced hazmat transportation officials who are used to the old "5,000 pound rule" of the past. The weight limit was lowered from 5,000 to 2,205 pounds in 1998.

Figure 5.28. Explosives 1.1, 1.2, and 1.3 placards (§172.522)
Except for size and color, the explosives 1.1, explosives 1.2, and explosives 1.3 placards must be as shown here. In addition to complying with §172.519 of this subpart, the background color on the explosives 1.1, explosives 1.2, and explosives 1.3 placards must be orange. The "" must be replaced with the appropriate division number, and when required, the appropriate compatibility group letter. The symbol, text, numerals, and inner border must be black.*

Figure 5.29. Explosives 1.4 placard (§172.523)
Except for size and color, the explosives 1.4 placard must be as shown here. In addition to complying with §172.519 of this subpart, the background color on the explosives 1.4 placard must be orange. The "*" must be replaced, when required, with the appropriate compatibility group letter. The division numeral "1.4" must measure at least 2.5 inches in height. The text, numerals and inner border must be black.

Figure 5.30. Explosives 1.5 placard (§172.524)
Except for size and color, the explosives 1.5 placard must be as shown here. In addition to complying with the §172.519 of this subpart, the background color on the explosives 1.5 placard must be orange. The "*" must be replaced, when required, with the appropriate compatibility group letter. The division numeral "1.5" must measure at least 2.5 inches in height. The text, numerals and inner border must be black.

Figure 5.31. Explosives 1.6 placard (§172.525)
Except for size and color, the explosives 1.6 placard must be as shown here. In addition to complying with §172.519 of this subpart, the background color on the explosives 1.6 placard must be orange. The "*" must be replaced, when required, with the appropriate compatibility group letter. The division numeral "1.6" must measure at least 2.5 inches in height. The text, numerals, and inner border must be black.

Labeling, Placarding, and Emergency Response

Figure 5.32. Square background placard (§172.527)
Except for size and color, the square background required by §172.510(a) for certain placards on rail cars, and §172.507 for placards on motor vehicles containing a package of highway route controlled quantity radioactive materials, must be as shown here. In addition to meeting the requirements of §172.5.19 for minimum durability and strength, the square background must consist of a white square measuring 14.25 inches on each side surrounded by a black border extending to 15.15 inches on each side.

Figure 5.33. Non-flammable gas placard (§172.528)
Except for size and color, the non-flammable gas placard must be as shown here. In addition to complying with §172.519, the background color on the non-flammable gas placard must be green. The letters in both words must be at least 1.5 inches high. The symbol, text, class number, and inner border must be white.

Figure 5.34. Oxygen placard (§172.530)
Except for size and color, the "oxygen" placard must be as shown here. In addition to complying with §172.519, the background color on the "oxygen" placard must be yellow. The symbol, text, class number, and inner border must be black.

Figure 5.35. Flammable gas placard (§172.532)
Except for size and color, the flammable gas placard must be as shown here. In addition to complying with §172.519, the background color on the flammable gas placard must be red. The symbol, text, class number and inner border must be white.

Figure 5.36. Poison gas placard (§172.540)
Except for size and color, the poison gas placard must be as shown here. In addition to complying with §172.519, the background on the poison gas placard and the symbol must be white. The background of the upper diamond must be black and the lower point of the upper diamond must be 2.625 inches above the horizontal center line. The text, class number, and inner border must be black.

Figure 5.37. Flammable placard (§172.542)
Except for size and color, the flammable placard must be as shown here. In addition to complying with §172.519, the background color on the flammable placard must be red. The symbol, text, class number, and inner border must be white. The word "gasoline" may be used in place of the word "flammable" on a placard that is displayed on a cargo tank or a portable tank being used to transport gasoline by highway. The word "gasoline" must be shown in white.

Labeling, Placarding, and Emergency Response

Figure 5.38. Combustible placard (§172.544)
Except for size and color, the combustible placard must be as shown here. In addition to complying with §172.519, the background color on the combustible placard must be red. The symbol, text, class number, and inner border must be white. On a combustible placard with a white bottom as prescribed by §172.332(c)(4), the class number must be red or black. The words "fuel oil" may be used in place of the word "combustible" on a placard that is displayed on a cargo tank or portable tank being used to transport by highway fuel oil that is not classed as a flammable liquid. The words "fuel oil" must be shown in white.

Figure 5.39. Flammable solid placard (§172.546)
Except for size and color, the flammable solid placard must be as shown here. In addition to complying with §172.519, the background on the flammable solid placard must be white with seven vertical red stripes. The stripes must be equally spaced, with one red stripe placed in the center of the label. Each red stripe and each white space between two red stripes must be 1.0 inch wide. The letters in the word "solid" must be at least 1.5 inches high. The symbol, text, class number, and inner border must be black.

Figure 5.40. Spontaneously combustible placard (§172.547)
Except for size and color, the spontaneously combustible placard must be as shown here. In addition to complying with §172.519, the background color on the spontaneously combustible placard must be red in the lower half and white in upper half. The letters in the word "spontaneously" must be at least 0.5 inch high. The symbol, text, class number, and inner border must be black.

Figure 5.41. Dangerous when wet placard (§172.548)
Except for size and color, the dangerous when wet placard must be as shown here. In addition to complying with §172.519, the background color on the dangerous when wet placard must be blue. The letters in the words "when wet" must be at least 1.0 inch high. The symbol, text, class number, and inner border must be white.

Figure 5.42. Oxidizer placard (§172.550)
Except for size and color, the oxidizer placard must be as shown here. In addition to complying with §172.519, the background color on the oxidizer placard must be yellow. The symbol, text, division number, and inner border must be black.

Figure 5.43. Organic peroxide placard (§172.552)
Except for size and color, the organic peroxide placard must be as shown here. In addition to complying with §172.519, the background color on the organic peroxide placard must be yellow. The symbol, text, division number, and inner border must be black.

Labeling, Placarding, and Emergency Response

Figure 5.44. Poison placard (§172.554)
Except for size and color, the poison placard must be as shown here. In addition to complying with §172.519, the background on the poison placard must be white. The symbol, text, class number, and inner border must be black. The word "toxic" may be used in lieu of the word "poison."

Figure 5.45. Poison inhalation hazard placard (§172.555)
Except for size and color, the poison inhalation hazard placard must be as shown here. In addition to complying with §172.519, the background on the poison inhalation hazard placard and the symbol must be white. The background of the upper diamond must be black and the lower point of the upper diamond must be 2.625 inches above the horizontal center line. The text, class number, and inner border must be black.

Figure 5.46. Radioactive placard (§172.556)
Except for size and color, the radioactive placard must be as shown here. In addition to complying with §172.519, the background color on the radioactive placard must be white in the lower portion with a yellow triangle in the upper portion. The base of the yellow triangle must be 1.1 inches ± 0.2 inches above the placard horizontal center line. The symbol, text, class number, and inner border must be black.

Figure 5.47. Corrosive placard (§172.558)
Except for size and color, the corrosive placard must be as shown here. In addition to complying with 519, the background color on the corrosive placard must be black in the lower portion with a white triangle in the upper portion. The base of the white triangle must be 1.5 inches ± 0.2 inches above the placard horizontal center. The text and class number must be white. The symbol and inner border must be black.

Figure 5.48. Class 9 (miscellaneous hazardous materials) placard (§172.560)
Except for size and color, the Class 9 (miscellaneous hazardous materials) placard must be as shown here. In addition to conformance with §172.519, the background on the Class 9 placard must be white with seven black vertical stripes on the top half extending from the top of the placard to one inch above the horizontal centerline. The black vertical stripes must be spaced so that visually they appear equal in width to the six white spaces between them. The space below the vertical lines must be white with the class number 9 underlined and centered at the bottom.

General Placarding Requirements

Each bulk packaging, freight container, unit load device, transport vehicle, or rail car containing any quantity of a hazardous material must be placarded on each side and each end with the type of placards specified in Tables 5.3 and 5.4.

A freight container, unit load device, transport vehicle, or rail car that contains non-bulk packagings with two or more categories of hazardous materials that require different placards specified in Table 5.4 may be placarded with "dangerous" placards instead of the separate placarding specified for each of the materials in Table 5.4. However, when 2,205 pounds aggregate gross weight or more of one category of material is loaded therein at one loading facility on a freight container, unit load device, transport vehicle, or rail car, the placard specified in Table 5.4 of Paragraph (e) of 49 CFR 172.504 for that category must be applied.

Labeling, Placarding, and Emergency Response

In 49 CFR 172.504, the two tables shown below (5.3 and 5.4) are Table 1 and Table 2. Table 1 (5.3) materials in non-bulk packages must always be placarded (unless excepted from placarding by the Hazardous Materials Regulations). Look for Table 2 (5.4) materials in non-bulk packages with a subsidiary hazard that requires placarding. Placards are specified for hazardous materials in accordance with the following tables:

Table 5.3. 49 CFR 172.504 Table 1

Category of Material (Hazard class or section reference division number and (§) additional description as appropriate)	Placard Name	Placard Design Section Reference (§)
1.1	EXPLOSIVES 1.1	172.522
1.2	EXPLOSIVES 1.2	172.522
1.3	EXPLOSIVES 1.3	172.522
2.3	POISON GAS	172.540
4.3	DANGEROUS WHEN WET	172.548
5.2 (Organic Peroxide, type B)	ORGANIC PEROXIDE	172.552
6.1 (Inhalation hazard, Zone A or B)	POISON INHALATION HAZARD	172.555
7 (Radioactive Yellow III label only)	RADIOACTIVE*	172.556

* Radioactive placard also required for exclusive use shipments of low specific activity material and surface contaminated objects transported in accordance with §173.427(a) of this subchapter.

> ✓ **Important Note:** Placarding is not required for Class 9 materials that are transported in the U.S. A bulk package of Class 9 hazardous materials is excepted from placarding, but it is still required to display the ID number according to the marking requirements.

Table 5.4. 49 CFR 172.504 Table 2

Category of Material (Hazard class or section division number and reference (§) additional description, as appropriate)	Placard Name	Placard Design Section Reference (§)
1.4	EXPLOSIVES 1.4	172.523
1.5	EXPLOSIVES 1.5	172.524
1.6	EXPLOSIVES 1.6	172.525
2.1	FLAMMABLE GAS	172.532
2.2	NON-FLAMMABLE GAS	172.528
3	FLAMMABLE	172.542
Combustible liquid	COMBUSTIBLE	172.544
4.1	FLAMMABLE SOLID	172.546
4.2	SPONTANEOUSLY COMBUSTIBLE	172.547
5.1	OXIDIZER	172.550
5.2 (Other than organic peroxide, type B)	ORGANIC PEROXIDE	172.552
6.1 (other than inhalation hazard, Zone A or B)	POISON	172.554
6.2	(None)	
8	CORROSIVE	172.558
9	CLASS 9	172.560
ORM-D	(None)	

Exceptions to the Placarding Requirements

- The restrictions of §172.502(a) do not apply to bulk packagings, freight containers, unit load devices, transport vehicles, or rail cars that are placarded in conformance with the TDG Regulations, the IMDG Code, or the UN Recommendations.

- The restrictions of §172.502(a)(2) do not apply to the display of an identification number on a white square-on-point configuration in accordance with §172.336(b) of this part.

- The restrictions in Paragraph (a)(2) of this section do not apply until October 1, 2001 to a safety sign or safety slogan (e.g., "Drive Safely" or "Drive Carefully"), which was permanently marked on a transport vehicle, bulk packaging, or freight container on or before August 21, 1997.

Except for bulk packagings and hazardous materials subject to §172.505, when hazardous materials covered by Table 5.4 of this section are transported by highway or rail, placards are not required on:

Labeling, Placarding, and Emergency Response

- A transport vehicle or freight container that contains less than 1001 pounds aggregate gross weight of hazardous materials covered by Table 5.4.

- A rail car loaded with transport vehicles or freight containers, none of which is required to be placarded.

> ✓ **Important Note:** The prohibitions provided in this section do not prohibit the display of placards in the manner prescribed in this subpart, if not otherwise prohibited (see §172.502), on transport vehicles or freight containers that are not required to be placarded.

A non-bulk packaging that contains only the residue of a hazardous material covered by Table 5.2 of Paragraph (e) of this section need not be included in determining placarding requirements.

When more than one division placard is required for Class 1 materials on a transport vehicle, rail car, freight container, or unit load device, only the placard representing the lowest division number must be displayed.

- A "flammable" placard may be used in place of a "combustible" placard on:
  - A cargo tank or portable tank
  - A compartmented tank car which contains both flammable and combustible liquids

- A "non-flammable gas" placard is not required on a transport vehicle that contains non-flammable gas if the transport vehicle also contains flammable gas or oxygen and it is placarded with "flammable gas" or "oxygen" placards, as required.

- "Oxidizer" placards are not required for Division 5.1 materials on freight containers, unit load devices, transport vehicles, or rail cars that also contain Division 1.1 or 1.2 materials and that are placarded with "explosives" 1.1 or 1.2 placards, as required.

- For transportation by transport vehicle or rail car only, an "oxidizer" placard is not required for Division 5.1 materials on a transport vehicle, rail car, or freight container that also contains Division 1.5 explosives and is placarded with "explosives" 1.5 placards, as required.

- The "explosive" 1.4 placard is not required for those Division 1.4 Compatibility Group S (1.4S) materials that are not required to be labeled 1.4S.

- For domestic transportation of oxygen, compressed or oxygen, refrigerated liquid, the "oxygen" placard in §172.530 of this subpart may be used in place of a "non-flammable gas" placard.

- Except for a material classed as a combustible liquid that also meets the definition of a Class 9 material, a "combustible" placard is not required for a material classed as a combustible liquid when transported in a non-bulk packaging classed as a bulk packaging.

- For domestic transportation, a Class 9 placard is not required. A bulk packaging containing a Class 9 material must be marked on each side and each end with the appropriate identification number displayed on an orange panel, or a white-square-on-point display configuration are required.

- For Division 6.1, PG III materials, a "poison" placard may be modified to display the text "PG III" below the mid-line of the placard.

- For domestic transportation, a "poison" placard is not required on a transport vehicle or freight container required to display a "poison inhalation hazard" or "poison gas" placard.

- For shipments of Class 1 (explosive) materials by aircraft or vessel, the applicable compatibility group letter must be displayed on the placards required by this section.

Permissive Placarding

Placards may be displayed for a hazardous material, even when not required, if the placarding otherwise conforms to the requirements of this subpart.

> ✓ **Important Note:** For procedures and limitations pertaining to the display of identification numbers on placards, see §172.334.

Prohibited Placarding (§172.502)

No person may affix or display on a packaging, freight container, unit load device, motor vehicle, or rail car:

- Any placard described in §172.502 unless:
  - The material being offered or transported is a hazardous material
  - The placard represents a hazard of the hazardous material being offered or transported
  - Any placarding conforms to the requirements of §172.502

Labeling, Placarding, and Emergency Response

- Any sign, advertisement, or other device that, by its color, design, shape or content could be confused with any placard prescribed §172.502.

Placarding for Subsidiary Hazards (§172.505)

a. Each transport vehicle, freight container, portable tank, unit load device, or rail car that contains a poisonous material subject to the "poison-inhalation hazard" shipping description of §172.203(m)(3) must be placarded with a "poison inhalation hazard" or "poison gas" placard, as appropriate, on each side and each end, in addition to any other placard required for that material in §172.504. Duplication of the poison inhalation hazard or poison gas placard is not required.

b. In addition to the "radioactive" placard that may be required by §172.504(e) of this subpart, each transport vehicle, portable tank or freight container that contains 1001 pounds or more gross weight of fissile or low specific activity uranium hexafluoride must be placarded with a "corrosive" placard on each side and each end.

c. Each transport vehicle, portable tank, freight container, or unit load device that contains a material that has a subsidiary hazard of being dangerous when wet (as defined in §173.124 of this subchapter) must be placarded with "dangerous when wet" placards, on each side and each end, in addition to the placards required by §172.504.

d. Hazardous materials that possess secondary hazards may exhibit subsidiary placards that correspond to the placards described in this part, even when not required by this part (see also §172.519(b)(4) of this subpart).

Providing and Affixing Placards: Highway (§172.506)

a. Each person offering a motor carrier a hazardous material for transportation by highway must provide to the motor carrier the required placards for the material being offered prior to or at the same time the material is offered for transportation, unless the carrier's motor vehicle is already placarded for the material as required.

> ✓ **Important Note:** Transport vehicles, freight containers, unit load devices, or rail cars are always placarded as described earlier, unless the placard(s) on the package are clearly visible from the exterior of the transportation device.

> ✓ **Important Point:** The individual who offers a bulk package of hazardous materials is responsible for affixing the required placards before it can be transported. Likewise, the individual offering a rail car, freight container, or unit load device is responsible for affixing the placards to these transportation devices.

Special Placarding Provisions: Highway (§172.507)

a. Each motor vehicle used to transport a package of highway controlled quantity Class 7 (radioactive) materials must have the required radioactive warning placard placed on a square background as described in §172.527.

b. A nurse tank meeting the provisions of §173.315(m) is not required to be placarded on an end containing valves, fittings, regulators, or gauges when those appurtenances prevent the markings and placard from being properly placed and visible.

Providing and Affixing Placards: Rail (§172.508)

a. Each person offering a hazardous material for transportation by rail must affix to the rail car containing the material the placards specified by this subpart. Placards displayed on motor vehicle transport containers or portable tanks may be used to satisfy this requirement if the placards otherwise conform to the provisions of this subpart.

b. No rail carrier may accept a rail car containing a hazardous material for transportation unless the placards for the hazardous material are affixed thereto as required by this subpart.

Special Placarding Provisions: Rail (§172.510)

a. **White square background.** The following must have the specified placards placed on a white square background, as described in §172.527:

- Division 1.1 and 1.2 (explosive) materials that require "explosives 1.1" or "explosives 1.2" placards affixed to the rail car

- Materials classed in Division 2.3 Hazard Zone A or 6.1 Packing Group I Hazard Zone A that require "poison gas" or "poison" placards affixed to the rail car, including tank cars containing only a residue of the material

- Class DOT 113 tank cars used to transport a Division 2.1 (flammable gas) material, including tank cars containing only a residue of the material.

b. **Chemical ammunition.** Each rail car containing Division 1.1 or 1.2 (explosive) ammunition that also meets the definition of a material poisonous by inhalation must be placarded "explosives 1.1" or "explosives 1.2" and "poison gas" or "poison inhalation hazard."

Freight Containers and Aircraft Unit Load Devices

a. **Capacity of 640 cubic feet or more.** Each person who offers for transportation, and each person who loads and transports, a hazardous material in a freight container or aircraft unit load device having a capacity of 640 cubic feet or more must affix to the freight container or aircraft unit load device the placards specified for the material in accordance with §172.504. However:

- The placarding exception provided in §172.504(c) applies to motor vehicles transporting freight containers and aircraft unit load devices.

- The placarding exception provided in §172.504(c) applies to each freight container and aircraft unit load device being transported for delivery to a consignee immediately following an air or water shipment

- Placarding is not required on a freight container or aircraft unit load device if it is only transported by air and is identified as containing a hazardous material in the manner provided in Part 5, Chapter 2, Section 2.7 of the ICAO Technical Instructions.

b. **Capacity less than 640 cubic feet.** Each person who offers for transportation by air, and each person who loads and transports by air, a hazardous material in a freight container or aircraft unit load device having a capacity of less than 640 cubic feet must affix one placard of the type specified by Paragraph (a) of this section unless the freight container or aircraft unit load device:

- Is labeled in accordance with Subpart E of this part, including §172.406(e);

- Contains radioactive materials requiring the Radioactive Yellow III label and is placarded with one Radioactive placard and is labeled in accordance with Subpart E of this part, including §172.406(e); or

- Is identified as containing a hazardous material in the manner provided in Part 5, Chapter 2, Section 2.7 of the ICAO Technical Instructions.

When hazardous materials are offered for transportation not involving air transportation in a freight container having a capacity of less than 640 cubic feet, the freight container need not be placarded. However, if not placarded, it must be labeled in accordance with Subpart E of this part.

(a) Notwithstanding paragraphs (a) and (b) of this section, packages containing hazardous materials (other than ORM-D) offered for transportation by air in freight containers are subject to the inspection requirements of §175.30.

Bulk Packagings (§172.514)

For hazardous materials transported in bulk packagings, placards (as specified in §§172.504 and 172.505) are generally used instead of labels to communicate the hazard(s).

Each bulk packaging that is required to be placarded when it contains a hazardous material must remain placarded when it is emptied, unless it is:

- Sufficiently cleaned of residue and purged of vapors to remove any potential hazard; or
- Refilled with a material requiring different placards or no placards, to such an extent that any residue remaining in the packaging is no longer hazardous.

> ✓ **Important Note:** Packagings that are "empty" but still contain hazardous material residue or vapors are transported as if they were full. The packaging is considered empty *only* if it is completely purged of residue or refilled with a non-hazardous material so that a hazard no longer exists. "Residue" placards may no longer be used.

Exceptions. The following packagings may be placarded on only two opposite sides, or alternatively, may be labeled instead of placarded in accordance with Subpart E of this part:

(1) A portable tank having a capacity of less than 1000 gallons

(2) A DOT 106 or 110 multi-unit tank car tank

(3) A bulk packaging other than a portable tank, cargo tank, or tank car (for example, a bulk bag or box) with a volumetric capacity of less than 640 cubic feet

(4) An intermediate bulk container

Visibility and Display of Placards (§172.516)

Each placard on a motor vehicle and each placard on a rail car must be readily visible from the direction it faces except from the direction of another motor vehicle or rail car to which the motor vehicle or rail car is coupled. This requirement may be met by the placards displayed on the freight containers or portable tanks loaded on a motor vehicle or rail car.

The required placarding of the front of a motor vehicle may be on the front of a truck-tractor, instead of or in addition to the placarding on the front of the cargo body to which a truck-tractor is attached.

Labeling, Placarding, and Emergency Response

Each placard on a transport vehicle, bulk packaging, freight container, or aircraft unit load device must:

- Be securely attached or affixed thereto or placed in a holder thereon.
- Be located clear of appurtenances and devices such as ladders, pipes, doors, and tarpaulins.
- So far as practicable, be located so that dirt or water is not directed to it from the wheels of the transport vehicle;
- Be located away from any marking (such as advertising) that could substantially reduce its effectiveness, and in any case at least 3 inches away from such marking;
- Have the words or identification number (when authorized) printed on it displayed horizontally, reading from left to right;
- Be maintained by the carrier in a condition so that the format, legibility, color, and visibility of the placard will not be substantially reduced due to damage, deterioration, or obscurement by dirt or other matter;
- Be affixed to a background of contrasting color, or have a dotted or solid line outer border that contrasts with the background color.

Recommended specifications for a placard holder are set forth in Appendix C of this part. Except for a placard holder similar to that contained in Appendix C to this part, the means used to attach a placard may not obscure any part of its surface other than the borders.

A placard or placard holder may be hinged provided that the required format, color, and legibility of the placard are maintained.

> ✓ **Important Note:** A placard may be made of any plastic, metal, or other material capable of withstanding, without deterioration or a substantial reduction in effectiveness, a 30-day exposure to open weather conditions (§172.519(a)(1)).

Types of Placards

As with hazardous material labels and markings, several different placards are mandated for use in identifying specific materials being shipped. The primary purpose of these placards is twofold: (1) to alert the public to the potential dangers of hazardous materials; and (2) to guide emergency response personnel in their actions during a hazmat incident.

Shipping placards must conform to specific size specifications. For example, each placard must measure at least 10.8 inches on each side and must have a solid line inner border approximately 0.5 inches from each edge.

Hazard class or division number (as appropriate) must be shown in numbers measuring at least 1.6 inches in height.

When text is used on a placard to indicate a hazard, the printing must be in letters measuring at least 1.6 inches in height.

DOT Emergency Response (Subpart G—§172.600)

With billions of tons of hazardous material and waste transported each year throughout the U.S., and with truck transport accounting for approximately half of the hazardous material shipments per day, accidents happen. The most common hazmat accident is relatively minor (for example, a package becomes punctured or crushed, and a minor leak or spill results). However, no matter how minor the incident, hazmat employers and employees must be trained to properly respond.

Proper preparation for mitigation of hazmat incidents begins with having a written plan. The second step is to make sure that the plan is accessible to all who may have to employ it during an actual incident. The third step is to make sure that those who use the plan are properly trained—not only to mitigate the situation but to do it safely.

When we speak of "safely," we are speaking of a three-branched process. One branch requires us to ensure the responder's safety. Another branch requires us to ensure the safety of the public. The final branch of this process is to ensure the safety of the environment (air, water, and soil).

The safe handling and transport of hazardous materials, which include explosives, flammables, corrosive or toxic chemicals, spent reactor fuel, low-level radioactive wastes, and disease-causing biological agents, are major concerns to those agencies and organizations responsible for responding to incidents involving them.

What happens in the first few minutes of any hazardous material incident can determine its outcome. The actions of the first responder on the scene may therefore be crucial. It does not take a rocket scientist to determine who is most likely to be the first responder. In a trucking incident, if the driver is still able to respond, it will probably be the driver—a hazmat employee.

If the incident is responded to by professional firefighters, law enforcement officers, emergency medical services personnel, and transporters of hazardous materials, one thing is certain: They all face immediate challenges at the scene of any accident that involves or could involve hazardous materials. Thus their training is of primary importance. Indeed, an untrained responder who is overcome by the effects of the hazardous materials will simply have to be rescued along with other victims on the scene. This is a result that the DOT Hazardous Materials Regulations are trying to prevent.

Labeling, Placarding, and Emergency Response

DOT's Emergency Response Information is contained in Subpart G §172.600, 172.602, 172.604, and 172.606. We present the basic tenets of each of these sections below.

The standard general precautions the hazmat employee should take any time a hazmat incident occurs include:

1. Always protect yourself. Use protective equipment at all times.

2. Identify the hazard, both to protect yourself and to provide guidance for an appropriate response.

3. Secure the scene immediately. Keep bystanders away and cordon off the area.

4. Request assistance. Contact your dispatcher or CHEMTREC directly to request that proper authorities be notified of the incident.

> ✓ **Important Note:** CHEMTREC, the Chemical Transportation Emergency Center, is operated by the Chemical Manufacturers Association. CHEMTREC is available 24 hours a day, seven days a week. Responders can dial CHEMTREC's toll free number (1-800-424-9300) from anywhere in the U.S. CHEMTREC provides immediate advice at the scene of the emergency.

5. CHEMTREC communicators can help more efficiently if they receive the basic information needed. If the vehicle is placarded, provide the ID number from the placard or orange panel on the vehicle, and the name of the product from the shipping papers, if they are available. The shipping papers also contain information about the shipper and the point of origin and destination.

The dispatcher selects a response team based on the information gathered at the scene by the first responder and additional information and assistance provided by CHEMTREC and the manufacturer.

> ✓ **Important Note:** Emergency response information must be printed in English and offered with most hazmat shipments in one of three ways:
> - On a shipping paper
> - In a separate document that includes both the proper shipping name and technical name of the material (e.g., MSDS)
> - In a separate document that cross-references the description of the material on the shipping paper with the response information in the document (e.g., the North American Emergency Response Guidebook)

Using the *North American Emergency Response Guidebook*

The *North American Emergency Response Guidebook* is published every 3-4 years by DOT's Research and Special Programs Administration (RSPA). RSPA is responsible for making and revising the Hazardous Materials Regulations.

The purpose of the *Guidebook* is to assist responders in making initial decisions upon arriving at the scene of a dangerous goods incident. In addition, the *Guidebook* is designed to assist the driver in relaying information, if necessary, about the dangerous goods he or she is transporting to the emergency responders.

If you are a carrier, you should keep a copy of the *Guidebook* close at hand in the cab of your vehicle at all times. You will find the phone numbers of CHEMTREC (for United States), SETIQ (Mexico), and CANUTEC (Canada) in the guide, along with the phone number for the National Response Center (CHEMTREC and the National Response Center (NRC) work cooperatively to provide 24-hour assistance to emergency responders, carriers, shippers, and anyone handling hazardous materials).

If the need arises to call any of these emergency information centers, try to provide as much information as possible about the incident. Five basic pieces of information that should be provided are:

- Caller name and call back number
- Name of carrier, shipper/manufacturer or facility operator, and responsible party
- Nature, location, and time of the incident
- Name of material released or any identifying information
- Container type, railcar/truck number, vessel name, or other identifying information

Table of Placards and Initial Response Guides

The Table of Placards and Initial Response Guides to Use On-Scene (pages 14-15 in the *North American Emergency Response Guidebook*) should only be used if materials can't be specifically identified by using shipping papers, numbered placards, or the orange panel number. The placards themselves are referred to as guides. Each guide has an identifying number. For example, guide 153 on this table shows the placard for a shipment of corrosives. The number at the apex of the guide number (number "8") refers to the hazard class or division (corrosives). Page 11 of the *Guidebook* lists the hazard classes. These guides are referenced later in the *Guidebook*.

Guidebook *Yellow Pages*

The yellow section of the *Guidebook* provides a listing of chemicals chronologically in the order of their UN shipping number. For example:

Labeling, Placarding, and Emergency Response 111

ID No.	GUIDE No.	Name of Material
1590	153	Dichloroanilines

The number 1590 is the UN shipping number designation for the material dichloroanilines, which must appear on the shipping manifest. Turning back to the Table of Placards, we see that guide 153 is a placard showing the skull and crossbones (Note: Guide 153 is also used for class 8 corrosives), and the hazard class or division is 6. Class 6 is a poison.

Guidebook *Blue Pages*

The blue section contains the same information on materials found in the yellow section, but listed alphabetically.

Example:

Look up "kerosene" and obtain

1) the UN shipping number

2) the guide number you should refer to

3) the hazard class or division

From the *Guidebook* blue section we find: UN 1223, guide 128, and hazard class 3 (fuel oil, a flammable liquid).

Example:

Look up the material "london purple" and obtain

1) the UN shipping number

2) the guide number you should refer to

3) the hazard class or division

From the *Guidebook* blue section we find: UN 1621, guide 151, and hazard class 6 (Harmful: Stow Away from Foodstuffs).

Guidebook *Orange Pages*

The orange section provides specific information on the potential hazards and the emergency response actions to be taken in the event of a spill. For london purple, look up guide 151. Guide 151 provides us with information on how to handle this material in a spill or fire, as well as first aid information.

Guidebook *Green Pages*

The green section contains a Table of Initial Isolation and Protective Action Distances. This provides some guidance on safe distances to retreat in the event of a spill. As an example, if we have a hazardous material highway incident where coal gas is spilled (UN 1023), by examining the table we find that for a small spill, we would have to isolate a minimum of 100 feet in all directions and up to 0.1 mile downwind from the source. For a large spill, the initial isolation distance recommended from the tale is 100 feet in all directions, and a minimum of 0.2 miles downwind. Exactly what constitutes a large spill and a small spill is subjective and depends on personal judgement.

> ✓ **Important Note:** The *North American Emergency Response Guidebook* can be obtained From Labelmaster, 5724 N. Pulaski Road, Chicago, IL 60646-6797, 1-800-621-5808.

Summary

Training hazmat employees in the DOT labeling and placarding requirements will increase safety for all concerned. Labeling and placarding regulations go hand in hand with the regulations and training that cover the materials used to contain the hazardous materials to be transported. These materials are the hazardous materials packaging, and are the subject of Chapter 6.

Notes

[1] From *Hazardous Materials Response Handbook*. 1989. Quincy, Massachusetts: National Fire Protection Association. 1.

[2] 29 CFR 1910.1200 (F)(1)(i)(ii)(iii)

Chapter 6
Packaging and Shipping

While consulting the hazardous materials tables, 49 CFR Part 172 to determine proper classification and shipping descriptions, the shipper should also determine proper packaging. Among the factors influencing the choice of package are quantity of hazardous material per package, cushioning material, proper closure and reinforcement, and proper pressure. The mode of transport selected also influences the method of packaging.[1]

Introduction

After manufacture and processing, hazardous materials may be used on site for various applications, or, more likely, are shipped from the manufacturer and/or processor to another end-user. Because the materials we are discussing are "hazardous," their proper delivery is crucial to the safety of everyone involved in their handling and transport. To help ensure their safe handling and transport, DOT requires that hazardous materials be properly packaged.

> ✓ **Important Note:** *Packaging* refers to receptacles or containers that meet DOT requirements of 49 CFR.

To comply with DOT's packaging requirements, certain steps must be followed (see Figure 6.1). This chapter briefly discusses DOT's packaging requirements.

> ✓ **Important Note:** Specific instructions for all types of packagings and all modes of transportation can be found in the Hazardous Material Regulations, Part 173: Shippers General Requirements for Shipments and Packagings.

Figure 6.1. Steps to packaging compliance

Packagings and Packages: General Requirements (§173.24)

This subchapter applies to bulk and non-bulk packagings, both new and reused packagings, and specification and non-specification packagings.

Each package used for the shipment of hazardous materials must be designed, constructed, maintained, filled, its contents limited, and closed, so that under conditions normally incident to transportation:

1. No identifiable (without the use of instruments) release of hazardous materials to the environment can occur

2. The effectiveness of the package will not be substantially reduced (for example, impact resistance, strength, packaging compatibility, etc. must be maintained for the minimum and maximum temperatures encountered during transportation)

3. No mixture of gases or vapors in the package occur that could, through any credible spontaneous increase of heat or pressure, significantly reduce the effectiveness of the packing.

Authorized Packagings

The packaging prescribed or permitted for the hazardous material in a packing section is specified for that material in column 8 of the §172.101 table; the material must also conform to applicable requirements in the special provisions of column 7 of the §172.101 table.

> ✓ **Important Note:** Packing authorizations can be found in columns (8A)-(8C) of the Hazardous Materials Table.

Specification Packagings

A specification packaging, including a UN standard packaging manufactured in the United States, must conform in all details to the applicable specification or standard in Part 178 or Part 179.

A UN standard packaging manufactured outside the United States, in accordance with national or international regulations based on the UN Recommendations on the Transport of Dangerous Goods, may be incorporated and used and is considered to be an authorized packaging under the provisions of this section, subject to the following conditions and limitations:

1. The packaging fully conforms to applicable provisions in the UN Recommendations on the Transport of Dangerous Goods and the requirements of this subpart, including reuse provisions;

2. The packaging is capable of passing the prescribed tests in Part 178 of this subchapter applicable to that standard; and

3. The competent authority of the country of manufacture provides reciprocal treatment for UN standard packagings manufactured in the United States.

Compatibility

The person offering a hazardous material for transportation has the responsibility of ensuring that such packagings are compatible with their lading. This particularly applies to corrosivity, permeability, softening, premature aging, and embrittlement.

Packaging materials and contents must be designed and manufactured to allow no significant chemical or galvanic reaction between the materials and contents of the package.

Plastic used in packagings and receptacles must be of a type compatible with the lading and may not be permeable to the extent that a hazardous condition is likely to occur during transportation, handling, or refilling.

Hazardous materials may not be mixed or packed together in the same outer packaging with other hazardous or nonhazardous materials if such materials are capable of reacting dangerously with each other and causing:

- combustion or dangerous evolution of heat
- evolution of flammable, poisonous, or asphyxiant gases
- formation of unstable or corrosive materials.

Packagings used for solids that may become liquid at temperatures likely to be encountered during transportation must be capable of containing the hazardous materials in the liquid state.

Standard Packaging

Standard hazardous materials packaging consists of outer packaging and inner packaging, or combination packaging. Combination packaging is comprised of one or more inner packagings used in combination with a non-bulk outer packaging (not including a composite packaging).

The outermost packaging is a combination or composite package used with cushioning or absorbent materials designed to protect and contain the inner packagings.

We are concerned with three types of packagings: single packaging, composite packaging, and overpacks.

Single packaging is usually a non-bulk packaging other than a combination packaging. The package is simply a single receptacle into which the material is directly loaded (for example, a single drum).

Composite packaging is a combination outer packaging + an inner receptacle. Both the inner receptacle and outer packaging form an integral packaging (for example, a drum with an inner lining).

Overpacks are enclosures used by a single consignor to provide either protection or convenience in the handling operation, or to consolidate two or more single packages. To be in compliance for overpacks, specific rules must be followed:

- Each package in an overpack must be in full compliance with all applicable packaging regulations. Each package must be properly marked and labeled as per §173.25.
- The markings and labels on each of the packages overpacked **must** be reproduced on the outside of the overpack. An exception to this rule is that if the overpacked packages are visible from the outside of the overpack, remarking and labeling is not necessary (§173.25(a)(2)).

Packaging and Shipping

- Packages are subject to orientation markings as per §173.312 (§173.(a)(3)).

- Overpacks of specification packagings *must* be marked with an indication that the overpacked packages inside comply with prescribed specifications (§173.25(a)(5)).

Performance-Oriented Packaging (POP)

The *Performance-Oriented Packaging (POP) Standard* was developed by the UN Committee of Experts on the Transport of Dangerous Goods. It provides for packaging safety while facilitating the free flow of these packagings within international commerce. These standards can be found in the *"UN Recommendations for the Transport of Dangerous Goods"* (known as the "Orange Book"), which serves as the basis and authority for the performance-oriented packaging requirements. It also serves as the basis for the performance-oriented packaging standards adopted by individual countries. The 49 CFR authorizes the use of the UN Performance-Oriented Packaging Standards for non-bulk packagings for most domestic transportation of hazardous materials §171.14(a). Unless excepted, the use of UN Performance-Oriented Packaging Standards is mandatory. UN POP standards apply to non-bulk packagings; they do not apply to cylinders, radioactive materials, and bulk packagings.

The UN Performance Oriented Packaging Standard is based on the anticipated performance of the packaging within the transportation system, rather than the detailed design and construction of packaging. The objective of POP standards is to secure the level of safety desired by ensuring the ability of a given package type to successfully pass a series of performance tests designed to test its ability to withstand the normal conditions incident to transportation, rather than its design or construction. POP tests are generally based on the packing group for which the packaging is being tested or qualified. Remember, the packing group gives the degree of the hazard, while the hazard class gives the type of hazard.

POP Design Tests

Recall that the performance criteria in the Hazardous Material Regulations are based on the UN Recommendation for the Transport of Dangerous Goods. In light of this, each POP packaging design must pass specific tests, including:

Drop—to contain and protect the dangerous goods if the package is dropped

Leakproofness—to prevent leakage of liquids under conditions of normal transport

Hydrostatic pressure—to prevent leakage of liquids under pressure

Stacking—to maintain stability within a stack while stacked with similar packages

> ✓ **Important Note:** In addition to the standard tests listed, hazardous material packagings transported in the U.S. must be capable of withstanding the vibration standard.

In the actual testing process, the Hazardous Material Regulation specifies how many samples must be tested, how often testing is required, and how each test must be conducted. Packaging manufacturers are required to mark each package that meets the POP standard.

> ✓ **Important Note:** POP marking is made up of a series of symbols, numbers, and letters that indicate specific characteristics of the package (see next section, "POP Marking Codes").

POP Marking Codes

Markings must be stamped, embossed, printed, burned, or otherwise marked on the packaging in a way that provides adequate accessibility, permanency, legibility, and contrast (§178.3(a)(3)).

All POP markings must contain codes for the following information, in the order listed below.

1. UN symbol or letters ("u" over the top of the letter "n" inside a circle) must be on embossed metal packagings.

2. Packaging code: the type of packaging, the material of construction, and the category of packaging (§178.503(a)(2)). The most commonly used codes are found in Table 6.1.

Table 6.1 Most Commonly Used Packaging Codes

Type	Material	Category
1 Drums	A Steel	A, B, or H Drums-Jerricans
2 Barrels	B Aluminum	1 Closed Head 2 Open Head
3 Jerricans	C Natural Wood	A or B Boxes
4 Boxes	D Plywood	1 Ordinary A or B
5 Bags	F Reconstituted Wood	2 A or B w/inner liner or coating
6 Composite Packagings	G Fiber	C Boxes
	H Plastic	1 Ordinary 2 w/sht proof walls
7 Pressure Receptacle	L Textile	H Boxes
	M Paper, multiwall	1 Expanded Plastic 2 Solid Plastic
	N Metal other than steel or aluminum	L Bags 2 Sift proof
	P Glass Porcelain, or stoneware	M Bags
		2 Multi wall, water resistant

Packaging and Shipping

Examples: From Table 6.1, we see that the outer packaging code 4D is for a plywood box; 4H2 is for a solid plastic box; 1G is for a fiber drum, and so forth.

3. Performance: an upper case X, Y, or Z indicates the performance standard or level for which the packaging has been successfully tested. The following codes should be committed to memory:

X   Meets Packing Group I, II, and III

Y   Meets Packing Group II, and III

Z   Meets Packing Group III only

4. Specific gravity (relative density or gross mass): a number designating either the specific gravity or gross mass, as appropriate, for which the packing has been successfully tested should come right after the performance standard code (§178.503(a)(4)).

> ✓ **Important Note:** Packagings without inner packagings intended to contain liquids must be marked with the specific gravity rounded down to the first decimal place for which the packaging has been successfully tested. The specific gravity is not required when the specific gravity does not exceed 1.2.

5. Hydrostatic test pressure: for liquids or gases or "S" for solids or inner packagings.

6. Year the package was manufactured.

7. Country in which the package was manufactured.

8. Name and address or registration number of manufacturer.

9. Minimum thickness (for reusable or reconditioned packagings). Note: Additional information is required for reconditioned containers.

The following is an example of the amount of information that can be obtained from the manufacturer's marking on the package.

U
    4G/Y145/S/94USA/VL824
N

"4G" means fiberboard box
"Y" means it meets Packing Groups II and III tests
"145" means gross weight of solid
"S" means solids
"94" is the year of manufacture
"USA" country the package was manufactured in
"VL824" is the code of the manufacturer

Shipping Papers (§172.200)

Because time is critical in an emergency, information on hazardous materials must be capable of being easily and quickly found on the shipping paper. Therefore, each person who offers a hazardous material for transportation must properly describe the material on the shipping paper. DOT defines a *shipping paper* as a bill of lading (the straight bill of lading is the most commonly used), a shipping order, manifest, or other shipping document serving a similar purpose and containing the required information.

When hazardous materials are the shipment or part of the shipment, the shipping paper must contain certain specific items of information. The DOT basic shipping description consists of the following parts, all found in the Hazardous Materials Table and required by §172.202:

- Proper shipping name
- Hazard class or division number
- UN (United Nations) or NA (North America) identification (ID) number
- Packing Group (PG) number, in Roman numerals. [Note: a few materials do not have a Packing Group number].

When a description of a hazardous material must be included on a shipping paper, that description must conform to the following requirements (§172.201):

- When a hazardous material and a material not subject to the requirements of this subchapter are described on the same shipping paper, the hazardous material description entries required by §172.202 and those additional entries that may be required by §172.203:
    - Must be entered first, or
    - Must be entered in a color that clearly contrasts with any description on the shipping paper of a material not subject to the requirements of this subchapter, except that a description on a reproduction of a shipping paper may be highlighted, rather than printed, in a contrasting color, or
    - Must be identified by the entry of an "X" placed before the proper shipping name in a column captioned "HM." (Note: the "X" may be replaced by "RQ" (reportable quantity), if appropriate).
- The regulations require that the basic description must be legibly printed (manually or mechanically), and spelled correctly in English.
- Unless it is specifically authorized or required, the required shipping description may not contain any code or abbreviation.
- A shipping paper may contain additional information concerning the material, provided that the information is not inconsistent with the required

description. Unless otherwise permitted or required, additional information must be placed after the basic description required by §172.202.

- A shipping paper may consist of more than one page, if each page is consecutively numbered and the first page bears a notation specifying the total number of pages included in the shipping paper. For example, "Page 1 of 4 pages."

- A shipping paper must contain an emergency response telephone number. The contact information of a knowledgeable person or someone who has immediate access to such a person must be on the shipping papers.

- The quantity of material shipped, by weight, volume, etc.

In addition to the information required on shipping papers discussed to this point, certain additional information may also be required on shipping papers. For example,

- When certain N.O.S. (not otherwise specified) or other shipping names are used that do not specifically identify the hazardous material or hazardous ingredient, the technical name of the material or ingredient must be entered in parentheses in association with the basic description. Usually this means after the proper shipping name or after the packing group number. If the material is a mixture containing two or more hazardous material components, the technical names of at least two components most predominately contributing to the hazard must be entered.

- If the material is a hazardous substance, the letters "RQ" (reportable quantity) must be entered either before or after the basic description. When a package has been emptied but contains a residue that meets the hazardous substance definition, the letters "RQ" must be entered.

- In addition, if the proper shipping name does not identify the hazardous substance(s) by name (as shown in Appendix A to the Hazardous Materials Table), the name(s) of the hazardous substance component(s) must be shown in parentheses in association with the basic description. The word "contains" may precede the component names if desired.

- If the material is a marine pollutant and is transported in bulk packaging via air, rail, or truck, or in any size packaging by vessel, the words "marine pollutant" must be added in association with the basic description. If the proper shipping name does not identify the marine pollutant component by means as shown in Appendix B to the Hazardous Materials Table, the name of the component(s) must be entered in parentheses in association with the basic description.

- The words "residue: last contained" must precede the proper shipping name for tank cars that have been unloaded but still contain the residue of a hazardous material, and for **any** type of packaging that contains a residue that is a hazardous substance.

- Unless already included in the proper shipping name for waste materials, the word "waste" must precede the proper shipping name for a hazardous waste.

- If a liquid material in a package meets the definition of an elevated temperature material (other than molten sulfur or molten aluminum) and that fact is not disclosed in the proper shipping name, the word "hot" must immediately precede the proper shipping name.

- In the case that a material is a poison (as defined by Division 6.1, Packing Group I or II), and is not disclosed as such in the shipping name or class entry, the word "poison" must be entered on the shipping paper in association with the shipping description. If the material is poisonous by inhalation, the words "poison - inhalation hazard" followed by the hazard zone designation must be entered immediately following the basic description. The word "poison" need not be repeated if it otherwise appears in the shipping description.

- As specified in 49 CFR 172.203 (d), additional information is required for radioactive materials.

- The notation "dangerous when wet" must be entered in association with the basic description for any material meeting the Class 4.3 definition.

- The notation "limited quantity" or "LTD QTY" must be added following the basic description for any shipment moving under a limited quantity exception.

- For shipments moving under a DOT exemption, the notation "DOT-E" must be added, followed by the exemption number assigned, and so located that it is clearly associated with the description to which it applies.

> ✓ **Important Note:** Although not required, other information **may**, at the shipper's option, be included on the shipping paper (for example, flashpoint, trade name, subsidiary hazard class, etc.).

Summary

Packaging is the first line of defense against spills during handling and transportation, ensuring safety for the transporter, the population, and the environment. Shipping papers with the proper information will save time in the event of an emergency. It is important to train employees on how to properly package hazardous materials and prepare shipping papers with complete information.

Notes

[1] From Kenworthy, W. E. 1992. *Transportation of Hazardous Materials*, 2nd ed. Rockville, Maryland: Government Institutes, Inc. 116.

Chapter 7
Function-Specific Training

Introduction

So far in this text, we have stressed the importance (and the requirement under §177.800(c)) of providing the hazmat employee with safety, general awareness, and familiarization training. General awareness/familiarization training teaches hazmat employees the requirements of the Hazardous Materials Regulations and enables them to recognize and identify hazardous materials using uniform procedures found in the these regulations. General awareness training includes the basic concepts associated with:

- Classification
- The Hazardous Materials Table
- Labeling
- Marking
- Placarding
- Emergency Response
- Packaging

But the above list does not cover everything that hazmat employees need to know to perform their hazmat jobs safely. A couple of key ingredients are still required to ensure that safe material handling and transportation occurs. Specifically, we have not discussed hazmat duties beyond those of the manufacturer (hazmat employer), shipper, and transporter. We have said little about those additional hazmat employees who must be trained. These employees include personnel involved with handling and storage, clerical personnel, dispatch, sales, laboratory/research and development, training/safety personnel, administrative/management, mechanics, production personnel, and purchasing agents.

Function-specific training is intended to teach the hazmat employee the specific tasks of the job. The purpose of function-specific training is to increase a hazmat employee's awareness of safety considerations and regulatory requirements, thereby reducing the occurrence of hazardous material incidents caused by human error. Included in this training is job-specific information that is not covered under general awareness/familiarization training. Because it is wide-ranging, function-specific training can cover the use of specific packagings, tools, procedures, materials, or any other variations required to help determine what additional training may be needed. Function-specific training focuses on each hazmat job, and covers any information necessary to do the job correctly according to the guidelines provided in the Hazardous Material Regulations. To determine which employees need function-specific training, employers analyze the functions. Does the job require driving, packaging, loading, or auxiliary functions such as preparing manifests, using MSDSs, or labeling cargo?

That an organizational purchasing agent must receive function-specific training may come as a surprise to many readers. Some might ask: "Why does someone involved in purchasing need to have hazmat training or function-specific training?" Purchasing agents are typically tasked with writing specifications for hazardous materials packaging. Thus, the purchasing agent must know the ins and outs of both hazmat packagings and hazardous materials.

In the example used above, purchasing agents who write specifications for hazardous material packaging (or other requirements) must actually receive training over and above that provided in the general awareness and familiarization training. For example, the function-specific training that purchasing agents should receive must include a thorough introduction to and an understanding of the Hazardous Materials Table. Shipping papers are also obviously important to purchasing agents; they must be very familiar with these standard shipping documents. An understanding of packaging and package marking, labeling, placarding, and loading and proper stowage are also required. Purchasing agents should also be familiar with the organization's emergency response procedure.

This chapter on function-specific training provides information that these additional personnel should have to work with and around hazardous materials as per regulatory requirements. After we discuss function-specific training requirements, we briefly discuss proper loading and storage requirements. Keep in mind that drivers must also receive function-specific training. Function-specific training for drivers must include all procedures not directly connected to driving (for example, proper loading and unloading techniques, or how to inspect valves on cargo tanks). Chapter 8 covers the training information necessary for hazmat drivers.

> ✓ **Important Note:** Keep in mind that 800(c) specifically states that a carrier may not transport a hazardous material by motor vehicle unless each of its hazmat employees involved in that transportation (whatever the involvement might be) is trained as required by this part and Subpart H of Part 172 of this subchapter.

Loading/Unloading (§177.834) and Storage

One of the primary function-specific duties for hazmats is loading and storing hazardous materials. Specific information on loading and storing hazardous material containers for all modes of transportation is located in the Hazardous Materials Regulations. Specifically, Part 177, Carriage by Public Highway, Subpart B: Loading and Unloading, stipulates loading and storing requirements for motor vehicles.

General Loading and Unloading Requirements

DOT stipulates that hazmat employees be made aware of the following loading and unloading requirements:

1. **Packages secured in a vehicle.** Any tank, barrel, drum, cylinder, or other packaging, not permanently attached to a motor vehicle, which contains any Class 3 (flammable liquid), Class 2 (gases), Class 8 (corrosive), Division 6.1 (poisonous), or Class 7 (radioactive) material must be secured against movement within the vehicle on which it is being transported, under conditions normally incident to transportation.

2. **No smoking while loading or unloading.** Smoking on or about any motor vehicle while loading or unloading any Class 1 (explosive), Class 3 (flammable liquid), Class 4 (flammable solid), Class 5 (oxidizing), or Division 2.1 (flammable gas) materials is forbidden.

3. **Keep fire away, loading and unloading.** Extreme care must be taken in the loading or unloading of any Class 1 (explosive), Class 3 (flammable liquid), Class 4 (flammable solid), Class 5 (oxidizing), or Division 2.1 (flammable gas) materials into or from any motor vehicle to keep fire away and to prevent persons in the vicinity from smoking, lighting matches, or carrying any flame or lighted cigar, pipe, or cigarette.

4. **Handbrake set while loading and unloading.** No hazardous material shall be loaded into or on, or unloaded from, any motor vehicle unless the handbrake be securely set and all other reasonable precautions be taken to prevent motion of the motor vehicle during such loading and unloading process.

5. **Use of tools, loading and unloading.** No tools that are likely to damage the effectiveness of the closure of any package or other container or likely adversely to affect such package or container, must be used for the loading or unloading of any Class 1 (explosive) material or other dangerous article.

6. **Prevent relative motion between containers.** Containers of Class 1 (explosive), Class 3 (flammable liquid), Class 4 (flammable solid), Class 5 (oxidizing), Class 8 (corrosive), Class 2 (gases) and Division 6.1 (poisonous) materials must be so braced as to prevent motion thereof relative to the vehicle while in transit. Containers having valves or other fittings must be so loaded that there will be the minimum likelihood of damage thereto during transportation.

7. **Precautions concerning containers in transit; fueling road units.** Reasonable care should be taken to prevent undue rise in temperature of containers and their contents during transit. There must be no tampering with such container or the contents thereof nor any discharge of the contents of any container between point of origin and point of billed destination. Discharge of contents of any container, other than a cargo tank or IM portable tank, must not be made prior to removal from the motor vehicle. Nothing contained in this paragraph shall be so construed as to prohibit the fueling of machinery or vehicles used in road construction or maintenance.

8. **Attendance requirements. (1) Loading.** A cargo tank must be attended by a qualified person at all times when it is being loaded. The person who is responsible for loading the cargo tank is also responsible for ensuring that it is so attended. **(2) Unloading.** A motor carrier who transports hazardous materials by a cargo tank must ensure that the cargo tank is attended by a qualified person at all times during unloading. However, the carrier's obligation to ensure attendance during unloading ceases when:

    a. The carrier's obligation for transporting the materials is fulfilled

    b. The cargo tank has been placed upon the consignee's premises

    c. The motive power has been removed from the cargo tank and removed from the premises

Except for unloading operations subject to §§177.840(p) or (q), a qualified person "attends" the loading or unloading of a cargo tank if, throughout the process, he is alert and is within 25 feet of the cargo tank. The qualified person attending the unloading of a cargo tank must have an unobstructed view of the cargo tank and delivery hose to the maximum extent practicable during the unloading operation.

A person is "qualified" if he or she has been made aware of the nature of the hazardous material that is to be loaded or unloaded, has been instructed on the procedures to be followed in emergencies, is authorized to move the cargo tank, and has the means to do so.

Function-Specific Training 127

9. **Manholes and valves closed.** A person may not drive a cargo tank and a motor carrier may not permit a person to drive a cargo tank motor vehicle containing a hazardous material regardless of quantity unless:

   a. All manhole closures are closed and secured

   b. All valves and other closures in liquid discharge systems are closed and free of leaks.

10. **Use of cargo heaters when transporting certain hazardous material.** Transportation includes loading, carrying, and unloading.

    a. **When Transporting Class 1 (explosive) Materials.** A motor vehicle equipped with a cargo heater of any type may transport Class 1 (explosive) materials only if the cargo heater is rendered inoperable by: (1) draining or removing the cargo heater fuel tank; and (2) disconnecting the heater's power source.

    b. **When Transporting Certain Flammable Material—(1) Use of combustion cargo heaters.** A motor vehicle equipped with a combustion cargo heater may be used to transport Class 3 (flammable liquid) or Division 2.1 (flammable gas) materials only if each of the following requirements are met:

    - It is a catalytic heater

    - The heater's surface temperature cannot exceed 54°C (130°F) either on a thermostatically controlled heater or on a heater without thermostatic control when the outside or ambient temperature is 16°C (61°F) or less

    - The heater is not ignited in a loaded vehicle

    - There is no flame, either on the catalyst or anywhere in the heater

    - The manufacturer has certified that the heater meets the requirements under Paragraph (l)(2)(i) of this section by permanently marking the heater "**meets dot requirements for catalytic heaters used with flammable liquid and gas.**"

    - The heater is marked "**do not load into or use in cargo compartments containing flammable liquid or gas if flame is visible on catalyst or in heater.**"

    - Heater requirements under §393.77 of this title are complied with.

11. **Tanks constructed and maintained in compliance with Spec 106A or 110A.** Tanks authorized for the shipment of hazardous materials by highway in Part 173 must be carried in accordance with the following requirements:

    a. Tanks must be securely chocked or clamped on vehicles to prevent any shifting

b. Equipment suitable for handling a tank must be provided at any point where a tank is to be loaded upon or removed from a vehicle

   c. No more than two cargo carrying vehicles may be in the same combination of vehicles.

12. **Specification 56, 57, IM 101, and IM 102 portable tanks.** When loaded, portable tanks may not be stacked on each other nor placed under other freight during transportation by motor vehicle.

13. **Unloading of IM portable tanks.** An IM portable tank may be unloaded while remaining on a transport vehicle with the power unit attached if the tank meets the outlet requirements in §178.345-11 of this subchapter and the tank is attended by a qualified person during the unloading in accordance with the requirements in Paragraph (i) of this section.

Separation of Hazardous Materials (177.842)

Separation distances are established for transporting radioactive materials and are also required for people and cargo compartment dividing petitions.

Specifically, §177.842 states the packages of Class 7 (radioactive) material bearing "radioactive yellow-II" or "radioactive yellow-III" labels may not be placed in a transport vehicle, storage location, or any other place closer than the distances shown in Table 7.1 to any area which may be continuously occupied by any passenger, employee, or animal, nor closer than the distances shown in Table 7.1 to any package containing undeveloped film (if so marked), and must conform to the following conditions:

- If more than one of these packages is present, the distance must be computed from Table 7.1 on the basis of the total transport index number determined by adding together the transport index number on the labels on the individual packages and overpacks in the vehicle or storeroom.

- Where more than one group of packages is present in any single storage location, a single group may not have a total transport index greater than 50. Each group of packages must be handled and stowed no closer than 20 feet (measured edge to edge) to any other group.

Function-Specific Training

Table 7.1. Separation Distance Table

Total Transport Index	Minimum separation distance in feet to nearest undeveloped film for various time of transit					Minimum distance in feet to area of persons, or minimum distance in feet from dividing partition of cargo compartments
	2 hours	2-4 hours	4-8 hours	8-12 hours	Over 12 hours	
None	0	0	0	0	0	0
0.1 to 1.0	1	2	3	4	5	4
1.1 to 5.0	3	4	6	8	11	15
5.1 to 10.1	3	4	6	8	11	2
10.1 to 20.0	5	8	12	16	22	4
20.1 to 30.0	7	10	15	20	29	5
30.1 to 40.0	8	11	17	22	33	6
40.1 to 50.0	9	12	19	24	36	7

Segregation of Hazardous Materials (§177.848)

Certain hazardous materials can't be carried on the same load. For example, acids can't be transported with cyanides or cyanide mixtures. A segregation table (see Table 7.2) provides a reference for segregating certain hazardous materials.

Specifically, §177.848 states that it applies to materials that meet one or more of the hazard classes and are:

- In packages that require labels in accordance with Part 172
- In a compartment within a multi-compartmented cargo tank subject to the restrictions in §173.33
- In a portable tank loaded in a transport vehicle or freight container

When a transport vehicle is to be transported by vessel, other than a ferry vessel, hazardous materials on or within that vehicle must be stored and segregated in accordance with §176.83(b).

Hazardous materials may not be loaded, transported, or stored together, except as provided in this section, and in accordance with Table 7.2.

Table 7.2. Segregation Table for Hazardous Materials

Class or division	Notes	1.1 1.2	1.3	1.4	1.5	1.6	2.1	2.2	2.3 gas Zone A	2.3 gas Zone B	3	4.1	4.2	4.3	5.1	5.2	6.1 liquids PGI Zone A	7	8 liquids only
Explosives 1.1 and 1.2	A	*	*	*	*	*	X	X	X	X	X	X	X	X	X	X	X	X	X
Explosives 1.3		*	*	*	*	*	X	X	X	X	X	X	X	X	X	X	X	X	X
Explosives 1.4		*	*	*	*	*	O		O	O	O	X	O	X	X	X	O		O
Very insensitive explosives 1.5	A	*	*	*	*	*	X	X	X	X	X	X	X	X	X	X	X	X	X
Extremely insensitive explosives 1.6		*	*	*	*	*													
Flammable gases 2.1		X	X	O	X			X	O								O		
Non-toxic, non-flammable gases 2.2		X			X														
Poisonous gas Zone A 2.3		X	X	O	X		X				X	X	X	X	X	X	X		X
Poisonous gas Zone B 2.3		X	X	O	X		O				O	O	O	O	O	O			O
Flammable liquids 3		X	X	O	X				X	O			O		O		X		O
Flammable solids 4.1		X			X				X	O							X		X
Spontaneously combustible materials 4.2		X	X	O	X				X	O							X		O
Dangerous when wet materials 4.3		X	X		X				X	O							X		O
Oxidizers 5.1	A	X	X	O	X		O		X	O	O						X		O
Organic Peroxides 5.2		X	X	O	X		O		X	O	X	X	X	X	X		X		O
Poisonous liquids PG I Zone A 6.1		X	X		X														X
Radioactive materials 7		X			X														
Corrosive liquids 8		X	X	O	X				X	O		O	X	O	O	O	X		

Using the Segregation Table

Instructions for using the segregation table for hazardous materials are as follows:

1. The absence of any hazard class or division or a blank space in the table indicates that no restrictions apply.

2. The letter "X" in the table indicates that these materials may not be loaded, transported, or stored together in the same transport vehicle or storage facility during the course of transportation.

3. The letter "O" in the table indicates that these materials may not be loaded, transported, or stored together in the same transport vehicle or storage facility during the course of transportation unless separated in a manner that, in the event of leakage from packages under conditions normally incident to transportation, commingling of hazardous materials would not occur. Notwithstanding the methods of separation employed, Class 8 (corrosive) liquids may not be loaded above or adjacent to Class 4 (flammable) or Class 5 (oxidizing) materials; except that shippers may load truckload shipments of such materials together when it is known that the mixture of contents would not cause a fire or a dangerous evolution of heat or gas.

4. The "*" in the table indicates that segregation among different Class 1 (explosive) materials is governed by the compatibility table in Paragraph (f) of this section.

5. The note "A" in the second column of the table means that, notwithstanding the requirements of the letter "X," ammonium nitrate (UN1942) and ammonium nitrate fertilizer may be loaded or stored with Division 1.1 (Class A explosive) or Division 1.5 (blasting agents) materials.

Summary

General awareness training and function-specific training are critically important to the safe handling and transportation of hazardous materials. Employees whose jobs require loading or unloading hazardous materials must have proper training in these procedures. Knowledge of proper separation and segregation of hazardous materials will protect employees working with hazardous materials. Chapter 8 covers driver training, another essential type of function-specific training and one that is essential to safe hazmat handling and transportation.

Chapter 8
Driver Training

Driver training constitutes ... [a] grey area, which has plagued the trucking industry and government regulators ... involving the safety issue. Realizing that no federal or state requirements exist for tractor-trailer drivers and no federal standards were provided, the FHA, in 1984, developed proposed minimum standards for training programs. It followed a year later with a model curriculum for training tractor-trailer drivers.[1]

Introduction

When hazardous materials are properly and safely manufactured, processed, packaged, labeled, and handled, hazmat employers and employees, the public, and the environment all win. They are unaffected by the potentially dangerous properties of the hazardous materials involved.

However, tragic consequences can only be avoided if the final major step in the process is conducted as safely as the others—transportation of the hazardous materials.

At any given point of any given day, various hazardous materials are crisscrossing this country (and others) from point of manufacture to shipper to end-user. While rail, water, and air transport considerable quantities of hazardous materials, vehicles using the same highways the rest of us use transport the majority of finished products containing hazardous materials.

Through experience, DOT and other regulatory agencies have come to recognize that the transport of hazardous materials is the weakest, most dangerous link in the hazmat chain. In light of this, and because so much of the total bulk of hazardous material is transported by truck on common highways, measures had to be taken to ensure safe delivery of the hazardous loads.

To shore up this weak transportation link in the hazardous material delivery chain, the Federal government requires hazardous material drivers to be hazmat trained (on general awareness, Safety and function-specific topics) and trained on the safe operation of the motor vehicle they intend to operate, as well as the requirements of the Federal Motor Carrier Safety Regulations (49 CFR 390-397).

The Federal Motor Carrier Safety Regulations include guidelines on:

- General issues
- Qualifications of drivers
- Driving of motor vehicles
- Parts and accessories necessary for safe operation
- Hours of service of drivers
- Inspection, repair, and maintenance
- Transportation of hazardous materials; driving and parking rules

This chapter focuses on driver training and what that training entails. As we progress through this important information, keep in mind that the rules and requirements included herein, along with the DOT's Substance Abuse Guidelines (see Chapter 9), were designed to prevent the type of tragic incident that we described in the prelude to this text.

DOT-Required Driver Training

Hazmat drivers must be trained on several DOT-required safety elements, including:

- Pre-trip safety inspection
- Use of vehicle controls and equipment, including use of emergency equipment
- Operation of the vehicle
- Procedures for maneuvering about tunnels, bridges, and railroad crossings
- Requirements on vehicle attendance, parking, smoking, routing, and incident reporting
- Loading and unloading of materials
- Specialized training for drivers who operate a cargo tank or vehicle with a portable tank with a capacity of 1,000 gallons or more

Driver Training

Pre-Trip Safety Inspection (§392.7)

This section of the Motor Carrier Safety Regulations specifies that no commercial motor vehicle is to be driven unless the driver has satisfied himself/herself that the following parts and accessories are in good working order, nor shall any driver fail to use or make use of such parts and accessories when and as needed:

- Service brakes, including trailer brake connections
- Parking (hand) brake
- Steering mechanism
- Lighting devices and reflectors
- Tires
- Horn
- Windshield wiper or wipers
- Rear-vision mirror or mirrors
- Coupling devices

Figure 8.1 illustrates a standard pre-trip vehicle inspection form.

Use of Vehicle Controls and Equipment, Including Operation of Emergency Equipment (§392.7, §192.8, and §393.95)

In addition to knowing how perform a pre-trip inspection and what equipment to inspect, the driver must also know how to properly operate the equipment listed in the section above.

In addition, the driver must ensure that the emergency equipment required by §393.95 is in place and ready for use. Drivers must also make use of the equipment listed in the next section when and as needed.

Emergency Equipment

Except for lightweight vehicles, every bus, truck, and truck-tractor, and every driven vehicle in driveaway-towaway operations must be equipped with the following emergency equipment:

- **Fire extinguisher.** Every power unit must be equipped with a fire extinguisher that is properly filled and located so that it is readily accessible for use. The fire extinguisher must be securely mounted on the vehicle. The fire extinguisher must be designed, constructed, and maintained to permit visual determination of whether it is fully charged. The fire extinguisher must have an extinguishing agent that does not need protection from freezing. The fire extinguisher must not use a vaporizing liquid that gives off vapors more toxic than

DRIVER'S VEHICLE INSPECTION REPORT
AS REQUIRED BY THE D.O.T FEDERAL MOTOR CARRIER SAFETY REGULATIONS

CARRIER:_____
ADDRESS:_____
DATE:_____ TIME:_____ A.M._____ P.M.

CHECK ANY DEFECTIVE ITEM AND GIVE DETAILS UNDER "REMARKS"

TRACTOR/TRUCK NO._____ ODOMETER READING_____

- ☐ Air Compressor
- ☐ Air Lines
- ☐ Battery
- ☐ Brake Accessories
- ☐ Brakes, Parking
- ☐ Brakes, Service
- ☐ Clutch
- ☐ Coupling Devices
- ☐ Defroster/Heater
- ☐ Drive Line
- ☐ Engine
- ☐ Fifth Wheel
- ☐ Frame and Assembly
- ☐ Front Axle
- ☐ Fuel Tanks
- ☐ Generator

- ☐ Horn
- ☐ Lights
  - Head – Stop
  - Tail – Dash
  - Turn Indicators
- ☐ Mirrors
- ☐ Muffler
- ☐ Oil Pressure
- ☐ Radiator
- ☐ Rear End
- ☐ Reflectors
- ☐ Safety Equipment
  - Fire Extinguisher
  - Reflective Triangles
  - Flags – Flares - Fuses
  - Spare Bulbs & Fuses
  - Spare Seal Beam

- ☐ Suspension System
- ☐ Starter
- ☐ Steering
- ☐ Tachograph
- ☐ Tires
- ☐ Tire Chains
- ☐ Transmission
- ☐ Wheels and Rims
- ☐ Windshield Wipers
- ☐ other

TRAILER(S) NO.(S)_____

- ☐ Brake Connections
- ☐ Brakes
- ☐ Coupling Devices
- ☐ Coupling (King) Pin
- ☐ Doors
- ☐ Hitch
- ☐ Landing Gear

- ☐ Lights – All
- ☐ Roof
- ☐ Suspension System
- ☐ Tarpaulin
- ☐ Tires
- ☐ Wheels and Rims
- ☐ Other

Remarks:_____
_____
_____
_____
_____
_____
_____
_____

☐ CONDITION OF THE ABOVE VEHICLE IS SATISFACTORY
DRIVER'S SIGNATURE_____

☐ ABOVE DEFECTS CORRECTED
☐ ABOVE DEFECTS NEED NOT BE CORRECTED FOR SAFE OPERATION OF VEHICLE.
MECHANIC'S SIGNTURE:_____ DATE:_____
DRIVER'S SIGNATURE:_____ DATE:_____

Figure 8.1. Driver's vehicle inspection report

those produced by the substance shown as having a toxicity rate of 5 or 6 in the Underwriters' Laboratories *Classification of Comparative Life Hazard of Gases and Vapors.*[2]

- **Spare fuses.** Every vehicle must be equipped with at least one spare fuse or other overload protective devise, if the devices used are not of a reset type for each kind and size used.

- **Warning devices for stopped vehicles.** One of the following warning devices must be carried on board the vehicle:

    Vehicles equipped with warning devices before January 1, 1974

    - Three liquid-burning emergency flares that satisfy the requirements of SAFE Standard J596, "Liquid Burning Emergency Flares," and three fusees and two red flags; or

    - Three electric emergency lanterns that satisfy the requirements of SAE Standard J596, "Electric Emergency Lanterns," and two red flags; or

    - Three red emergency reflectors that satisfy the requirements of Paragraph (i) of this section, and two red flags; or

    - Three red emergency reflective triangles that satisfy the requirements of Paragraph (h) of this section; or

    - Three bi-directional emergency reflective triangles that conform to the requirements of Federal Motor Vehicle Safety Standard No 125, §571.125 of this title.

    Vehicles equipped with warning devices on and after January 1, 1974

    - Three bi-directional emergency reflective triangles that conform to the requirements of Federal Motor Vehicle Safety Standard No. 125 §571.125 of this title; or

    - At least six fusees or three liquid-burning flares. The vehicle must have as many additional fusees or liquid-burning flares as are necessary to satisfy the requirements of §392.22.

Operation of the Vehicle

For operation of the vehicle, driver training must include:

- Proper turning techniques
- Proper backing
- Proper parking
- Proper handling

- Information concerning vehicle characteristics that affect vehicle stability (such as the effects of curves and braking)
- Effects of speed on vehicle control
- Dangers associated with weather and road conditions that the driver may experience
- High center of gravity

Procedures for Maneuvering about Tunnels, Bridges, and Railroad Crossings

The required procedures for maneuvering about tunnels, bridges, and railroad crossings are covered by the Hazardous Materials Regulations. Briefly, hazmat drivers cannot park within 300 feet of a bridge, tunnel, dwelling, or place where people park. Drivers must stop within 15 feet of railroad tracks, but not any further than 50 feet. While crossing the tracks, the driver is prohibited from shifting gears. Exemptions to the not shifting gears while crossing railroad tracks rule are for green traffic signals, directed traffic, industrial switching tracks, streetcar tracks, and tracks abandoned or marked "exempt." Drivers must check company policy regarding procedures. Keep in mind that the above stipulations are the minimum requirements mandated by DOT; company policies may be more stringent.

> ✓ **Important Note:** Unless no alternative route exists, tunnels must be avoided when carrying hazardous materials.

Requirements for Vehicle Attendance, Parking, Smoking, Routing, and Incident Reporting

In addition to the following requirements, hazmat drivers must be informed of company guidelines for attendance of vehicle, parking, smoking, routing, and incident reporting.

Attendance of Vehicle

When fueling, the driver must ensure that the engine is turned off, and must remain in attendance. Occupying the sleeper berth is not considered in attendance. Simply, hazardous loads must be attended at all times by drivers. Drivers should check with company attendance polices to ensure compliance.

> ✓ **Important Note:** During breaks, the driver must be within 100 feet of the vehicle and have an unobstructed view of it.

Driver Training

Parking

Hazmat drivers must not park within five (5) feet of the traveled portion of a roadway, except when a vehicle problem requires the driver to stop and other parking isn't available. Drivers should check company policy regarding parking.

Smoking

Drivers are not allowed to smoke within 25 feet of their vehicles if they are carrying explosives, oxidizers, or flammables on board. Drivers should check company policy regarding smoking.

Routing

Drivers carrying explosives must have a written route plan. Drivers should be alert for special bypasses designated for hazardous loads. Drivers should avoid alleys, stadiums, churches, schools, shopping centers, or other places where large numbers of people may be found. Drivers should check company policy regarding routing.

Incident Reporting

Trainers must ensure that drivers are fully aware of company procedures and forms for incident reporting. Such procedures and forms typically include guidelines on obtaining witness(es) statement(s) and filling out a Hazardous Materials Incident Information Form.

The Hazardous Materials Incident Information Form should include the following information:

- Driver's name
- Location
- Telephone number
- Location of incident
- Carrier involved
- Placard/labels applied
- Name of commodity
- Shipper
- Accessibility
- Local population information
- Topographical features
- Weather conditions
- Availability of water

Loading/Unloading of Materials

Loading/unloading and securing practices are important topics with which the driver must be familiar. Drivers should check company policy regarding loading/unloading practices.

Drivers need to be informed of the following guidelines:

- During loading and unloading of hazardous materials, the vehicle engine must be turned off
- All packages must be secure against movement
- Pole trailers cannot be used for transporting hazardous materials
- Smoking is not allowed on or near a vehicle during loading or unloading of Class 1 (explosives), Class 3 (flammable liquids), Class 4 (flammable solids), Class 5 (oxidizers), and Class 2-Division 2.1 (flammable gas) materials
- Handbrake must be set and wheels chocked to prevent vehicle movement during loading and unloading
- Caution must be used whenever tools are employed to prevent damage to any package or container
- Class 1 (explosives), Class 3 (flammable liquids), Class 4 (flammable solids), Class 5 (oxidizers), Class 8 (corrosives), Class 2 (gas), and Class 6-Division 6.1 (poisonous) materials must be braced to prevent motion, and loaded in such a manner that valves and other fittings are protected from damage
- Protect containers in transit from an undue rise in temperature
- A "qualified" person must be in attendance at all times during the loading and unloading of a cargo tank
- Guidelines for proper separation distances for transporting radioactive materials must be followed
- Guidelines for proper segregation requirements must be followed
- Containers and their contents are not to be tampered with between point of origin and point of billed destination

> ✓ **Important Note:** Drivers and others involved with loading/unloading hazardous materials should be alert to broken, severely dented, leaking, or improper containers, which must be rejected.

Cargo Tank and Portable Tank Operators

Drivers who operate a cargo tank or vehicle with a portable tank with a capacity of 1,000 gallons or more must receive hazmat training (properties and hazards of the material transported), have the appropriate state-issued CDL (with proper endorsement), and receive specialized training.

In addition to training on proper loading and unloading procedures and retest and inspection requirements for cargo tanks, the specialized training must include:

- The operation of emergency control features of the cargo tank or portable tank
- Special vehicle handling characteristics including:
  - high center of gravity
  - fluid-load subject to surge
  - effects of fluid-load surge on braking
  - characteristic differences in stability among baffled, unbaffled, and multi-compartmented tanks
  - effects of partial loads on vehicle stability

Summary

Transportation is the weakest link in the hazardous material delivery chain. Driver training is essential to safe transportation of these hazardous materials, and therefore the Federal government requires hazardous material drivers to be hazmat trained. However, though essential, training isn't the entire answer to safe hazmat transportation.

Notes

[1] GAO. August 1989. Truck Safety: Information on Driver Training. *GAO/RCED-89-163*. 1-7, 12-23.

[2] Copies of the Classification can be obtained by writing to Underwriters' Laboratories, Inc., 205 East Ohio Street, Chicago, IL, 60611.

Chapter 9
Controlled Substances and Alcohol Use and Testing

The U.S. Department of Transportation set out to strengthen the nation's program, subjecting 8.3 million transportation workers to drug and alcohol testing, and also to make the rules easier for all parties to understand. By all accounts, the revisions unveiled last December [1999] achieve both goals.[1]

Introduction

Since their enactment, Part 382 and Part 40 of 49 CFR have received mixed reviews from drivers, from employers required to test drivers, and from drug testing laboratories. From the driver's point of view, many feel that the controlled substances and alcohol use and testing requirements mandated for drivers are unnecessary and unwarranted invasions of privacy.

However, experience (backed by many records demonstrating that on-the-road highway accidents related to drug/alcohol abuse are frequent) shows that when drivers use drugs or alcohol to push their bodies beyond the normal limits, the results can be disastrous. Drivers may think they are working faster and/or more efficiently, but the truth is that they are more likely to make mistakes. As the drug or alcohol wears off, drivers work more slowly and sluggishly. Since driving is both a mental and reflex task, being slow and sluggish can have terrible consequences.

Consider one common example of drug abuse. Illegal use of cocaine may give drivers the idea that they can do anything, but this is a false belief that could cause them to take unnecessary risks on the highway—risks that could put their lives and the lives of others in great danger. Add to that the risk inherent in the action of the drug itself.

Cocaine takes a terrible toll on even normal, healthy hearts—just one line can aggravate a hidden heart defect or trigger a reaction to an unknown allergy (both of which can result in death). Until only a few years ago, the popular perception of cocaine was that it was a "safe" drug, and not even physically addictive. In reality, this drug is anything but safe. In fact, doctors consider it three times more lethal than heroin.

When used, those under the influence of these commonly abused substances may exhibit the symptoms as follows:

Alcohol

    slowed reaction time

    poor hand-eye coordination

    impaired judgement

    poor concentration and short-term memory

Marijuana

    impaired coordination

    impaired sense of time and distance

    inability to retain information

Cocaine

    strong mood swings

    false belief that the drug enhances physical ability

    tendency to work too quickly, performing jobs carelessly and superficially

Crack

    confusion

    inability to concentrate

    difficulty making decisions

    distrust

    delusions and hallucinations

Negative reactions to substance abuse testing from drivers range from anger at the invasion of privacy to irritation at the hassle and inconvenience. However, those drivers who do use drugs or alcohol while driving put us all at risk. From the above list of possible symptoms from various substances used, it should be clear that our highways must be used only drivers that have full control of their faculties.

Negative reactions to substance abuse testing from the employer's point of view are also prevalent. Some see DOT's substance abuse prevention requirements as wasteful, expensive, and a hassle. However, since employers are usually directly or indirectly responsible for their employees' performance on the job, they have a very real personal stake in ensuring that the drivers are alcohol and drug free.

Drug testing laboratories are in business to perform certain analytical procedures for profit. The possibility that a for-profit testing facility might have difficulty with testing for alcohol/drug abuse by DOT-sanctioned drivers might seem strange. But this is the case, to a degree. When testing procedures are vague and ambiguous (as many federal requirements are in an attempt to cover as many contingencies as possible), attempting to perform analytical procedures in full compliance with mandated regulations is not always straightforward. In fact, because of vagaries and ambiguities, DOT's 49 CFR Part 40 *Drug Testing*, originally enacted in 1989, was just recently revised (December 1999) to make requirements easier to follow and implement. Despite these concerns, drug and alcohol testing of DOT drivers is done to prevent the type of incident discussed in the prelude to this text.

Key Terms (§382.107)

Air blank	A reading by an evidential breath testing device (EBT) of ambient air containing no alcohol. (In EBTs using gas chromatography technology, a reading of the driver's internal standard.)
Alcohol	The intoxicating agent in beverage alcohol, ethyl alcohol, or other low molecular weight alcohols including methyl and isopropyl alcohol.
Alcohol concentration (or content)	The alcohol in a volume of breath expressed in terms of grams of alcohol per 210 liters of breath as indicated by an evidential breath test under this part.
Alcohol use	The consumption of any beverage, mixture, or preparation, including any medication, containing alcohol.
Aliquot	A portion of a specimen used for testing.
Blind sample or blind performance test specimen	A urine specimen submitted to a laboratory for quality control testing purposes, with a fictitious identifier, so that the laboratory cannot distinguish it from employee specimens, and which is spiked with known quantities of specific drugs or which is blank, containing no drugs.
Breath Alcohol Technician (BAT)	An individual who instructs and assists individuals in the alcohol testing process and operates an EBT.
Canceled or invalid test	In drug testing, a drug test that has been declared invalid by a Medical Review Officer (MRO). A canceled test is neither a positive nor a negative test. For purposes of this part, a sample that has been rejected for testing by a laboratory is treated the same as a canceled test. In alcohol testing, a test that is deemed invalid under §40.79. It is neither a positive nor a negative test.
Chain of custody	Procedures to account for the integrity of each urine or blood specimen by tracking its handling and storage from point of specimen collection to final disposition of the specimen. With respect to drug testing, these procedures shall require that an appropriate drug testing custody form (see §40.23(a)) be used from time of collection to receipt by the laboratory and that upon receipt by the laboratory, an appropriate laboratory chain of custody form(s) account(s) for the sample or sample aliquots within the laboratory.
Collection container	A container into which the employee urinates to provide the urine sample used for a drug test.
Collection site	A place designated by the employer where individuals present themselves for providing a specimen of their urine to be analyzed for the presence of drugs.

Collection site person	A person who instructs and assists individuals at a collection site and who receives and makes a screening examination of the urine specimen provided by those individuals.
Commerce	Means: 1. Any trade, traffic, or transportation within the jurisdiction of the U.S. between a place in a state and a place outside of such state, including a place outside the U.S. and 2. Trade, traffic, and transportation in the U.S. that affects any trade, traffic, and transportation described in Paragraph (1) of this definition.
Commercial Motor Vehicle	A motor vehicle or combination of motor vehicles used in commerce to transport passengers or property if the motor vehicle: 1. has a gross combination weight rating of 11,794 or more kilograms (26,001 or more pounds) inclusive of a towed unit with a gross vehicle weight rating of more than 4,536 kilograms (10,000 pounds); or 2. has a gross vehicle weight rating of 11,794 or more kilograms (26,001 or more pounds); or 3. is designed to transport 16 or more passengers, including the driver; or 4. is of any size and is used in the transportation of materials found to be hazardous for the purposes of the Hazardous Materials Transportation Act and which require the motor vehicle to be placarded under the Hazardous Materials Regulations (49 CFR Part 172, Subpart F).
Confirmation test for alcohol testing	A second test following a screening test with a result of 0.02 or greater that provides quantitative data of alcohol concentration. For controlled substances, testing requires a second analytical procedure to identify the presence of a specific drug or metabolite, independent of the screen test, one that uses a different technique and chemical principle from that of the screen test to ensure reliability and accuracy. (Gas chromatography/mass spectrometry (GC/MS) is the only authorized confirmation method for cocaine, marijuana, opiates, amphetamines, and phycyclidine).
Consortium	An entity, including a group or association of employers or contractors, which provides alcohol or controlled substances testing as required by this part, or other DOT alcohol or controlled substances testing rules, and that acts on behalf of the employers.
Controlled substances	Those substances identified in §40.21(a) of this title.
DHHS	The Department of Health and Human Services or any designee of the Secretary, Department of Health and Human Services.

Disabling damage	Damage that precludes departure of a motor vehicle from the scene of the accident in its usual manner in daylight after simple repairs.
	
	1. **Inclusions.** Damage to motor vehicles that could have been driven, but would have been further damaged if so driven.
	
	2. **Exclusions**
	
	   a. Damage that can be remedied temporarily at the scene of the accident without special tools or parts.
	
	   b. Tire disablement without other damage even if no spare tire is available.
	
	   c. Headlight or taillight damage.
	
	   d. Damage to turn signals, horn, or windshield wipers that make them inoperative.

DOT Agency	An agency (or "operating administration") of the USDOT administering regulations requiring alcohol and/or drug testing, in accordance with Part 40 of this title.
Driver	Any person who operates a commercial motor vehicle. This includes, but is not limited to: full time, regularly employed drivers; casual, intermittent, or occasional drivers; and leased drivers and independent, owner-operator contractors who are either directly employed by or under lease to an employer or who operate a commercial motor vehicle at the direction of or with the consent of an employer.

Controlled Substances and Alcohol Use and Testing

EBT (or evidential breath testing device)	An EBT approved by the National Highway Traffic Safety Administration (NHTSA) for the evidential testing of breath and placed on NHTSA's "Conforming Products List of Evidential Breath Measurement Devices" (CPL, and identified on the CPL as conforming with the model specifications available from the National Highway Traffic Safety Administration, Office of Alcohol and State Programs).
Employee	An individual designated in a DOT agency regulation as subject to drug testing and/or alcohol testing. As used in this part, "employee" includes an applicant for employment. "Employee" and "individual" or "individual to be tested" have the same meaning for purposes of this part.
Employer	Any person (including the United States, a state, District of Columbia, tribal government, or a political subdivision of a state) who owns or leases a commercial motor vehicle or assigns persons to operate such a vehicle. The term employer includes an employer's agents, officers, and representatives.
Licensed medical practitioner	A person licensed, certified, and/or registered in accordance with applicable Federal, State, local, or foreign laws and regulations to prescribe controlled substances and other drugs.
Medical Review Office (MRO)	A licensed physician (medical doctor or doctor of osteopathy) responsible for receiving laboratory results generated by an employer's drug testing program who has knowledge of substance abuse disorders and has appropriate medical training to interpret and evaluate an individual's confirmed positive test result together with his or her medical history and any other relevant biomedical information. [Note: A Doctor of Chiropractic holding a Certified Addiction Professional degree is not considered a licensed medical doctor or doctor of osteopathy and therefore cannot serve as an MRO].
Performing (a safety-sensitive function)	A driver is considered to be performing a safety-sensitive function during any period in which he or she is actually performing, ready to perform, or immediately available to perform any safety-sensitive functions.
Positive rate	The number of positive results for random controlled substances tests conducted under this part, plus the number of refusals of random controlled substances tests required by this part, divided by the total of random controlled substances tests conducted under this part, plus the number of refusals or random tests required by this part.
Refuse to submit (to alcohol or controlled substances test)	A driver who: • fails to provide adequate breath for alcohol testing as required by Part 40 of this title, without a valid medical explanation, after he or she has received notice of the requirement for breath testing in accordance with the provisions of this part; • fails to provide an adequate urine sample for controlled substances testing as required by Part 40 of this title, without a genuine inability to provide a specimen (as determined by a medical evaluation), after he or she has received notice of the requirement for urine testing in accordance with the provisions of this part; or • engages in conduct that clearly obstructs the testing process.
Safety-sensitive function	All time from the time a driver begins to work or is required to be in readiness to work until the time he/she is relieved from work and all responsibility for performing work. Safety-sensitive functions include: • all time at an employer or shipper plant, terminal, facility, or other property, or on any public property, waiting to be dispatched, unless the driver has been relieved from duty by the employer; • all time inspecting equipment as required by §§392.7 and 392.8 of this subchapter or otherwise inspecting, servicing, or conditioning any commercial motor vehicle at any time; • all time spent at the driving controls of a commercial motor vehicle in operation; • all time, other than driving time, in or upon any commercial motor vehicle except time spent resting in a sleeper berth (a berth conforming to the requirements of §393.76 of this subchapter).

- all time loading or unloading a vehicle, supervising, or assisting in the loading or unloading, attending a vehicle being loaded or unloaded, remaining in readiness to operate the vehicle, or in giving or receiving receipts for shipments loaded or unloaded; and

- all time repairing, obtaining assistance, or remaining in attendance upon a disabled vehicle.

Screening test (also known as initial test)	In alcohol testing, an analytical procedure to determine whether a driver may have a prohibited concentration of alcohol in his or her system. In controlled substance testing, an immunoassay screen to eliminate "negative" urine specimens from further consideration.
Secretary	The Secretary of Transportation or the Secretary's designee.
Shipping container	A container capable of being secured with a tamper-evident seal used for transfer of one or more urine specimen bottle(s) and associated documentation from the collection site to the laboratory.
Specimen bottle	The bottle that after being labeled and sealed according to the procedures in Part 40 is used to transmit a urine sample to the laboratory.
Substance Abuse Professional (SAP)	A licensed physician (Medical Doctor or Doctor of Osteopathy); or a licensed or certified psychologist, social worker, or employee assistance professional; or an addiction counselor (certified by the National Association of Alcoholism and Drug Abuse Counselors Certification Commission or by the International Certification Reciprocity Consortium/ Alcohol & Other Drug Abuse). All must have knowledge of and clinical experience in the diagnosis and treatment of alcohol and controlled substances-related disorders.
Trip lease driver	A driver generally employed by one motor carrier, but who is temporarily leased to another motor carrier for one or more trips generally for a time period less than 30 days; would also apply to volunteer organizations that use loaned drivers.
Violation rate	The number of drivers (as reported under §382.305 of this part) found during random tests given under this part to have an alcohol concentration of 0.04 or greater, plus the number of drivers who refuse a random test required by this part, divided by the total reported number of drivers in the industry given random alcohol tests under this part, plus the total reported number of drivers in the industry who refuse a random test required by this part.

Controlled Substances and Alcohol Use and Testing (Part 382)

The following material is largely derived from 49 CFR Part 382. The numbering and lettering of paragraphs and subparagraphs are shown as they appear in this Part.

Subpart A—General

In Subpart A (General), DOT states that the purpose of this part is to establish programs designed to help prevent accidents and injuries resulting from the misuse of alcohol or use of controlled substances by drivers of commercial motor vehicles.

- §382.103 (a) (Applicability) - applies to every person and to all employers of such persons who operate a commercial motor vehicle in commerce in any state, and is subject to:

    (1) The commercial driver's license requirements of Part 383 of this subchapter;

(2) The Licencia Federal de Conductor (Mexico) requirements; or

(3) The commercial driver's license requirements of the Canadian National Safety Code.

- **§382.103 (b)** - an employer who employs himself/herself as a driver must comply with both the requirements in this part that apply to employers and the requirements in this part that apply to drivers. An employer who employs only himself/herself as a driver is required to implement a random alcohol and controlled substances testing program of two or more covered employees in the random testing selection pool.

- **§382.103 (c)** - The exceptions contained in §390.3(f) of this subchapter do not apply to this part. The employers and drivers identified in §390.3(f) must comply with the requirements of this part, unless otherwise specifically provided in Paragraph (d) of this section.

- **§382.103 (d) Exceptions** - This part lists the following exceptions for employers and drivers:

(1) Required to comply with the alcohol and/or controlled substances testing requirements of parts 653 and 654 of this title (Federal Transit Administration alcohol and controlled substances testing regulations); or

(2) Who a State must waive from the requirements of Part 383 of this subchapter. These individuals include active duty military personnel; members of the reserves; and members of the national guard on active duty, including personnel on full-time national guard duty, personnel on part-time national guard training and national guard military technicians (civilians who are required to wear military uniforms), and active duty U.S. Coast Guard personnel; or

(3) Who a State has, at its discretion, exempted from the requirements of Part 383 of this subchapter. These individuals may be:

(i) Operators of a farm vehicle which is:

(A) Controlled and operated by a farmer;

(B) Used to transport either agricultural products, farm machinery, farm supplies, or both to or from a farm;

(C) Not used in the operations of a common or contract motor carrier; and

(D) Used within 241 kilometers (150 miles) of the farmer's farm.

(ii) Firefighters or other persons who operate commercial motor vehicles that are necessary for the preservation of life or property or the execution of emergency governmental functions, are equipped with audible and visual signals, and are not subject to normal traffic regulations.

Intrastate drivers of commercial motor vehicles (CMVs) who are required to obtain commercial driver's licenses (CDLs) are required to be alcohol and drug tested by their employer (49 U.S.C. Section 31301). Students who will be trained to be motor vehicle operators are subject to alcohol and drug testing, and are required to obtain a CDL to operate training vehicles provided by the school.

> ✓ **Important Note:** A foreign resident driver operating between the U.S. and a foreign country from a U.S. terminal for a U.S.-based employer is subject to the FHWA alcohol and controlled substances testing regulations.

- **§382.105 Testing Procedures**—each employer must ensure that all alcohol or controlled substances testing conducted under this part complies with the procedures set forth in Part 40 of this title. The provisions of Part 40 of this title that address alcohol or controlled substances testing are made applicable to employers by this part.

- **§382.109 Preemption of State and local laws**

  (a) Except as provided in Paragraph (b) of this section, this part preempts any State or local law, rule, regulation, or order to the extent that:

  (1) Compliance with both the State or local requirement and this part is not possible; or

  (2) Compliance with the State or local requirement is an obstacle to the accomplishment and execution of any requirement in this part.

  (b) This part can not be construed to preempt provisions of State criminal law that impose sanctions for reckless conduct leading to actual loss of life, injury, or damage to property, whether the provisions apply specifically to transportation employees, employers, or the general public.

- **§382.111 Other requirements imposed by employers** - Except as expressly provided in this part, nothing in this part is to be construed to affect the authority of employers, or the rights of drivers, with respect to the use of alcohol, or the use of controlled substances, including authority and rights with respect to testing and rehabilitation.

- **§382.113 Requirement for notice** - Before performing an alcohol or controlled substances test under this part, each employer must notify the driver that the alcohol or controlled substances test is required by this part. No employer is allowed to falsely represent that a test is administered under this part.

> ✓ **Important Note:** A driver must be notified before submitting to each test required by Part 382. This notification can be provided to the driver either verbally or in writing.

- **§382.115 Starting date for testing programs.**

    (a) **All domestic employers.** Each domestic-domiciled employer that begins commercial motor vehicle operations is required to implement the requirements of this part on the date the employer begins such operations.

    (b) **Large foreign employers.** Each foreign-domiciled employer with fifty or more drivers assigned to operate commercial motor vehicles in North America on December 17, 1995, is required to implement the requirements of this part beginning on July 1, 1996.

    (c) **Small foreign employers.** Each foreign-domiciled employer with less than fifty drivers assigned to operate commercial motor vehicles in North America on December 17, 1995, must implement the requirements of this part beginning on July 1, 1997.

    (d) **All foreign employers.** Each foreign-domiciled employer that begins commercial motor vehicle operations in the U.S. after December 17, 1995, but before July 1, 1997, must implement the requirements of this part beginning on July 1, 1997. A foreign employer that begins commercial motor vehicle operations in the U.S. on or after July 1, 1997, must implement the requirements of this part on the date the foreign employer begins such operations.

> ✓ **Important Note:** Any governmental entity, or a subunit of it that controls CMVs and the day-to-day operations of its drivers, may be considered the employer for purposes of Part 382. For example, a city government divided into various departments, such as parks and public works, could consider the departments as separate employers if the CMV operations are separately controlled. The city also has the option of deeming the city as the employer of all of the drivers of the various departments.

Subpart B—Prohibitions

- **§382.201 Alcohol concentration** - Drivers are strictly prohibited from reporting for duty or remaining on duty requiring the performance of safety-sensitive functions while having an alcohol concentration of 0.04 or greater. No employer having actual knowledge that a driver has an alcohol concentration of 0.04 or greater shall permit the driver to perform or continue to perform safety-sensitive functions.

- **§382.205 On-duty use** - No driver is allowed to use alcohol while performing safety-sensitive functions. No employer having actual knowledge [Note: The form of "actual knowledge" is not specified, but may result from the employer's direct observation of the employee, the driver's previous employer(s), the employee's admission of alcohol use, or other occurrence] that a driver is using alcohol while performing safety-sensitive functions shall permit the driver to perform or continue to perform safety-sensitive functions.

- **§382.207 Pre-duty use** - No driver is allowed to perform safety-sensitive functions within four (4) hours after using alcohol. No employer having actual knowledge that a driver has used alcohol within four (4) hours shall permit a driver to perform or continue to perform safety-sensitive functions.

- **§382.209 Use following an accident** - No driver required to take a post-accident alcohol test under §382.303 of this part shall use alcohol for eight (8) hours following the accident, or until he/she undergoes a post-accident alcohol test, whichever occurs first.

- **§382.211 Refusal to submit to a required alcohol or controlled substances test** - No driver shall refuse to submit to a post-accident alcohol or controlled substances test required under §382.303, a random alcohol or controlled substances test required under §382.305, a reasonable suspicion alcohol or controlled substances test required under §382.307, or a follow-up alcohol or controlled substances test required under §382.311. No employer shall permit a driver who refuses to such tests to perform or continue to perform safety-sensitive functions.

- §382.213 **Controlled substances use.**

    (a) No driver shall report for duty or remain on duty requiring the performance of safety-sensitive functions when the driver uses any controlled substance, except when the use is pursuant to the instructions of a licensed medical practitioner, as defined in §382.107 of this part, who has advised the driver that the substance will not adversely affect the driver's ability to safely operate a CMV.

    (b) No employer having actual knowledge that a driver has used a controlled substance shall permit the driver to perform or continue to perform a safety-sensitive function.

    (c) An employer may require a driver to inform the employer of any therapeutic drug use.

Controlled Substances and Alcohol Use and Testing

> ✓ **Important Note:** A physician, and only a physician—not a pharmacist or anyone else—must specifically advise the driver that the substances in a prescription will not adversely affect the driver's ability to safely operate a CMV.

- §382.215 **Controlled substances testing** - No driver shall report for duty, remain on duty or perform a safety-sensitive function, if the driver tests positive for controlled substances. [**Note:** The Federal Highway Administration considers test results to be complete for the calendar year in which the MRO makes a final determination of the test results, regardless of the date the specimen was collected]. No employer having actual knowledge that a driver has tested positive for controlled substances shall permit the driver to perform or continue to perform safety-sensitive functions.

Subpart C—Tests Required

- §382.301 **Pre-employment Testing**

    (a) Prior to the first time a driver performs safety-sensitive functions for an employer, the driver shall undergo testing for alcohol and controlled substances as a condition prior to being used, unless the employer uses the exception in paragraphs (c) and (d) of this section. No employer shall allow a driver, who the employer intends to hire or use, to perform safety-sensitive functions unless the driver has been administered an alcohol test with a result indicating an alcohol concentration less than 0.04, and has received a controlled substances test result from the Medical Review Officer (MRO) indicating a verified negative test result. If pre-employment alcohol test result under this section indicates an alcohol content of 0.02 or greater but less than 0.04, the provision of §382.505 shall apply.

    (b) **Exception for pre-employment alcohol testing** - an employer is not required to administer an alcohol test required by Paragraph (a) of this section if:

    (1) The driver has undergone an alcohol test required by this section or the alcohol misuse rule of another DOT agency under Part 40 of this title within the previous six months, with a result indicating an alcohol concentration less than 0.04; and

    (2) The employer ensures that no prior employer of the driver of whom the employer has knowledge has records of a violation of this part or the alcohol misuse rule of another DOT agency within the previous six months.

(c) **Exception for pre-employment controlled substances testing**—an employer is not required to administer a controlled substances test required by Paragraph (a) of this section if:

   (1) The driver has participated in a controlled substances testing program that meets the requirements of this part within the previous 30 days; and

   (2) While participating in that program, either

      (i) Was tested for controlled substances within the past six months (from the date of application with the employer) or

      (ii) Participated in the random controlled substances testing program for the previous 12 months (from the date of application with the employer); and

   (3) The employer ensures that no prior employer of the driver of whom the employer has knowledge has records of a violation of this part or the controlled substances use rule of another DOT agency within the previous six months.

(d) (1) An employer who exercises the exception in either Paragraph (b) or (c) of this section shall contact the alcohol and/or controlled substances testing program(s) in which the driver participates or participated and shall obtain and retain from the testing program(s) the following information:

      (i) Name(s) and address(es) of the program(s).

      (ii) Verification that the driver participates or participated in the program(s).

      (iii) Verification that the program(s) conforms to Part 40 of this title.

      (iv) Verification that the driver is qualified under the rules of this part, including that the driver has not refused to be tested for controlled substances.

      (v) The date the driver was last tested for alcohol or controlled substances.

      (vi) The results of any tests taken within the previous six months and any other violations of Subpart B of this part.

   (2) An employer who uses, but does not employ [i.e., employs "trip lease" drivers involved in intestate commerce or this statement applies to volunteer organizations that use loaned drivers], a driver more than once a year to operate commercial motor vehicles must obtain the information in Paragraph (d)(1) of this section at least once every six months. The records prepared under this paragraph shall be maintained

Controlled Substances and Alcohol Use and Testing

in accordance with §382.401. If the employer cannot verify that the driver is participating in a controlled substances testing program in accordance with this part and Part 40, the employer shall conduct a pre-employment alcohol and/or controlled substances test.

(e) Notwithstanding any other provisions of this subpart, all provisions and requirements in this section pertaining to pre-employment testing for alcohol are vacated as of May 1, 1995.

- §382.303 Post-accident testing

(a) As soon as practicable following an occurrence involving a commercial motor vehicle operating on a public road in commerce, each employer shall test for alcohol and controlled substances each surviving driver:

(1) Who was performing safety-sensitive functions with respect to the vehicle, if the accident involved the loss of human life; or

(2) Who receives a citation under State or local law for a moving traffic violation arising from the accident, if the accident involved:

(i) Bodily injury to any person who, as a result of the injury, immediately receives medical treatment away from the scene of the accident; or

(ii) One or more motor vehicles incurring disabling damage as a result of the accident, requiring the motor vehicle to be transported away from the scene by a tow truck or other motor vehicle.

Table 9.1 notes when a post-accident test is required.

Table 9.1. Requirements for a Post-Accident Test

Type of Accident Involved	Citation issued to the CMV	Test must be performed by driver employer
Human fatality	YES	YES
	NO	YES
Bodily injury with immediate medical treatment away from the scene	YES	YES
	NO	NO
Disabling damage to any motor vehicle requiring tow away	YES	YES
	NO	NO

(b) (1) **Alcohol tests.** If a test required by this section is not administered within two hours following the accident, the employer shall prepare and maintain on file a record stating the reasons the test was not promptly administered. If a test required by this section is not administered within eight hours following the accident, the employer shall cease attempts to administer an alcohol test and shall prepare and maintain the same record. Records shall be submitted to the FHWA upon request of the Associate Administer.

(2) For the years stated in this paragraph, employers who submit MIS (Management Information System) reports shall submit to the FHWA each record of a test required by this section that is not completed within eight hours. The employer's records of tests that are not completed within eight hours shall be submitted to the FHWA by March 15, 1996; March 15, 1997, and March 15, 1998, for calendar years 1995, 1996, and 1997, respectively. Employers shall append these records to their MIS submissions. Each record shall include the following information:

   (i) Type of test (reasonable suspicion/post-accident);

   (ii) Triggering event (including date, time, and location);

   (iii) Reason(s) test could not be completed within eight hours:

   (iv) If blood alcohol testing could have been completed within eight hours, the name, address, and telephone number of the testing site where blood testing could have occurred; and

(3) Records of alcohol tests that could not be completed in eight hours shall be submitted to the FHWA at the following address: Attn: Alcohol Testing Program, Office of Motor Carrier Research and Standards (HCS-1), Federal Highway Administration, 400 Seventh Street, SW., Washington, DC 20590.

(4) **Controlled substances tests.** If a test required by this section is not administered within 32 hours following the accident, the employer shall cease attempts to administer controlled substances test, and prepare and maintain on file a record stating the reasons the test was not promptly administered. Records shall be submitted to the FHWA upon request upon request of the Associate Administrator.

(c) A driver who is subject to post-accident testing shall remain readily available for such testing or may be deemed by the employer to have refused to submit to testing. Nothing in this section shall be construed to require the delay of necessary medical attention for injured people following an accident or to prohibit a driver from leaving the scene of an accident for the period necessary to obtain assistance in responding to the accident, or to obtain necessary emergency medical care.

(d) An employer shall provide drivers with necessary post-accident information, procedures and instructions, prior to the driver operating a commercial motor vehicle, so that drivers will be able to comply with the requirements of this section.

(e) (1) The results of a breath or blood test for the use of alcohol, conducted by Federal, State, or local officials having independent authority for the test, shall be considered to meet the requirements of this section, provided such tests conform to the applicable Federal, State or local alcohol testing requirements, and that the results of the tests are obtained by the employer.

(2) The results of a urine test for the use of controlled substances, conducted by Federal, State, or local officials having independent authority for the test, shall be considered to meet the requirements of this section, provided such tests conform to the applicable Federal, State or local controlled substances testing requirements, and that the results of the tests are obtained by the employer.

(f) **Exception.** This section does not apply to:

(1) An occurrence involving only boarding or alighting from a stationary vehicle; or

(2) An occurrence involving only the loading or unloading of cargo; or

(3) An occurrence in the course of the operation of a passenger car or a multipurpose passenger vehicle by an employer unless the motor vehicle is transporting passengers for hire or hazardous materials of a type and quantity that require the motor vehicle to be marked or placarded in accordance with §177.823 of this title.

✓ **Important Note:** An employer may allow a driver, subject to post-accident controlled substance testing, to continue to drive pending receipt of the results of the controlled substances test, so long as no other restrictions are imposed by §382.307 or by law enforcement officials.

✓ **Important Note:** U.S. employers are responsible for ensuring that drivers who have an accident (as defined by §390.5) in a foreign country are post-accident alcohol and drug tested in conformance with the requirements of 49 CFR parts 40 and 382. If the test(s) cannot be administered within the required 8 or 32 hours, the employer shall prepare and maintain a record stating the reasons the test(s) was not administered.

- **§382.305 Random testing**

    (a) Every employer shall comply with the requirements of this section. Every driver shall submit to random alcohol and controlled substance testing as required in this section.

    (b)(1) Except as provided in paragraphs (c) through (e) of this section, the minimum annual percentage rate for random alcohol testing shall be 25 percent (10 percent for calendar year 1998) of the average number of driver positions.

    (2) Except as provided in paragraphs (f) through (h) of this section, the minimum annual percentage rate for random controlled substances testing shall be 50 percent of the average number of driver positions.

    (c) The FHWA Administrator's decision to increase or decrease the minimum annual percentage rate for alcohol testing is based on the reported violation rate for the entire industry. All information used for this determination is drawn from the alcohol management information system reports required by §382.403 of this part. To ensure reliability of the data, the FHWA Administrator considers the quality and completeness of the reported data, may obtain additional information or reports from employers, and may make appropriate modifications in calculating the industry violation rate. Each year, the FHWA Administrator will publish in the *Federal Register* the minimum annual percentage rate for random alcohol testing of drivers (e.g., as pointed out earlier, the random testing rate was reduced to 10 percent for calendar year 1998). The new minimum annual percentage rate for random alcohol testing will be applicable starting January 1 of the calendar year following publication.

    (d)(1) When the minimum annual percentage rate for random alcohol testing is 25 percent or more, the FHWA Administrator may lower this rate to 10 percent of all driver positions if the FHWA Administrator determines that the data received under the reporting requirements of §382.403 for two consecutive calendar years indicates that the violation rate is less than 0.5 percent.

    (2) When the minimum annual percentage rate for random alcohol testing is 50 percent, the FHWA Administrator may lower this rate to 25 percent of all driving positions if the FHWA Administrator determines that the data received under the reporting requirements of §382.403 for two consecutive calendar years indicate that the violation rate is less than 1.0 percent but equal to or greater than 0.5 percent.

    (e)(1) When the minimum annual percentage rate for random alcohol testing is 10 percent, and the data received under the reporting requirements of §382.403 for that calendar year indicate that the violation rate is equal to or greater than 0.5 percent, but less than 1.0 percent, the FHWA Administrator will increase the minimum annual percentage rate for random alcohol testing to 25 percent for all driver positions.

(2) When the minimum annual percentage rate for random alcohol testing is 25 percent or less, and the data received under the reporting requirements of §382.403 for that calendar year indicate that the violation rate is equal to or greater than 1.0 percent, the FHWA Administrator will increase the minimum annual percentage rate for random alcohol testing to 50 percent testing for all driver positions.

(f) The FHWA Administrator's decision to increase or decrease the minimum annual percentage rate for controlled substances testing is based on the reported positive rate for the entire industry. All information used for this determination is drawn from the controlled substances management information system reports required by §382.403 of this part. To ensure reliability of the data, the FHWA Administrator considers the quality and completeness of the reported data, may obtain additional information or reports from employers, and may make appropriate modifications in calculating the industry positive rate. Each year, the FHWA Administrator will publish in the *Federal Register* the minimum annual percentage rate for random controlled substances testing of drivers. The new minimum annual percentage rate for random controlled substances testing will be applicable starting January 1 of the calendar year following publication.

(g) When the minimum annual percentage rate for random controlled substances testing is 50 percent, the FHWA Administrator may lower this rate to 25 percent of all driver positions if the FHWA Administrator determines that the data received under the reporting requirements of §382.403 for two consecutive calendar years indicate that the positive rate is less than 1.0 percent. However, after the initial two years of random testing by large employers and the initial first year of testing by small employers under this section, the FHWA Administrator may lower the rate the following calendar year, if the combined positive testing rate is less than 1.0 percent, and if it would be in the interest of safety.

(h) When the minimum annual percentage rate for random controlled substances testing is 25 percent, and the data received under the reporting requirements of §382.403 for any calendar year indicate that the reported positive rate is equal to or greater than 1.0 percent, the FHWA Administrator will increase the minimum annual percentage rate for random controlled substances testing to 50 percent of all driver positions.

(i) The selection of drivers for random alcohol and controlled substances testing shall be made by a scientifically valid method, such as a random number table or a computer-based random number generator that is matched with drivers' Social Security numbers, payroll identification numbers, or other comparable identifying numbers. Under the selection process used, each driver shall have an equal chance of being tested each time selections are made.

(j) The employer shall randomly select a sufficient number of drivers for testing during each calendar year to equal an annual rate not less than the minimum annual percentage rate for random alcohol and controlled substances testing determined by the FHWA Administrator. If the employer conducts random testing for alcohol and/or controlled substances through a consortium, the number of drivers to be tested may be calculated for each individual employer or may be based on the total number of drivers covered by the consortium who are subject to random alcohol and/or controlled substances testing at the same minimum annual percentage rate under this part or any DOT alcohol or controlled substances random testing rule.

(k) Each employer shall ensure that random alcohol and controlled substances tests conducted under this part are unannounced and that the dates for administering random alcohol and controlled substances tests are spread reasonably throughout the calendar year.

(l) Each employer shall require that each driver who is notified of selection for random alcohol and/or controlled substances testing proceeds to the test site immediately; provided, however, that if the driver is performing a safety-sensitive function other than driving a commercial motor vehicle at the time of notification, the employer shall instead ensure that the driver ceases to perform the safety-sensitive function and proceeds to the testing site as soon as possible.

(m) A driver shall only be tested for alcohol while the driver is performing safety-sensitive functions, or just after the driver has ceased performing such functions.

(n) If a given driver is subject to random alcohol or controlled substances testing under the random alcohol or controlled substances testing rules of more than one DOT agency for the same employer, the drive shall be subject to random alcohol and/or controlled substances testing at the annual percentage rate established for the calendar year by the DOT agency regulating more than 50 percent of the driver's functions.

(o) If an employer is required to conduct random alcohol or controlled substances testing under the alcohol or controlled substances testing rules of more than one DOT agency, the employer may—

   (1) Establish separate pools for random selection, with each pool containing the DOT-covered employees who are subject to testing at the same required minimum annual percentage rate; or

(2) Randomly select such employees for testing at the highest minimum annual percentage rate established for the calendar year by any DOT agency to which the employer is subject.

A driver must be about to perform, or immediately available to perform, a safety-sensitive function to be considered subject to random alcohol testing. A supervisor, mechanic, or clerk, etc., who is on call to perform safety-sensitive functions may be tested at any time they are on call, ready to be dispatched while on-duty. For the employee who does not drive as part of the employee's usual job functions, but who holds a CDL and may be called upon at any time, on an occasional or emergency basis to drive, the employer's obligations, in terms of random testing, must ensure such an employee is included in the random testing pool, like a fulltime driver. A drug test must be administered each time the employee's name is selected from the pool. However, alcohol testing may only be conducted just before, during, or just after the performance of safety-sensitive functions. If an employee is off work due to temporary layoff, illness, injury or vacation, the individual's name should not be removed from the random pool, so long as there is a reasonable expectation of the employee's return. Once an employee is randomly tested during a calendar year, he or she must be returned to the pool for each new selection. Each driver must be subject to an equal chance of being tested during each selection process.

> ✓ **Important Note:** Driver positions that are vacant for a testing cycle are not to be included in the determination of how many random tests must be conducted.

> ✓ **Important Note:** When a driver works for two or more employers, the driver must be in the pool of each employer for which the driver works.

- **§382.307 Reasonable suspicion testing**

    (a) An employer shall require a driver to submit to an alcohol test when the employer has reasonable suspicion to believe that the driver has violated the prohibitions of Subpart B of this part concerning alcohol. The employer's determination that reasonable suspicion exists to require the driver to undergo an alcohol test must be based on specific, contemporaneous, articulable observations concerning the appearance, behavior, speech or body odors of the driver.

    (b) An employer shall require a driver to submit to a controlled substances test when the employer has reasonable suspicion to believe that the driver has violated the prohibitions of Subpart B of this part concerning controlled substances. The employer's determination that reasonable suspicion exists to require the driver to undergo a controlled substances test must be based on specific, contemporaneous, articulable observations concerning the appearance, behavior, speech or body odors of the driver. The observations may include indications of the chronic and withdrawal effects of controlled substances.

(c) The required observations for alcohol and/or controlled substances reasonable suspicion testing shall be made by a supervisor or company official who is trained in accordance with §382.603 of this part. The person who makes the determination that reasonable suspicion exists to conduct an alcohol test shall not conduct the alcohol test of the driver.

(d) Alcohol testing is authorized by this section only if the observations required by Paragraph (a) of this section are made during, just preceding, or just after the period of the work day that the driver is required to be in compliance with this part. A driver may be directed by the employer to only undergo reasonable suspicion testing while the driver is performing safety-sensitive functions, just before the driver is to perform safety-sensitive functions, or just after the driver has ceased performing such functions.

(e) (1) If an alcohol test required by this section is not administered within two hours following the determination under Paragraph (a) of this section, the employer shall prepare and maintain on file a record stating the reasons the alcohol test was not promptly administered. If an alcohol test required by this section is not administered within eight hours following the determination under Paragraph (a) of this section, the employer shall cease attempts to administer an alcohol test and shall state in the record the reasons for not administering the test.

(2) For the years stated in this paragraph, employers who submit MIS reports shall submit to the FHWA each record of a test required by this section that is not completed within eight hours. The employer's records of tests that could not be completed within eight hours shall be submitted to the FHWA by March 15, 1996; March 15, 1997; and March 15, 1998; for calendar years 1995, 1996, and 1997, respectively. Employers shall append these records to their MIS submissions. Each record shall include the following information:

(i) Type of test (reasonable suspicion/post accident);

(ii) Triggering event (including date, time, and location);

(iii) Reason(s) test could not be completed within eight hours; and

Controlled Substances and Alcohol Use and Testing 163

> (iv) If blood alcohol testing could have been completed within eight hours, the name, address, and telephone number of the testing site where blood testing could have occurred.
>
> (3) Records of tests that could not be completed in eight hours shall be submitted to the FHWA at the following address: Attn.: Alcohol Testing program, Office of Motor Carrier Research and Standards (HCS-1), Federal Highway Administration, 400 Seventh Street, SW., Washington, DC 20590.
>
> (4) Notwithstanding the absence of a reasonable suspicion alcohol test under this section, no driver shall report for duty or remain on duty requiring the performance of safety-sensitive functions while the driver is under the influence of or impaired by alcohol, as shown by the behavioral, speech, and performance indicators of alcohol misuse, nor shall an employer permit the driver to perform or continue to perform safety-sensitive functions, until:
>
>> (i) An alcohol test is administered and the driver's alcohol concentration measures less than 0.02; or
>>
>> (ii) Twenty four hours have elapsed following the determination under Paragraph (a) of this section that there is reasonable suspicion to believe that the driver has violated the prohibitions in this part concerning the use of alcohol.
>
> (5) Except as provided in Paragraph (e)(2) of this section, no employer shall take any action under this part against a driver based solely on the driver's behavior and appearance, with respect to alcohol use, in the absence of an alcohol test. This does not prohibit an employer with independent authority of this part from taking any action otherwise consistent with law.
>
> (f) A written record shall be made of the observations leading to a controlled substance reasonable suspicion test, and signed by the supervisor or company official who made the observations, within 24 hours of the observed behavior or before the results of the controlled substances test are released, whichever is earlier.

A reasonable suspicion alcohol test can't be based upon any information or observations of alcohol use or possession, other than a supervisor's actual knowledge. In other words, information conveyed by third parties of a driver's alcohol use may not be the only determining factor used to conduct a reasonable suspicion test. Only a trained supervisor may conduct a reasonable suspicion test based on observed specific, contemporaneous, articulable appearance. Alcohol is generally a legal substance; thus only its use or presence in sufficient concentrations while operating a CMV is a violation of FHWA regulation. Alcohol withdrawal effects, standing alone, do not indicate that a driver has used alcohol in violation of the regulations, and would not constitute reasonable suspicion to believe so.

> ✓ **Important Note:** The requirements of §§382.307 and 382.603 are not applicable to owner-operators in non-supervisory positions.

- **§382.309 Return-to-duty testing**

  (a) Each employer shall ensure that before a driver returns to duty requiring the performance of a safety-sensitive function after engaging in conduct prohibited by Subpart B of this part concerning alcohol, the driver shall undergo a return-to-duty alcohol test with a result indicating an alcohol concentration of less than 0.02.

  (b) Each employer shall ensure that before a driver returns to duty requiring the performance of a safety-sensitive function after engaging in the conduct prohibited by Subpart B of this part concerning controlled substances, the driver shall undergo a return-to-duty controlled substances test with a result indicating a verified negative result for controlled substances use.

- **§382.311 Follow-up testing.**

  (a) Following a determination under §382.605(b) that a driver is in need of assistance in resolving problems associated with alcohol misuse and/or use of controlled substances, each employer shall ensure that the driver is subject to unannounced follow-up alcohol and/or controlled substances testing as directed by a substance abuse professional in accordance with the provisions of §382.605(c)(2)(ii).

  (b) Follow-up alcohol testing shall be conducted only when the driver is performing safety-sensitive functions, just before the driver is to perform safety-sensitive functions, or just after the driver has ceased performing safety-sensitive functions.

Subpart D—Handling of Test Results, Record Retention and Confidentiality

- **§382.401 Retention of Records**

  (a) **General requirements.** Each employer shall maintain records of its alcohol misuse and controlled substances use prevention programs as provided in this section. The records shall be maintained in a secure location with controlled access.

(b) **Period of retention.** Each employer shall maintain the records in accordance with the following schedule:

(1) **Five years.** The following records shall be maintained for a minimum of five years:

(i) Records of driver alcohol test results indicating an alcohol concentration of 0.02 or greater.

(ii) Records of driver verified positive controlled substances test results.

(iii) Documentation of refusals to take required alcohol and/or controlled substances tests,

(iv) Driver evaluation and referrals,

(v) Calibration documentation.

(vi) Records related to the administration of the alcohol and controlled substances testing programs, and

(vii) A copy of each annual calendar year summary required by §382.403.

(2) **Two years.** Records related to the alcohol and controlled substances collection process (except calibration of evidential breath testing devices).

(3) **One year.** Records of negative and canceled controlled substances test results (as defined in Part 40 of this title) and alcohol test results with a concentration of less than 0.02 shall be maintained for a minimum of one year.

(4) **Indefinite period.** Records related to the education and training of breath alcohol technicians, screening test technicians, supervisors, and drivers shall be maintained by the employer while the individual performs the functions that require the training and for two years after ceasing to perform those functions.

(c) **Types of records.** The following specific types of records shall be maintained. "Documents generated" are documents that may have to be prepared under a requirement of this part. If the record is required to be prepared, it must be maintained.

(1) Records related to the collection process:

(i) Collection logbooks, if used;

(ii) Documents relating to the random selection process;

(iii) Calibration documentation for evidential breath testing devices;

(iv) Documentation of breath alcohol technician training;

(v) Documents generated in connection with decisions to administer reasonable suspicion alcohol or controlled substances tests;

(vi) Documents generated in connection with decisions on post-accident tests;

(vii) Documents verifying existence of a medical explanation of the inability of a driver to provide adequate breath or to provide a urine specimen for testing; and

(viii) Consolidated annual calendar year summaries as required by §382.403.

(2) Records related to a driver's test results:

(i) The employer's copy of the alcohol test form, including the results of the test;

(ii) The employer's copy of the controlled substances test chain of custody and control form;

(iii) Documents sent by the MRO to the employer, including those required by §382.407(a).

(iv) Documents related to the refusal of any driver to submit to an alcohol or controlled substances test required by this part; and

(v) Documents presented by a driver to dispute the result of an alcohol or controlled substances test administered under this part.

(vi) Documents generated in connection with verifications of prior employers' alcohol or controlled substances test results that the employer;

(A) Must obtain in connection with the exception contained in §382.301 of this part, and

(B) Records related to other violations of this part.

(4) Records related to evaluations:

(i) Records pertaining to a determination by a substance abuse professional concerning a driver's need for assistance; and

(ii) Records concerning a driver's compliance with recommendations of the substance abuse professional.

(5) Records related to education and training:

(i) Materials on alcohol misuse and controlled substance use awareness, including a copy of the employer's policy on alcohol misuse and controlled substance use;

(ii) Documentation of compliance with the requirements of §382.601, including the driver's signal receipt of education materials;

(iii) Documentation of training provided to supervisors for the purpose of qualifying the supervisors to make a determination concerning the need for alcohol and/or controlled substances testing based on reasonable suspicion;

(iv) documentation of training for breath alcohol technicians as required by §40.51(a) of this title, and

(v) Certification that any training conducted under this part complies with the requirements for such training.

(6) Administrative records related to alcohol and controlled substances testing:

(i) Agreements with collection site facilities, laboratories, breath alcohol technicians, screening test technicians, medical review officers, consortia, and third party service providers;

(ii) Names and positions of officials and their role in the employer's alcohol and controlled substances testing program(s);

(iii) Quarterly laboratory statistical summaries of urinalysis required by §40.29(g)(6) of this title; and

(iv) The employer's alcohol and controlled substances testing policy and procedures.

(d) **Location of records.** All records required by this part shall be maintained as required by §390.31 of this subchapter and shall be made available for inspection at the employer's principal place of business within two business days after a request has been made by an authorized representative of the Federal Highway Administration.

(e) (1) **OMB control number.** The information collection requirements of this part have been reviewed by the Office of Management and Budget pursuant to the Paperwork Reduction Act of 1995 and have been assigned OMB control number 2125-0543.

✓ **Important Note:** Employers may use agents to maintain records, as long as they are in a secure location with controlled access. The employer must also make all records available for inspection at the employer's principal place of business within two business days of when an FHWA representative made a request.

- **§382.403 Reporting of results in a management information system**

  (a) An employer shall prepare and maintain a summary of the results of its alcohol and controlled substances testing programs performed under this part during the previous calendar year, when requested by the Secretary of Transportation, any DOT agency, or any State or local officials with regulatory authority over the employer or any of its drivers.

  (b) If an employer is notified, during the month of January, of a request by the Federal Highway Administration to report the employer's annual calendar year summary information, the employer shall prepare and submit the report to the Federal Highway Administration by March 15 of that year. The employer shall ensure that the annual summary report is accurate and received by March 15 at the location that the Federal Highway Administration specifies in its request. The report shall be in the form and manner prescribed by the Federal Highway Administration in its request. When the report is submitted to the Federal Highway Administration by mail or electronic transmission, the information requested shall be typed, except for the signature of the certifying official. Each employer shall ensure the accuracy and timeliness of each report submitted by the employer or a consortium.

  (c) **Detailed summary.** Each annual calendar year summary that contains information on a verified positive controlled substances test result, an alcohol screening test result of 0.02 or greater, or any other violation of the alcohol misuse provisions of Subpart B of this part shall include the following informational elements:

  (1) Number of drivers subject to Part 382;

  (2) Number of drivers subject to testing under the alcohol misuse or controlled substances use rules of more than one DOT agency, identified by each agency;

  (3) Number of urine specimens collected by type of test (e.g., pre-employment, random, reasonable suspicion, post-accident);

  (4) Number of positives verified by a MRO by type of test, and type of controlled substance;

  (5) Number of negative controlled substance tests verified by a MRO by type of test;

  (6) Number of persons denied a position as a driver following a pre-employment verified positive controlled substances test and/or a pre-employment alcohol test that indicates an alcohol concentration of 0.04 or greater;

(7) Number of drivers with tests verified positive by a medical review officer for multiple controlled substances;

(8) Number of drivers who refused to submit to an alcohol or controlled substances test required under this subpart;

(9) (i) Number of supervisors who have received required alcohol training during the reporting period; and

(ii) Number of supervisors who have received required controlled substances training during the reporting period;

(10) (i) Number of screening alcohol tests by type of test; and

(ii) Number of confirmation alcohol tests, by type of test;

(11) Number of confirmation alcohol tests indicating an alcohol concentration of 0.02 or greater but less than 0.04, by type of test;

(12) Number of confirmation alcohol tests indicating an alcohol concentration of 0.04 or greater, by type of test;

(13) Number of drivers who were returned to duty (having complied with the recommendations of a substance abuse professional as described in §382.503 and §382.605), in this reporting period, who previously:

(i) Had a verified positive controlled substance test result, or

(ii) Engaged in prohibited alcohol misuse under the provisions of this part;

(14) Number of drivers who were administered alcohol and drug tests at the same time, with both a verified positive drug test result and an alcohol test result indicating an alcohol concentration of 0.04 or greater; and

(15) Number of drivers who were found to have violated any non-testing prohibitions of Subpart B of this part, and any action taken in response to the violation.

(d) **Short summary.** Each employer's annual calendar year summary that contains only negative controlled substance test results, alcohol screening test results of less than 0.02, and does not contain any other violations of Subpart B of this part, may prepare and submit, as required by Paragraph (b) of this section, either a standard report form containing all the information elements specified in Paragraph (c) of this section, or an "EZ" report form. The "EZ" report shall include the following information elements:

(1) Number of drivers subject to this Part 382;

(2) Number of drivers subject to testing under the alcohol misuse or controlled substance use rules of more than one DOT agency, identified by each agency;

(3) Number of urine specimens collected by type of test (pre-employment, random, reasonable suspicion, post-accident);

(4) Number of negatives verified by a medical review officer by type of test;

(5) Number of drivers who refused to submit to an alcohol or controlled substances test required under this subpart;

(6) (i) Number of supervisors who have received required alcohol training during the reporting period; and

    (ii) Number of supervisors who have received required controlled substances training during the reporting period;

(7) Number of screen alcohol tests by type of test; and

(8) Number of drivers who were returned to duty (having complied with the recommendations of a substance abuse professional as described in §§382.503 and 382.605), in this reporting period, who previously:

    (i) Had a verified positive controlled substance test result, or

    (ii) Engaged in prohibited alcohol misuse under the provisions of this part.

(e) Each employer that is subject to more than one DOT agency alcohol or controlled substances rule shall identify each driver covered by the regulations of more than one DOT agency. The identification will be by the total number of covered functions. Prior to conducting any alcohol or controlled substances test on a driver subject to the rules of more than one DOT agency, the employer shall determine which DOT agency rule or rules authorizes or requires the test. The test result information shall be directed to the appropriate DOT agency or agencies.

(f) A consortium may prepare annual calendar year summaries and reports on behalf of individual employers for purposes of compliance with this section. However, each employer shall sign and submit such a report and shall remain responsible for ensuring the accuracy and timeliness of each report prepared on its behalf by a consortium.

- **§382.405 Access to facilities and records**

(a) Except as required by law (i.e., federal statutes or an order of a competent Federal jurisdiction, such as an administrative subpoena) or expressly authorized or required in this section, no employer shall release driver information that is contained in records required to be maintained under §382.401.

(b) A driver is entitled, upon written request, to obtain copies of any records pertaining to the driver's use of alcohol or controlled substances, including any records pertaining to his or her alcohol or controlled substances tests. The employer shall promptly provide the records requested by the

Controlled Substances and Alcohol Use and Testing

driver. Access to a driver's records shall not be contingent upon payment for records other than those specifically requested.

(c) Each employer shall permit access to all facilities utilized in complying with the requirements of this part to the Secretary of transportation, any DOT agency, or any State or local officials with regulatory authority over the employer or any of its drivers.

(d) Each employer shall make available copies of all results for employer alcohol and/or controlled substances testing conducted under this part and any other information pertaining to the employer's alcohol misuse and/or controlled substances use prevention program, when requested by the Secretary of transportation, any DOT agency, or any State or local officials with regularity authority over the employer or any of its drivers.

(e) When requested by the National Transportation Safety Board as part of an accident investigation, employers shall disclose information related to the employer's administration of a post-accident alcohol and/or controlled substance test administered following the accident under investigation.

(f) Records shall be made available to a subsequent employer upon receipt of a written request from a driver. Disclosure by the subsequent employer is permitted only as expressly authorized by the terms of the driver's request.

(g) An employer may disclose information required to be maintained under this part pertaining to a driver, the decision-maker in a lawsuit, grievance, or other proceeding initiated by or on behalf of the individual, and arising from the results of an alcohol and/or controlled substance test administered under this part, or from the employer's determination that the driver engaged in conduct prohibited by Subpart B of this part (including, but not limited to, a worker's compensation, unemployment compensation, or other proceeding relating to a benefit sought by the driver).

(h) An employer shall release information regarding a driver's records as directed by the specific, written consent of the driver authorizing release of the information to an identified person. Release of such information by the person reviewing the information is permitted only in accordance with the terms of the employee's consent.

> ✓ **Important Note:** Employers who are subject to other Federal agencies' regulations, such as the Nuclear Regulatory Commission, Department of Energy, Department of Defense, etc., may allow those agencies access to view or have access to test records required to be prepared and maintained by parts 40 and/or 382 only when a specific, contemporaneous authorization for release of the test records is allowed by the driver.

A motor carrier must respond to a third-party request (as directed by the specific, written consent of the driver authorizing release of the information on behalf of an entity such as a motor carrier) to release driver information that is contained in records required to be maintained under §382.401. However, the third-party administrator must comply with the conditions established concerning confidentiality, test results, and record keeping as stipulated in the "Notice: Guidance on the role of Consortia and Third-Party Administrators in DOT Drug and Alcohol Testing Programs." An employer (motor carrier) may disclose information required to be maintained under 49 CFR Part 382 (pertaining to the driver) to the driver or the decision maker in a lawsuit, grievance, or other proceeding (including, but not limited to, worker's compensation, unemployment compensation) initiated by or on behalf of the driver concerning prohibited conduct under 49 CFR Part 382.

Also an employer (motor carrier) may be required to provide the test result information pursuant to other Federal statues or an order of a competent federal jurisdiction, such as an administrative subpoena, as allowed by §382.405(a) without the driver's written consent.

- **§382.407 Medical review officer notifications to the employer.**

    (a) The medical review officer may report to the employer using any communications device, but in all instances a signed, written notification must be forwarded within three business days of completion of the medical review officer's review, pursuant to Part 40 of this title. A legible photocopy of the fourth copy of Part 40 Appendix A subtitled COPY 4—SEND DIRECTLY TO MEDICAL REVIEW OFFICER—DO NOT SEND TO LABORATORY of the *Federal Custody and Control Form OMB Number 9999-0023* may be used to make the signed, written notification to the employer for all test results (positive, negative, canceled, etc.), provided that the controlled substance(s) verified as positive, and the MRO's signature, shall be legibly noted in the remarks section of step 8 of the form completed by the medical review officer. The MRO must sign all verified positive test results. An MRO may sign or rubber-stamp negative test results. An MRO's staff may rubber-stamp negative test results under written authorization of the MRO. In no event shall an MRO, or his/her staff, use electronic signature technology to comply with this section. All reports, both oral and in writing, from the medical review officer to an employer shall clearly include:

    (1) A statement that the controlled substances test being reported was in accordance with Part 40 of this title and this part, except for legible photocopies of Copy 4 of the Federal Custody and Control Form;

    (2) The full name of the driver for whom the test results are being reported;

    (3) The type of test indicated on the custody and control form (i.e., random, post-accident, follow-up);

Controlled Substances and Alcohol Use and Testing 173

(4) The date and location of the test collection;

(5) The identities of the persons or entities performing the collection, analyzing the specimens, and serving as the medical review officer for the specific test;

(6) The results of the controlled substances test, positive, negative, test canceled, or test not performed, and if positive, the identity of the controlled substance(s) for which the test was verified positive.

(b) A medical review officer shall report to the employer that the medical review officer has made all reasonable efforts to contact the driver as provided in §40.33(c) of this title. The employer shall, as soon as practicable, request that the driver contacts the medical review officer prior to dispatching the driver or within 24 hours, whichever is earlier.

- **§382.409 Medical review officer record retention for controlled substances**

    (a) A medical review officer shall maintain all dated records and notifications, identified by individual, for a minimum of five years for verified positive controlled substances test results.

    (b) A medical review officer shall maintain all dated records and notifications, identified by individual, for a minimum of one year for negative and canceled controlled substances test results.

    (c) No person may obtain the individual controlled substances test results retained by a medical review officer, and no medical review officer shall release the individual controlled substances test results of any driver to any person, without first obtaining a specific, written authorization from the tested driver. Nothing in this authorization shall prohibit a medical review officer from releasing, to the employer or to officials of the Secretary of Transportation, and DOT agency, or any State or local officials with regulatory authority over the controlled substances testing program under this part, the information delineated in §382.407(a) of this subpart.

- **§382.411 Employer notifications**

    (a) An employer shall notify a driver of the results of a pre-employment controlled substance test conducted under this part, if the driver requests such results within 60 calendar days of being notified of the disposition of the employment application. An employer shall notify a driver of the results of random, reasonable suspicion and post-accident tests for controlled substances conducted under this part if the test results are verified positive. The employer shall also inform the driver which controlled substance or substances were verified as positive.

(b) The designated management official shall make reasonable efforts to contact and request each driver who submitted a specimen under the employer's program, regardless of the driver's employment status, to contact and discuss the results of the controlled substances test with a medical review officer who has been unable to contact the driver.

(c) The designated management official shall immediately notify the medical review officer that the driver has been notified to contact the medical review officer within 24 hours.

- **§382.413 Inquiries for alcohol and controlled substances information from previous employers**

    (a) (1) An employer shall, pursuant to the driver's written authorization, inquire about the following information on a driver from the driver's previous employers, during the preceding two years from the date of application, which are maintained by the driver's previous employers under §382.401(b)(1)(i) through (iii) of this subpart:

    (i) Alcohol tests with a result of 0.04 alcohol concentration or greater;

    (ii) Verified positive controlled substances test results; and

    (iii) Refusals to be tested.

    (2) The information obtained from a previous employer may contain any alcohol and drug information the previous employer obtained from other previous employers under Paragraph (a)(1) of this section.

    (b) If feasible, the information in Paragraph (a) of this section must be obtained and reviewed by the employer prior to the first time a driver performs safety-sensitive functions for the employer. If not feasible, the information must be obtained and reviewed as soon as possible, but no later than 14 calendar days after the first time a driver performs safety-sensitive functions for the employer. An employer may not permit a driver to perform safety-sensitive functions after 14 days without having made a good faith effort to obtain the information as soon as possible. If a driver hired or used by the employer ceases performing safety-sensitive functions for the employer before expiration of the 14-day period or before the employer has obtained the information in Paragraph (a) of this section, the employer must still make a good faith effort to obtain this information.

    (c) An employer must maintain a written, confidential record of the information obtained under Paragraph (a) or (f) of this section. If, after making a good faith effort, an employer is unable to obtain the information from a previous employer, a record must be made of the efforts to obtain the information and retained in the driver's qualification file.

Controlled Substances and Alcohol Use and Testing

(d) The prospective employer must provide to each of the driver's previous employers the driver's specific, written authorization for release of the information in Paragraph (a) of this section.

(e) The release of any information under this section may take the form of personal interviews, telephone interviews, letters, or any other method of transmitting information that ensures confidentiality.

(f) The information in Paragraph (a) of this section may be provided directly to the prospective employer by the driver, provided the employer assures itself that the information is true and accurate.

(g) An employer may not use a driver to perform safety-sensitive functions if the employer obtains information on a violation of the prohibitions in Subpart B of this part by the driver, without obtaining information on subsequent compliance with the referral and rehabilitation requirements of §382.605 of this part.

(h) Employers need not obtain information under Paragraph (a) of this section generated by previous employers prior to the starting dates in §382.115 of this part.

> ✓ **Important Note:** If a previous employer refuses to make a driver's records available, in violation of §382.405, pursuant to the new employer's and driver's request, the new employer should note the attempt to obtain the information and place the note with the driver's other testing information.

Subpart E—Consequences for Drivers Engaging in Substance Use-Related Conduct

- **§382.501 Removal from safety-sensitive function**

    (a) Except as provided in Subpart F of this part, no driver shall perform safety-sensitive functions, including driving a commercial motor vehicle, if the driver has engaged in conduct prohibited by Subpart B of this part or an alcohol or controlled substances rule of another DOT agency.

    (b) No employer shall permit any driver to perform safety-sensitive functions, including driving a commercial motor vehicle, if the employer has determined that the driver has violated this section.

    (c) For purposes of this subpart, commercial motor vehicle means a commercial motor vehicle in commerce as defined in §382.107, and a commercial motor vehicle in interstate commerce as defined in Part 390 of this subchapter.

> ✓ **Important Note:** If a driver violates the prohibitions in Subpart B, he/she is prohibited from performing safety-sensitive functions. However, a driver who has violated the prohibitions of Subpart B may perform any duties for an employer that are not considered "safety-sensitive functions."

- **§382.503 Required evaluation and testing**

No driver who has engaged in conduct prohibited by Subpart B of this part shall perform safety-sensitive functions, including driving a commercial motor vehicle, unless the driver has met the requirements of §382.605. No employer shall permit a driver who has engaged in conduct prohibited by Subpart B of this part to perform safety-sensitive functions, including driving a commercial motor vehicle, unless the driver has met the requirements of §382.605.

- **§382.505 Other alcohol-related conduct**

    (a) No driver tested under the provisions of Subpart C of this part who is found of have an alcohol concentration of 0.02 or greater but less than 0.04 shall perform or continue to perform safety-sensitive functions for an employer, including driving a commercial motor vehicle, nor shall an employer permit the driver to perform or continue to perform safety-sensitive functions, until the start of the driver's next regularly scheduled duty period, but not less than 24 hours following administration of the test.

    (b) Except as provided in Paragraph (a) of this section, no employer shall take any action under this part against a driver based solely on test results shown an alcohol concentration less than 0.04. This does not prohibit an employer with authority independent of this part from taking any action otherwise consistent with law.

- **§382.507 Penalties**

Any employer or driver who violates the requirements of this part shall be subject to the penalty provisions of 49 U.S.C. Section 521(b).

> ✓ **Important Note:** Title 49 U.S.C. 521(b)(2)(A) provides for civil penalties not to exceed $500 for each instance of refusing or failing of provide the information required by §382.405. Criminal penalties may also be imposed under 49 U.S.C. 521(b)(6).

Subpart F—Alcohol Misuse and Controlled Substances Use Information, Training, and Referral

- **§382.601 Employer obligation to promulgate a policy on the misuse of alcohol and use of controlled substances**

  (a) **General requirements.** Each employer shall provide educational materials that explain the requirements of this part and the employer's policies and procedures with respect to meeting these requirements.

  (1) The employer shall ensure that a copy of these materials is distributed to each driver prior to the start of alcohol and controlled substances testing under this part and to each driver subsequently hired or transferred into a position requiring driving a commercial motor vehicle.

  (2) Each employer shall provide written notice to representatives of employee organizations of the availability of this information.

  (b) **Required content.** The materials to be made available to drivers shall include detailed discussion of at least the following:

  (1) The identity of the person designated by the employer to answer driver questions about the materials;

  (2) The categories of drivers who are subject to the provisions of this part;

  (3) Sufficient information about the safety-sensitive functions performed by those drivers to make clear what period of the workday the driver is required to comply with this part.

  (4) Specific information concerning driver conduct that is prohibited by this part;

  (5) The circumstances under which a driver will be tested for alcohol and/or controlled substances under this part, including post-accident testing under §382.303(d);

  (6) The procedures that will be used to test for the presence of alcohol and controlled substances, protect the driver and the integrity of the testing processes, safeguard the validity of the test results, and ensure that those results are attributed to the correct driver, including post-accident information, procedures and instructions required by §382.303(d) of this part;

  (7) The requirement that a driver submit to alcohol and controlled substances tests administered in accordance with this part;

  (8) An explanation of what constitutes a refusal to submit to an alcohol or controlled substances test and the attendant consequences;

(9) The consequences for drivers found to have violated Subpart B of this part, including the requirement that the driver be removed immediately from safety-sensitive functions, and the procedures under §382.605;

(10) The consequences for drivers found to have an alcohol concentration of 0.02 or greater but less than 0.04;

(11) Information concerning the effects of alcohol and controlled substances use on an individual's health, work, and personal life; signs and symptoms of an alcohol or a controlled substances problem (the driver's or a coworker's); and available methods of intervening when an alcohol or a controlled substances problem is suspected, including confrontation, referral to any employee assistance program, and/or referral to management.

(c) **Optional provision.** The materials supplied to drivers may also include information on additional employer policies with respect to the use of alcohol or controlled substances, including any consequences for a driver found to have a specified alcohol or controlled substances level, that are based on the employer's authority independent of this part. Any such additional policies or consequences must be clearly and obviously described as being based on independent authority.

(d) **Certificate of receipt.** Each employer shall ensure that each driver is required to sign a statement certifying that he or she has received a copy of these materials described in this section. Each employer shall maintain the original of the signed certificate and may provide a copy of the certificate to the driver.

Because the employer is responsible, the employer would be in violation if using a driver who refuses to comply with §382.601 to perform any safety sensitive function, if the driver refuses to sign a statement certifying that he/she has received a copy of the educational materials required in §382.601. However, the employee would not be in violation if he or she drove without signing for the receipt of the policy.

- **§382.603 Training for supervisors.**

Each employer shall ensure that all persons designated to supervise drivers receive at least 60 minutes of training on alcohol misuse and receive at least an additional 60 minutes of training on controlled substances use. The training will be used by the supervisors to determine whether reasonable suspicion exists to require a driver to undergo testing under §382.307. The training shall include the physical, behavioral, speech, and performance indicators of probable alcohol misuse and use of controlled substances.

Controlled Substances and Alcohol Use and Testing 179

> ✓ **Important Note:** §382.603 does not require employers to provide recurrent training to supervisory personnel.

> ✓ **Important Note:** An employer may accept proof of supervisory training from a supervisor for another employer.

- **§382.605 Referral, evaluation, and treatment.**

  (a) Each driver who has engaged in conduct prohibited by Subpart B of this part shall be advised by the employer of the resources available to the driver in evaluating and resolving problems associated with the misuse of alcohol and use of controlled substances, including the names, addresses, and telephone numbers of substance abuse professionals (SAPS) and counseling and treatment programs.

  (b) Each driver who engages in conduct prohibited by Subpart B of this part shall be evaluated by a SAP who shall determine what assistance, if any, the employee needs in resolving problems associated with alcohol misuse and controlled substance use. [**Note:** The DOT rules define the SAP to be a licensed physician (medical doctor or doctor of osteopathy), a licensed or certified psychologist, a licensed or certified social worker, or a licensed or certified employee assistance professional. DOT does not certify the SAP, but expects the SAP to have knowledge of and clinical experience in the diagnosis and treatment of substances abuse-related disorders (the degrees and certificates alone do not confer this knowledge). Alcohol and drug abuse counselors certified by the National Association of Alcoholism and Drug Abuse Counselors Certification Commission, a national organization that imposes qualification standards for treatment of alcohol-related disorders, are included in the SAP definition].

  (c) (1) Before a driver returns to duty requiring the performance of a safety-sensitive function after engaging in conduct prohibited by Subpart B of this part, the driver shall undergo a return-to-duty alcohol test with a result indicating an alcohol concentration of less than 0.02 if the conduct involved alcohol, or a controlled substance.

  (2) In addition, each driver identified as needing assistance in resolving problems associated with alcohol misuse or controlled substances use:

  (i) Shall be evaluated by a SAP abuse professional to determine that the driver has properly followed any rehabilitation program prescribed under Paragraph (b) of this section, and

(ii) Shall be subject to unannounced follow-up alcohol and controlled substances tests administered by the employer following the driver's return to duty. The number and frequency of such follow-up testing shall be as directed by the substance abuse professional, and must consist of at least 6 tests in the first 12 months following the driver's return to duty. The employer may direct the driver to undergo return-to-duty and follow-up testing for both alcohol and controlled substances, if the SAP determines that return-to-duty and follow-up testing for both alcohol and controlled substances is necessary for that particular driver. Any such testing shall be performed in accordance with the requirements of 49 CFR 40. Follow-up testing shall not exceed 60 months from the date of the driver's return to duty. The SAP may terminate the requirement for follow-up testing at any time after the first six tests have been administered, if the SAP determines that such testing is no longer necessary.

(d) Evaluation and rehabilitation may be provided by the employer, by a SAP under contract with the employer, or by a SAP not affiliated with the employer. The choice of substance abuse professional and assignment of costs shall be made in accordance with employer/driver agreements and employer policies.

(e) The employer shall ensure that a SAP who determines that a driver requires assistance in resolving problems with alcohol misuse or controlled substances use does not refer the driver to the substance abuse professional's private practice or to a person or organization from which the SAP receives remuneration or in which the SAP has a financial interest. This paragraph does not prohibit a substance abuse professional from referring a driver for assistance provide through—

(1) A public agency, such as a State, county, or municipality;

(2) The employer or a person under contract to provide treatment for alcohol or controlled substance problems on behalf of the employer;

(3) The sole source of therapeutically appropriate treatment under the driver's health insurance program; or

(4) The sole source of therapeutically appropriate treatment reasonably assessable to the driver.

(f) The requirements of this section with respect to referral, evaluation and rehabilitation do not apply to applicants who refuse to submit to a pre-employment alcohol or controlled substances test or who have a pre-employment alcohol test with a result indicating an alcohol concentration of 0.04 or greater or a controlled substances test with a verified positive test result.

An employer who wants to continue to use or hire a driver who has violated the prohibitions in Subpart B in the past must ensure that a driver has complied with any SAP's recommended treatment prior to the driver returning to safety-sensitive functions. However, employers must only refer to an SAP drivers who have tested positive for controlled substances, tested positive for 0.04 or greater alcohol concentration, or have violated other prohibitions in Subpart B. An employee who has violated the rules is prohibited from performing any DOT safety-sensitive function until being evaluated by the SAP.

> ✓ **Important Note:** Self-help groups and community lectures qualify as education, but do not qualify as treatment.

> ✓ **Important Note:** An employer can't overrule an SAP treatment recommendation.

When an employer discharges an employee, the employer has no further obligation other than to provide a list of resources for evaluation and treatment numbers of SAPs and counseling and treatment programs to the employee. An employer is not obligated to return an employee to safety-sensitive duty following the SAP's finding during the follow-up evaluation that the employee has demonstrated successful compliance with the treatment recommendation. The SAP's findings and successful compliance with prescribed treatment and testing negative on the return-to-duty alcohol test and/or drug test are preconditions only. The employee must meet them to be considered for hiring or reinstatement to safety-sensitive duties by an employer.

> ✓ **Important Note:** The DOT rules do not affix responsibility for payment for SAP services upon any single party. The DOT has left discussion regarding payment to employer policies and to labor-management agreements.

If the SAP determines that an employee referred for alcohol misuse also uses drugs, or that an employee referred for drugs use also misuses alcohol, the SAP can require that the individual be tested for both substances. The SAP's decision to test for both can be based upon information gathered during the initial treatment program, and/or the information presented during the follow-up evaluation. In random testing and required follow-up testing, the two test types must be separate; one cannot be substituted for the other, or be conducted in lieu of the other. Follow-up testing is not to be conducted in a random way. An employee's follow-up testing program is to be individualized and designed to ensure that the employee is tested the appropriate number of times as directed by the SAP. If an employee tests positive while in follow-up testing, he/she is subject to the same specific DOT operating administration rules as if they tested positive on the initial test. The employee is also subject to employer policies related to second violation of DOT rules. Follow-up testing must be conducted a minimum of six times during the first twelve months following the employee's return to safety-sensitive functions. The intent of this requirement is that testing be spread throughout the 12-month period and not grouped into a shorter interval.

> ✓ **Important Note:** An employee who tests positive may be retained in a non-driving capacity.

> ✓ **Important Note:** Both the initial and follow-up SAP evaluations are clinical processes that must be conducted face-to-face.

A SAP's decision that an individual needs an education program constitutes a clinically based determination that the individual requires assistance in resolving problems with alcohol misuse and controlled substances use. Thus, the SAP is prohibited from referring the individual to his or her own practice for this recommended education unless exempted by DOT rules. SAPs are prohibited from referring an employee to themselves or to any program with which they are financially connected. SAP referrals to treatment programs must not give the impression of a conflict of interest. However, a SAP is not prohibited from referring an employee for assistance through a public agency; the employer or person under contract to provide treatment on behalf of the employer; the sole source of therapeutically appropriate treatment under the employee's health insurance program; or the sole source of therapeutically appropriate reasonably accessible to the employee.

> ✓ **Important Note:** Because evaluation by a qualified SAP rarely takes more than one diagnostic session, the requirement for an in-person evaluation is not unreasonable, even if it must be conducted some distance from the employee's home.

> ✓ **Important Note:** Foreign carriers are not required to have an employee assistance program (EAP).

Controlled Substances and Alcohol Use and Testing

> ✓ **Important Note:** Part 382 does not specify who is responsible (the employer or the driver) for paying for any testing under the alcohol and drug-testing program. The employer remains responsible at all times for ensuring compliance with the rule, regardless of who pays for testing.

> ✓ **Important Note:** An employer may use more than one Medical Review Officer, Breath Alcohol Technician, or Substance Abuse Professional.

> ✓ **Important Note:** Part 382 does not require a CMV driver to carry proof of compliance with Part 382 and Part 40; instead, the employer must maintain proof.

Procedures for Transportation Workplace Drug and Alcohol Testing Programs (Part 40)

As the title indicates, 49 CFR 40 deals specifically with procedures for conducting DOT drug and alcohol testing. In this section, we briefly cover the main sections and provide a listing of various paragraphs and subparagraphs contained within Part 40. For those seeking a more in-depth treatment of this Part, we refer you to the Code of Federal Regulations, 49 CFR Part 40.

Applicability (§40.1)

This part applies, through regulations that reference it issued by agencies of the DOT, to transportation employers (including self-employed individuals) required to conduct drug and/or alcohol testing programs by DOT agency regulations and to such transportation employers' officers, employees, agents, and contractors (including, but not limited to, consortia). Employers are responsible for the compliance of their officers, employees, agents, consortia, and/or contractors within the requirements of this part.

Subpart B—Drug Testing

- §40.21 The drugs

    (a) DOT agency drug testing programs require that employers test for marijuana, cocaine, opiates, amphetamines, and phencyclidine.

    (b) An employer may include in its testing protocols other controlled substances or alcohol only pursuant to a DOT agency approval, if testing for those substances is authorized under agency regulations and if the DHHS has established an approved testing protocol and positive threshold for each such substance.

(c) Urine specimens collected under DOT agency regulations requiring compliance with this part may only be used to test for controlled substances designated or approved for testing as described in this section and shall not be used to conduct any other analysis or test unless otherwise specifically authorized by DOT agency regulations.

(d) This section does not prohibit procedures reasonably incident to analysis of the specimen for controlled substance (e.g., determination of pH or tests for specific gravity, creatinine concentration or presence of adulterants).

> ✓ **Important Note:** Part 40 specifies that an employer must test for five drugs: marijuana, cocaine, amphetamines, opiates, and phencyclidine; however, an employer may not test for any other substances under DOT authority. Employers are not prohibited, however, from testing for other controlled substances as long as that testing remains under the authority of the employer.

Employers in the transportation industry who establish a drug testing program that tests beyond the five drugs currently required by Part 40 must also make clear to their employees what testing is required by DOT authority and what testing is required by the company. Additionally, employers must ensure that DOT urine specimens are collected in accordance with the provisions outlined in Part 40 and that a separate specimen collection process (including a separate act of urination) is used to obtain specimens for company testing programs.

Even if the drug testing custody and control form fails to indicate what tests are to be performed (which constitutes a collection site error), Part 40 indicates that DOT agency drug testing programs require that employers test for the five specified drugs. All DOT specimens, therefore, must be tested for the five categories of drugs, even if the accompanying drug testing custody and control form fails to indicate this.

While the DOT does not view this type of collection site error as a fatal flaw, it nevertheless jeopardizes the integrity of the entire collection process and could lead to a challenge and subsequent third party review. These errors should be addressed with the site supervisor in the hope of preventing future mistakes.

Controlled Substances and Alcohol Use and Testing 185

Briefly, the following sections, paragraphs and subparagraphs are also covered under Part 40.

1. Preparation for testing (§40.23)
2. Specimen collection procedures (§40.25)
    - Designation of collection site
    - Security
    - Chain of custody
    - Access to authorized personnel only
    - Privacy
    - Integrity and identity of specimen
    - Collection control
    - Transportation to laboratory
    - Failure to cooperate
    - Employee requiring medical attention
    - Use of chain of custody form
3. Laboratory personnel (§40.27)
    - Day-to-day management
    - Test validation
    - Day-to-day operations and supervision of analysts
    - Other personnel
    - Training
    - Files
4. Laboratory analysis procedures (§40.29)
    - Security and chain of custody
    - Receiving
    - Short-term refrigerated storage
    - Specimen processing
    - Initial test
    - Confirmatory test
    - Reporting results

- Long-term storage
- Retesting specimens
- Subcontracting
- Laboratory facilities
- Inspections
- Documentation
- Additional requirements for certified laboratories including:
    - Procedure manual
    - Standards and controls
    - Instruments and equipment
    - Remedial actions
    - Personnel available to testify at procedures

5. Quality assurance and quality control (§40.31)
    - Laboratory quality control requirements for initial tests
    - Laboratory quality control requirements for confirmation tests
    - Employer blind performance test procedures
6. Reporting and review of results (§40.33)
    - Medical review officer shall review confirmed positive results
    - Medical review officer—qualifications and responsibilities
    - Positive test result
    - Verification for opiates; review for prescription medication
    - Disclosure of information
7. Protection of employee records (§40.35)
8. Individual access to test and laboratory certification results
9. Use of certified laboratories (§40.39)

Subpart C—Alcohol Testing

Subpart C of Part 40 details DOT's specific requirements related to alcohol testing of employees assigned to safety-sensitive positions. Specific sections include:

- The breath and alcohol technician (BAT) (§40.51)
- Devices to be used for breath alcohol tests (§40.53)

- Quality assurance plans for EBT's (§40.55)
- Locations for breath alcohol testing (§40.57)
- The breath alcohol testing form (§40.59)
- Preparation for breath alcohol testing (§40.61)
- Procedures for screening tests (§40.63)
- Procedures for confirmation tests (§40.65)
- Refusals to test and uncompleted tests (§40.67)
- Inability to provide an adequate amount of breath (§40.69)
- Invalid tests (§40.79)
- Availability and disclosure of alcohol testing information about individual employees
- Maintenance and disclosure of records concerning EBTs and BATs

Subpart D—Non-Evidential Alcohol Screening Devices

- Authorization for use of non-evidential alcohol screening devices (§40.91)
- The screening test technician (§40.93)
- Quality assurance plans for non-evidential screening devices (§40.95)
- Locations for non-evidential alcohol screening tests (§40.97)
- Testing forms (§40.99)
- Screening test procedure (§40.101)
- Refusals to test and uncompleted tests (§40.103)
- Inability to provide an adequate amount of breath or saliva (§40.105)
- Invalid tests (§40.107)
- Availability and disclosure of alcohol testing information about individual employees (§40.109)
- Maintenance and disclosure of records concerning non- evidential testing devices and STT (Screening Test Technician (§40.111).

Sample Substance Abuse Policy

We have found that reading various State and/or Federal regulations is one thing, but attempting to decipher their meaning is quite another. To aid in your quest for understanding, you may be looking for example programs or policies. In Appendix C, we provide a sample substance abuse policy to help show you the way. The beauty of the particular sample policy in Appendix C is that it has been tested, both in the workplace and in the courtroom. However, we must point out that this is just a sample. It may not perfectly fit your particular situation, but it may give you the direction you need for assembling, adapting, or improving your own program.

Summary

The preventive measures in place through DOT and OSHA that control how hazardous materials are transported serve a valuable purpose: they protect us from loss, damage, injury, and danger. The need for such measures is unfortunate, as are the measures in place—the different types of training, safe work practices, and drug and alcohol testing—to ensure that the regulations are followed.

Notes

[1] Laws, J. April 2000. Rewriting The Testing Rulebook. *Occupational Health & Safety.* 36-38.

Afterword: Taming the Suicide Strip

All throughout the United States, a roadside practice has grown, one that serves as a grim reminder of the need for regulation. On stretches of highway where fatal accidents have occurred, the families and friends of those who died plant crosses and leave flowers for the dead. This practice serves both as memorial and as warning to those who still drive that road. The dangers of driving and the dangerous stretches of roadway—the suicide strips—cannot be entirely eliminated. As we all know, accidents can occur anywhere. The suicide strips must be approached, as all driving and all roads should be approached, with caution, attention, thought, and responsibility. One set of laws—a set of natural regulations we call the laws of physics—is the reason why.

The laws of physics cannot be repealed. They are the one set of regulations that we must always obey. The dynamics of the behavior of bodies under the action of forces that produce change in motion—mass, friction, gravity, velocity, acceleration, momentum, impact, stored kinetic energy—are a constant presence in any vehicle. Any driver works to control a constantly changing set of variables that affect where, when, and at what speed his or her vehicle will go. Any out-of-control vehicle becomes a deadly, massive projectile, dangerous to anything in its path, and in the inexorable laws of physics, the bigger they come, the greater the damage upon impact. Add a cargo of hazardous materials to the equation, and the dangers that those physical laws predict are magnified by laws of chemistry as well.

All drivers have a burden of responsibility to those with whom they share the road. That responsibility includes actively controlling—and not adding to—the variables we face when we're behind the wheel. Inattention, distraction, incapacity, and impairment and driving not only don't mix, they are deadly, and for commercial drivers they are inexcusable, especially when conveying a hazardous load. Because it's human nature to push the limits, cut corners, and squeeze as much out of a job as we can, we create human laws to monitor and enforce our own behavior and help us protect each other from those inexorable physical laws.

DOT hazmat and Hazcom regulations work to keep these variables under control. They control how dangerous materials are tracked, stored, handled, labeled, packaged, loaded, transported, and in case of accident, remediated. And they control the conditions under which commercial drivers may take to the roads.

However, regulations are only effective if they are followed, and that means both adherence to regulation and enforcement. While those regulations cost us time, money, stress, and aggravation, the costs of ignoring them are much higher. We know that cutting corners, unsafe driving practices, careless material handling, and substance abuse can cost us monetarily, but these practices also cost us in terms of short and long term environmental damage, potential and actual physical injury, death, the grief of familial loss, and spiritual loss within communities.

Appendix A
Hazardous Materials Table (§172.101)

§ 172.101 Hazardous Materials Table

Symbols	Hazardous materials descriptions and proper shipping names	Hazard class or Division	Identification Numbers	PG	Label Codes	Special provisions	Packaging (§173.***)			Quantity limitations		Vessel stowage	
							Exceptions	Non-bulk	Bulk	Passenger aircraft/rail	Cargo aircraft only	Location	Other
(1)	(2)	(3)	(4)	(5)	(6)	(7)	(8A)	(8B)	(8C)	(9A)	(9B)	(10A)	(10B)
	Accellerene, see p-Nitrosodimethylaniline.												
	Accumulators, electric, see Batteries, wet etc.												
D	Accumulators, pressurized, pneumatic or hydraulic (containing non-flammable gas).	2.2	NA1956		2.2		306	306	None	No limit	No limit	A	
	Acetal	3	UN1088	II	3	T7	150	202	242	5 L	60 L	E	
	Acetaldehyde	3	UN1089	I	3	A3, B16, T20, T26, T29	None	201	243	Forbidden	30 L	E	
A	Acetaldehyde ammonia	9	UN1841	III	9		155	204	240	200 kg	200 kg	A	34
	Acetaldehyde oxime	3	UN2332	III	3	B1, T8	150	203	242	60 L	220 L	A	
	Acetic acid, glacial or Acetic acid solution, with more than 80 percent acid, by mass.	8	UN2789	II	8, 3	A3, A6, A7, A10, B2, T8	154	202	243	1 L	30 L	A	
	Acetic acid solution, not less than 50 percent but not more than 80 percent acid, by mass.	8	UN2790	II	8	A3, A6, A7, A10, B2, T8	154	202	242	1 L	30 L	A	
	Acetic acid solution, with more than 10 percent and less than 50 percent acid, by mass.	8	UN2790	III	8	T8	154	203	242	5 L	60 L	A	
	Acetic anhydride	8	UN1715	II	8, 3	A3, A6, A7, A10, B2, T8	154	202	243	1 L	30 L	A	40
	Acetone	3	UN1090	II	3	T8	150	202	242	5 L	60 L	B	
	Acetone cyanohydrin, stabilized	6.1	UN1541	I	6.1	2, A3, B9, B14, B32, B76, B77, N34, T38, T43, T45	None	227	244	Forbidden	30 L	D	25, 40, 49
	Acetone oils	3	UN1091	II	3	T7, T30	150	202	242	5 L	60 L	B	
	Acetonitrile	3	UN1648	II	3	T14	150	202	242	5L	60L	B	40
	Acetyl acetone peroxide with more than 9 percent by mass active oxygen.	Forbidden											
	Acetyl benzoyl peroxide, solid, or with more than 40 percent in solution.	Forbidden											
	Acetyl bromide	8	UN1716	II	8	B2, T12, T26	154	202	242	1 L	30 L	C	40
	Acetyl chloride	3	UN1717	II	3, 8	A3, A6, A7, B100, N34, T18, T26	None	202	243	1 L	5 L	B	40
	Acetyl cyclohexanesulfonyl peroxide, with more than 82 percent wetted with less than 12 percent water.	Forbidden											
	Acetyl iodide	8	UN1898	II	8	B2, B101, T9	154	202	242	1 L	30 L	C	40
	Acetyl methyl carbinol	3	UN2621	III	3	B1, T1	150	203	242	60 L	220 L	A	
	Acetyl peroxide, solid, or with more than 25 percent in solution.	Forbidden											
	Acetylene, dissolved	2.1	UN1001		2.1		None	303	None	Forbidden	15kg	D	25, 40, 57
	Acetylene (liquefied)	Forbidden											
	Acetylene silver nitrate	Forbidden											
	Acetylene tetrabromide, see Tetrabromoethane.												
	Acid butyl phosphate, see Butyl acid phosphate.												
	Acid, sludge, see Sludge acid												
	Acridine	6.1	UN2713	III	6.1		153	213	240	100 kg	200kg	A	
	Acrolein dimer, stabilized	3	UN2607	III	3	B1, T1	150	203	242	60 L	220 L	A	40
	Acrolein, inhibited	6.1	UN1092	I	6.1, 3	1, B9, B14, B30, B42, B72, B77, T38, T43, T44	None	226	244	Forbidden	Forbidden	D	40
	Acrylamide	6.1	UN2074	III	6.1	T8	153	213	240	100 kg	200 kg	A	12
	Acrylic acid, inhibited	8	UN2218	II	8, 3	B2, T8	154	202	243	1 L	30 L	C	25, 40
	Acrylonitrile, inhibited	3	UN1093	I	3, 6.1	B9, T18, T26	None	201	243	Forbidden	30 L	E	40
	Actuating cartridge, explosive, see Cartridges, power device.												
	Adhesives, containing a flammable liquid.	3	UN1133	I	3	B42, T7, T30	150	201	243	1 L	30 L	B	
				II	3	B52, T7, T30	150	173	242	5 L	60 L	B	
				III	3	B1, B52, T7, T30	150	173	242	60 L	220 L	A	
	Adiponitrile	6.1	UN2205	III	6.1	T1	153	203	241	60 L	220 L	A	
	Aerosols, corrosive, Packing Group II or III, (each not exceeding 1 L capacity).	2.2	UN1950		2.2, 8	A34	306	None	None	75 kg	150 kg	A	40, 48, 85
	Aerosols, flammable, (each not exceeding 1 L capacity).	2.1	UN1950		2.1	N82	306	None	None	75 kg	150 kg	A	40, 48, 85
	Aerosols, flammable, n.o.s. (engine starting fluid) (each not exceeding 1 L capacity).	2.1	UN1950		2.1	N82	306	None	None	Forbidden	150kg	A	40, 48, 85
	Aerosols, non-flammable, (each not exceeding 1 L capacity).	2.2	UN1950		2.2		306, 307.	None	None	75 kg	150 kg	A	48, 85
	Aerosols, poison, each not exceeding 1 L capacity.	2.2	UN1950		2.2		306	None	None	Forbidden	Forbidden	A	40, 48, 85

Appendix A

§ 172.101 Hazardous Materials Table—Continued

Symbols	Hazardous materials descriptions and proper shipping names	Hazard class or Division	Identification Numbers	PG	Label Codes	Special provisions	Packaging (§173.***)			Quantity limitations		Vessel stowage	
							Exceptions	Non-bulk	Bulk	Passenger aircraft/rail	Cargo aircraft only	Location	Other
(1)	(2)	(3)	(4)	(5)	(6)	(7)	(8A)	(8B)	(8C)	(9A)	(9B)	(10A)	(10B)
	Air bag inflators, compressed gas or Air bag modules, compressed gas or Seat-belt pretensioners, compressed gas.	2.2	UN3353		2.2	133	166	166	166	75 kg	150 kg	A	
	Air bag inflators, pyrotechnic or Air bag modules, pyrotechnic or Seat-belt pretensioner, pyrotechnic.	9	UN3268	III	9		166	166	166	25 kg	100 kg	A	
	Air, compressed	2.2	UN1002		2.2		306	302	302	75 kg	150 kg	A	
	Air, refrigerated liquid, (cryogenic liquid).	2.2	UN1003		2.2, 5.1		320	316	318, 319.	Forbidden	150 kg	D	51
	Air, refrigerated liquid, (cryogenic liquid) non-pressurized.	2.2	UN1003		2.2, 5.1		320	316	318, 319.	Forbidden	Forbidden	D	51
	Aircraft evacuation slides, see Life saving appliances etc.												
	Aircraft hydraulic power unit fuel tank (containing a mixture of anhydrous hydrazine and monomethyl hydrazine) (M86 fuel).	3	UN3165	I	3, 6.1, 8.		None	172	None	Forbidden	42 L	E	
	Aircraft survival kits, see Life saving appliances etc.												
G	Alcoholates solution, n.o.s., in alcohol.	3	UN3274	II	3, 8		None	202	243	1 L	5 L	B	
	Alcoholic beverages	3	UN3065	II	3	24, B1, T1	150	202	242	5 L	60 L	A	
				III	3	24, B1, N11, T1	150	203	242	60 L	220 L	A	
	Alcohols, n.o.s	3	UN1987	I	3	T8, T31	None	201	243	1 L	30 L	E	
				II	3	T8, T31	150	202	242	5 L	60 L	B	
				III	3	B1, T7, T30	150	203	242	60 L	220 L	A	
G	Alcohols, flammable, toxic, n.o.s	3	UN1986	I	3, 6.1	T8, T31	None	201	243	Forbidden	30 L	E	40
				II	3, 6.1	T8, T31	None	202	243	1 L	60 L	B	40
				III	3, 6.1	B1, T8, T31	None	203	243	60 L	220 L	A	
	Aldehydes, n.o.s	3	UN1989	I	3	T8, T31	None	201	243	1 L	30 L	E	
				II	3	T8, T31	150	202	242	5 L	60 L	B	
				III	3	B1, T7, T30	150	203	242	60 L	220 L	A	
G	Aldehydes, flammable, toxic, n.o.s.	3	UN1988	I	3, 6.1	T8, T31	None	201	243	Forbidden	30 L	E	40
				II	3, 6.1	T8, T31	None	202	243	1 L	60 L	B	40
				III	3, 6.1	B1, T8, T31	150	203	242	60 L	220 L	A	
	Aldol	6.1	UN2839	II	6.1	T8	None	202	243	5 L	60 L	A	12
D	Aldrin, liquid	6.1	NA2762	II	6.1		None	202	243	5 L	60 L	B	
D	Aldrin, solid	6.1	NA2761	II	6.1		None	212	242	25 kg	100 kg	A	40
G	Alkali metal alcoholates, self-heating, corrosive, n.o.s.	4.2	UN3206	II	4.2, 8	64	None	212	242	15 kg	50 kg	B	
				III	4.2, 8	64	None	213	242	25 kg	100 kg	B	
	Alkali metal alloys, liquid, n.o.s	4.3	UN1421	I	4.3	A2, A3, B48, N34	None	201	244	Forbidden	1 L	D	
	Alkali metal amalgam, liquid	4.3	UN1389	I	4.3	A2, A3, N34	None	201	244	Forbidden	1 L	D	40
	Alkali metal amalgam, solid	4.3	UN1389	I	4.3	B101, B106, N40	None	211	242	Forbidden	15 kg	D	
	Alkali metal amides	4.3	UN1390	II	4.3	A6, A7, A8, A19, A20, B106	151	212	241	15 kg	50 kg	E	40
	Alkali metal dispersions, or Alkaline earth metal dispersions.	4.3	UN1391	I	4.3	A2, A3	None	201	244	Forbidden	1 L	D	
	Alkaline corrosive liquids, n.o.s., see Caustic alkali liquids, n.o.s.												
G	Alkaline earth metal alcoholates, n.o.s.	4.2	UN3205	II	4.2	65	None	212	241	15kg	50kg	B	
				III	4.2	65	None	213	241	25 kg	100 kg	B	
	Alkaline earth metal alloys, n.o.s	4.3	UN1393	II	4.3	A19, B101, B106	151	212	241	15 kg	50 kg	E	
	Alkaline earth metal amalgams	4.3	UN1392	I	4.3	A19, B101, B106, N34, N40	None	211	242	Forbidden	15 kg	D	
G	Alkaloids, liquid, n.o.s., or Alkaloid salts, liquid, n.o.s.	6.1	UN3140	I	6.1	A4, T42	None	201	243	1 L	30 L	A	
				II	6.1	T14	None	202	243	5 L	60 L	A	
				III	6.1	T7	153	203	241	60 L	220 L	A	
G	Alkaloids, solid, n.o.s. or Alkaloid salts, solid, n.o.s. poisonous.	6.1	UN1544	I	6.1		None	211	242	5 kg	50 kg	A	
				II	6.1		None	212	242	25 kg	100 kg	A	
				III	6.1		153	213	240	100 kg	200 kg	A	
	Alkyl sulfonic acids, liquid or Aryl sulfonic acids, liquid with more than 5 percent free sulfuric acid.	8	UN2584	II	8	B2, T8, T27	154	202	242	1 L	30 L	B	
	Alkyl sulfonic acids, liquid or Aryl sulfonic acids, liquid with not more than 5 percent free sulfuric acid.	8	UN2586	III	8	T8	154	203	241	5 L	60 L	B	
	Alkyl sulfonic acids, solid or Aryl sulfonic acids, solid, with more than 5 percent free sulfuric acid.	8	UN2583	II	8		154	212	240	15 kg	50 kg	A	
	Alkyl sulfonic acids, solid or Aryl sulfonic acids, solid with not more than 5 percent free sulfuric acid.	8	UN2585	III	8		154	213	240	25 kg	100 kg	A	
	Alkylphenols, liquid, n.o.s. (including C2-C12 homologues).	8	UN3145	I	8	T8	None	201	243	0.5 L	2.5 L	B	

§ 172.101 Hazardous Materials Table—Continued

Symbols (1)	Hazardous materials descriptions and proper shipping names (2)	Hazard class or Division (3)	Identification Numbers (4)	PG (5)	Label Codes (6)	Special provisions (7)	Packaging (§173.***) (8)			Quantity limitations (9)		Vessel stowage (10)	
							Exceptions (8A)	Non-bulk (8B)	Bulk (8C)	Passenger aircraft/rail (9A)	Cargo aircraft only (9B)	Location (10A)	Other (10B)
				II	8	T8	154	202	242	1 L	30 L	B	
				III	8	T7	154	203	241	5 L	60 L	A	
	Alkylphenols, solid, n.o.s. (including C2-C12 homologues).	8	UN2430	I	8	T8	None	211	242	1 kg	25 kg	B	
				II	8	T8	154	212	240	15 kg	50 kg	B	
				III	8	T8	154	213	240	25 kg	100 kg	A	
	Alkylsulfuric acids	8	UN2571	II	8	B2, T9, T27	154	202	242	1 L	30 L	C	14
	Allethrin, see Pesticides, liquid, toxic, n.o.s.												
	Allyl acetate	3	UN2333	II	3, 6.1	T8	None	202	243	1 L	60 L	E	40
	Allyl alcohol	6.1	UN1098	I	6.1, 3	2, B9, B14, B32, B74, B77, T38, T43, T45	None	227	244	Forbidden	Forbidden	D	40
	Allyl bromide	3	UN1099	I	3, 6.1	T18	None	201	243	Forbidden	30 L	B	40
	Allyl chloride	3	UN1100	I	3, 6.1	T18, T26	None	201	243	Forbidden	30 L	E	40
	Allyl chlorocarbonate, see Allyl chloroformate.												
	Allyl chloroformate	6.1	UN1722	I	6.1, 3, 8.	2, A3, B9, B14, B32, B74, N41, T38, T43, T45	None	227	244	Forbidden	Forbidden	D	40
	Allyl ethyl ether	3	UN2335	II	3, 6.1	T8	None	202	243	1 L	60 L	E	40
	Allyl formate	3	UN2336	I	3, 6.1	T18, T26	None	201	243	Forbidden	30 L	E	40
	Allyl glycidyl ether	3	UN2219	III	3	B1, T7	150	203	242	60 L	220 L	A	
	Allyl iodide	3	UN1723	II	3, 8	A3, A6, B100, N34, T18	None	202	243	1 L	5 L	B	40
	Allyl isothiocyanate, stabilized	6.1	UN1545	II	6.1, 3	A3, A7	None	202	243	Forbidden	60 L	D	40
	Allylamine	6.1	UN2334	I	6.1, 3	2, B9, B14, B32, B74, T38, T43, T45	None	227	244	Forbidden	Forbidden	D	40
	Allyltrichlorosilane, stabilized	8	UN1724	II	8, 3	A7, B2, B6, N34, T8, T26	None	202	243	Forbidden	30 L	C	40
	Aluminum alkyl halides	4.2	UN3052	I	4.2, 4.3	B9, B11, T28, T29, T40	None	181	244	Forbidden	Forbidden	D	
	Aluminum alkyl hydrides	4.2	UN3076	I	4.2, 4.3	B9, B11, T28, T29, T40	None	181	244	Forbidden	Forbidden	D	
	Aluminum alkyls	4.2	UN3051	I	4.2, 4.3	B9, B11, T28, T29, T40	None	181	244	Forbidden	Forbidden	D	
	Aluminum borohydride or Aluminum borohydride in devices.	4.2	UN2870	I	4.2, 4.3	B11	None	181	244	Forbidden	Forbidden	D	
	Aluminum bromide, anhydrous	8	UN1725	II	8	B106	154	212	240	15 kg	50 kg	A	40
	Aluminum bromide, solution	8	UN2580	III	8	T8	154	203	241	5 L	60 L	A	
	Aluminum carbide	4.3	UN1394	II	4.3	A20, B101, B106, N41	151	212	242	15 kg	50 kg	A	
	Aluminum chloride, anhydrous	8	UN1726	II	8	B106	154	212	240	15 kg	50kg	A	40
	Aluminum chloride, solution	8	UN2581	III	8	T8	154	203	241	5 L	60 L	A	
	Aluminum dross, wet or hot	Forbidden											
	Aluminum ferrosilicon powder	4.3	UN1395	II	4.3, 6.1	A19, B106, B108	151	212	242	15 kg	50 kg	A	40, 85, 103
				III	4.3, 6.1	A19, A20, B106, B108	151	213	241	25 kg	100 kg	A	40, 85, 103
	Aluminum hydride	4.3	UN2463	I	4.3	A19, B100, N40	None	211	242	Forbidden	15 kg	E	
D	Aluminum, molten	9	NA9260	III	9		None	None	247	Forbidden	Forbidden	D	
	Aluminum nitrate	5.1	UN1438	III	5.1	A1, A29	152	213	240	25 kg	100 kg	A	
	Aluminum phosphate solution, see Corrosive liquids, etc.												
	Aluminum phosphide	4.3	UN1397	I	4.3, 6.1	A8, A19, B100, N40	None	211	242	Forbidden	15 kg	E	40, 85
	Aluminum phosphide pesticides	6.1	UN3048	I	6.1	A8	None	211	242	Forbidden	15 kg	E	40, 85
	Aluminum powder, coated	4.1	UN1309	II	4.1		151	212	240	15 kg	50 kg	A	13, 39, 101
				III	4.1		151	213	240	25 kg	100 kg	A	13, 39, 101
	Aluminum powder, uncoated	4.3	UN1396	II	4.3	A19, A20, B106, B108	151	212	242	15 kg	50 kg	A	39
				III	4.3	A19, A20, B106, B108	151	213	241	25 kg	100 kg	A	39
	Aluminum resinate	4.1	UN2715	III	4.1		151	213	240	25 kg	100 kg	A	
	Aluminum silicon powder, uncoated.	4.3	UN1398	III	4.3	A1, A19, B108	151	213	241	25 kg	100 kg	A	40, 85, 103
	Aluminum smelting by-products or Aluminum remelting by-products.	4.3	UN3170	II	4.3	128, B106, B115	None	212	242	15 kg	50 kg	B	85, 103
				III	4.3	128, B106, B115	None	213	241	25 kg	100 kg	B	85, 103
	Amatols, see Explosives, blasting, type B.												
G	Amines, flammable, corrosive, n.o.s. or Polyamines, flammable, corrosive, n.o.s..	3	UN2733	I	3, 8	T42	None	201	243	0.5 L	2.5 L	D	40
				II	3, 8	T8, T31	None	202	243	1 L	5 L	B	40
				III	3, 8	B1, T8, T31	150	203	242	5 L	60 L	A	40
G	Amines, liquid, corrosive, flammable, n.o.s. or Polyamines, liquid, corrosive, flammable, n.o.s.	8	UN2734	I	8, 3	A3, A6, N34, T8, T31	None	201	243	0.5 L	2.5 L	A	
				II	8, 3	T8, T31	None	202	243	1 L	30 L	A	
G	Amines, liquid, corrosive, n.o.s., or Polyamines, liquid, corrosive, n.o.s.	8	UN2735	I	8	A3, A6, B10, N34, T42	None	201	243	0.5 L	2.5 L	A	
				II	8	B2, T8	154	202	242	1 L	30 L	A	

Appendix A

§ 172.101 Hazardous Materials Table—Continued

Symbols	Hazardous materials descriptions and proper shipping names	Hazard class or Division	Identification Numbers	PG	Label Codes	Special provisions	(8) Packaging (§173.***)			(9) Quantity limitations		(10) Vessel stowage	
							Exceptions	Non-bulk	Bulk	Passenger aircraft/rail	Cargo aircraft only	Location	Other
(1)	(2)	(3)	(4)	(5)	(6)	(7)	(8A)	(8B)	(8C)	(9A)	(9B)	(10A)	(10B)
G	Amines, solid, corrosive, n.o.s., or Polyamines, solid, corrosive n.o.s.	8	UN3259	III	8	T8	154	203	241	5 L	60 L	A	
				I	8		None	211	242	1 kg	25 kg	A	
				II	8		154	212	240	15 kg	50 kg	A	
				III	8		154	213	240	25 kg	100 kg	A	
	2-Amino-4-chlorophenol	6.1	UN2673	II	6.1		None	212	242	25 kg	100 kg	A	
	2-Amino-5-diethylaminopentane	6.1	UN2946	III	6.1	T1	153	203	241	60 L	220 L	A	
	2-Amino-4,6-Dinitrophenol, wetted with not less than 20 percent water by mass.	4.1	UN3317	I	4.1	23, A8, A19, A20, N41	None	211	None	1 kg	15 kg	E	28, 36
	2-(2-Aminoethoxy) ethanol	8	UN3055	III	8	T2	154	203	241	5 L	60 L	A	
	N-Aminoethylpiperazine	8	UN2815	III	8	T7	154	203	241	5 L	60 L	A	12
+	Aminophenols (o-; m-; p-)	6.1	UN2512	III	6.1	T1	153	213	240	100 kg	200 kg	A	
	Aminopropyldiethanolamine, see Amines, etc.												
	n-Aminopropylmorpholine, see Amines, etc.												
	Aminopyridines (o-; m-; p-)	6.1	UN2671	II	6.1	T7	None	212	242	25 kg	100 kg	B	12, 40
I	Ammonia, anhydrous	2.3	UN1005		2.3, 8	4	None	304	314, 315.	Forbidden	25 kg	D	40, 57
D	Ammonia, anhydrous	2.2	UN1005		2.2	13	None	304	314, 315.	Forbidden	25 kg	D	40, 57
D	Ammonia solution, relative density less than 0.880 at 15 degrees C in water, with more than 50 percent ammonia.	2.2	UN3318		2.2	13	None	304	314, 315.	Forbidden	25 kg	D	40, 57
I	Ammonia solution, relative density less than 0.880 at 15 degrees C in water, with more than 50 percent ammonia.	2.3	UN3318		2.3, 8	4	None	304	314, 315.	Forbidden	25 kg	D	40, 57
	Ammonia solutions, relative density between 0.880 and 0.957 at 15 degrees C in water, with more than 10 percent but not more than 35 percent ammonia.	8	UN2672	III	8	T14	154	203	241	5 L	60 L	A	40, 85
	Ammonia solutions, relative density less than 0.880 at 15 degrees C in water, with more than 35 percent but not more than 50 percent ammonia.	2.2	UN2073		2.2		306	304	314, 315.	Forbidden	150 kg	E	40, 57
	Ammonium arsenate	6.1	UN1546	II	6.1		None	212	242	25 kg	100 kg	A	
	Ammonium azide	Forbidden											
	Ammonium bifluoride, solid, see Ammonium hydrogen difluoride, solid.												
	Ammonium bifluoride solution, see Ammonium hydrogen difluoride, solution.												
	Ammonium bromate	Forbidden											
	Ammonium chlorate	Forbidden											
	Ammonium dichromate	5.1	UN1439	II	5.1		152	212	242	5 kg	25 kg	A	
	Ammonium dinitro-o-cresolate	6.1	UN1843	II	6.1	T8	None	212	242	25 kg	100 kg	B	36, 65, 66, 77
	Ammonium fluoride	6.1	UN2505	III	6.1		153	213	240	100 kg	200 kg	A	26
	Ammonium fluorosilicate	6.1	UN2854	III	6.1		153	213	240	100 kg	200 kg	A	26
	Ammonium fulminate	Forbidden											
	Ammonium hydrogen sulfate	8	UN2506	II	8		154	212	240	15 kg	50 kg	A	40
	Ammonium hydrogendifluoride, solid.	8	UN1727	II	8	B106, N34	154	212	240	15 kg	50 kg	A	25, 26, 40
	Ammonium hydrogendifluoride, solution.	8	UN2817	II	8, 6.1	N34, T15	None	202	243	1 L	30 L	B	40
				III	8, 6.1	T8	154	203	241	5 L	60 L	B	40, 95
	Ammonium hydrosulfide, solution, see Ammonium sulfide solution.												
D	Ammonium hydroxide, see Ammonia solutions, etc.												
	Ammonium metavanadate	6.1	UN2859	II	6.1		None	212	242	25 kg	100 kg	A	
D	Ammonium nitrate fertilizers	5.1	NA2072	III	5.1	7	152	213	240	25 kg	100 kg	B	48, 59, 60, 117

§ 172.101 HAZARDOUS MATERIALS TABLE—Continued

Symbols	Hazardous materials descriptions and proper shipping names	Hazard class or Division	Identification Numbers	PG	Label Codes	Special provisions	Packaging (§173.***)			Quantity limitations		Vessel stowage	
							Exceptions	Non-bulk	Bulk	Passenger aircraft/rail	Cargo aircraft only	Location	Other
(1)	(2)	(3)	(4)	(5)	(6)	(7)	(8A)	(8B)	(8C)	(9A)	(9B)	(10A)	(10B)
	Ammonium nitrate fertilizers; uniform non-segregating mixtures of ammonium nitrate with added matter which is inorganic and chemically inert towards ammonium nitrate, with not less than 90 percent ammonium nitrate and not more than 0.2 percent combustible material (including organic material calculated as carbon), or with more than 70 percent but less than 90 percent ammonium nitrate and not more than 0.4 percent total combustible material.	5.1	UN2067	III	5.1	52	152	213	240	25 kg	100 kg	B	48, 59, 60, 117
A, W	Ammonium nitrate fertilizers: uniform non-segregating mixtures of nitrogen/phosphate or nitrogen/postash types or complete fertilizers of nitrogen/phosphate/postash type, with not more than 70 percent ammonium nitrate and not more than 0.4 percent total added combustible material or with not more than 45 percent ammonium nitrate with unrestricted combustible material.	9	UN2071	III	9	132	155	213	240	200 kg	200 kg	A	
D	Ammonium nitrate-fuel oil mixture containing only prilled ammonium nitrate and fuel oil.	1.5D	NA0331	II	1.5D		None	62	None	Forbidden	Forbidden	B	1E, 5E
	Ammonium nitrate, liquid (hot concentrated solution).	5.1	UN2426		5.1	B5, B100, T25	None	None	243	Forbidden	Forbidden	D	59, 60
D	Ammonium nitrate mixed fertilizers.	5.1	NA2069	III	5.1	10	152	213	240	25 kg	100 kg	B	48, 59, 60, 117
	Ammonium nitrate, with more than 0.2 percent combustible substances, including any organic substance calculated as carbon, to the exclusion of any other added substance.	1.1D	UN0222	II	1.1D		None	62	None	Forbidden	Forbidden	B	1E, 5E, 19E
	Ammonium nitrate, with not more than 0.2 percent of combustible substances, including any organic substance calculated as carbon, to the exclusion of any other added substance.	5.1	UN1942	III	5.1	A1, A29	152	213	240	25 kg	100 kg	A	48, 59, 60, 116
	Ammonium nitrite	Forbidden											
	Ammonium perchlorate	1.1D	UN0402	II	1.1D	107	None	62	None	Forbidden	Forbidden	B	1E, 5E, 19E
	Ammonium perchlorate	5.1	UN1442	II	5.1	107, A9	152	212	242	5 kg	25 kg	E	58, 69, 106
	Ammonium permanganate	Forbidden											
	Ammonium persulfate	5.1	UN1444	III	5.1	A1, A29	152	213	240	25 kg	100 kg	A	
	Ammonium picrate, dry or wetted with less than 10 percent water, by mass.	1.1D	UN0004	II	1.1D		None	62	None	Forbidden	Forbidden	B	1E, 5E, 19E
	Ammonium picrate, wetted with not less than 10 percent water, by mass.	4.1	UN1310	I	4.1	23, A2, N41	None	211	None	0.5 kg	0.5 kg	D	28, 36
	Ammonium polysulfide, solution	8	UN2818	II	8, 6.1	T14	None	202	243	1 L	30 L	B	12, 26, 40
				III	8, 6.1	T7	154	203	241	5 L	60 L	B	12, 26, 40
	Ammonium polyvanadate	6.1	UN2861	II	6.1		None	212	242	25 kg	100 kg	A	
	Ammonium silicofluoride, see Ammonium fluorosilicate.												
	Ammonium sulfide solution	8	UN2683	II	8, 6.1, 3.	T14	None	202	243	1 L	30 L	B	12, 22, 26, 100
	Ammunition, blank, see Cartridges for weapons, blank.												
	Ammunition, illuminating with or without burster, expelling charge or propelling charge.	1.2G	UN0171	II	1.2G			62	None	Forbidden	Forbidden	B	
	Ammunition, illuminating with or without burster, expelling charge or propelling charge.	1.3G	UN0254	II	1.3G			62	None	Forbidden	Forbidden	B	
	Ammunition, illuminating with or without burster, expelling charge or propelling charge.	1.4G	UN0297	II	1.4G			62	None	Forbidden	75 kg	A	24E
	Ammunition, incendiary liquid or gel, with burster, expelling charge or propelling charge.	1.3J	UN0247	II	1.3J			62	None	Forbidden	Forbidden	E	7E, 13E, 23E

Appendix A

§ 172.101 HAZARDOUS MATERIALS TABLE—Continued

Symbols	Hazardous materials descriptions and proper shipping names	Hazard class or Division	Identification Numbers	PG	Label Codes	Special provisions	Packaging (§173.***)			Quantity limitations		Vessel stowage	
							Exceptions	Non-bulk	Bulk	Passenger aircraft/rail	Cargo aircraft only	Location	Other
(1)	(2)	(3)	(4)	(5)	(6)	(7)	(8A)	(8B)	(8C)	(9A)	(9B)	(10A)	(10B)
	Ammunition, incendiary (water-activated contrivances) with burster, expelling charge or propelling charge, see Contrivances, water-activated, etc.												
	Ammunition, incendiary, white phosphorus, *with burster, expelling charge or propelling charge.*	1.2H	UN0243	II	1.2H ...			62	None	Forbidden	Forbidden	E	8E, 14E, 15E, 17E
	Ammunition, incendiary, white phosphorus, *with burster, expelling charge or propelling charge.*	1.3H	UN0244	II	1.3H ...			62	None	Forbidden	Forbidden	E	8E, 14E, 15E, 17E
	Ammunition, incendiary *with or without burster, expelling charge, or propelling charge.*	1.2G	UN0009	II	1.2G ...			62	None	Forbidden	Forbidden	B	
	Ammunition, incendiary *with or without burster, expelling charge, or propelling charge.*	1.3G	UN0010	II	1.3G ...			62	None	Forbidden	Forbidden	B	
	Ammunition, incendiary *with or without burster, expelling charge or propelling charge.*	1.4G	UN0300	II	1.4G ...			62	None	Forbidden	75 kg	A	24E
	Ammunition, practice	1.4G	UN0362	II	1.4G ...			62	None	Forbidden	75 kg	A	24E
	Ammunition, practice	1.3G	UN0488	II	1.3G ...			62	None	Forbidden	Forbidden	B	
	Ammunition, proof	1.4G	UN0363	II	1.4G ...			62	None	Forbidden	75kg	A	24E
	Ammunition, rocket, see Warheads, rocket etc.												
	Ammunition, SA (small arms), see Cartridges for weapons, etc.												
	Ammunition, smoke (water-activated contrivances), white phosphorus, with burster, expelling charge or propelling charge, see Contrivances, water-activated, etc. (UN 0248).												
	Ammunition, smoke (water activated contrivances), without white phosphorus or phosphides, with burster, expelling charge or propelling charge, see Contrivances, water-activated, etc. (UN 0249).												
	Ammunition smoke, white phosphorus *with burster, expelling charge, or propelling charge.*	1.2H	UN0245	II	1.2H ...			62	None	Forbidden	Forbidden	E	8E, 14E, 15E, 17E
	Ammunition, smoke, white phosphorus *with burster, expelling charge, or propelling charge.*	1.3H	UN0246	II	1.3H ...			62	None	Forbidden	Forbidden	E	8E, 14E, 15E, 17E
	Ammunition, smoke *with or without burster, expelling charge or propelling charge.*	1.2G	UN0015	II	1.2G, 8			62	None	Forbidden	Forbidden	E	17E, 20E
	Ammunition, smoke *with or without burster, expelling charge or propelling charge.*	1.3G	UN0016	II	1.3G, 8			62	None	Forbidden	Forbidden	E	17E, 20E
	Ammunition, smoke *with or without burster, expelling charge or propelling charge.*	1.4G	UN0303	II	1.4G, 8			62	None	Forbidden	75 kg	E	17E, 20E
	Ammunition, sporting, see Cartridges for weapons, etc. (UN 0012; UN 0328; UN 0339).												
	Ammunition, tear-producing, non-explosive, *without burster or expelling charge, non-fuzed.*	6.1	UN2017	II	6.1, 8		None	212	None	Forbidden	50 kg	E	13, 40
	Ammunition, tear-producing *with burster, expelling charge or propelling charge.*	1.2G	UN0018	II	1.2G, 8, 6.1.			62	None	Forbidden	Forbidden	E	20E
	Ammunition, tear-producing *with burster, expelling charge or propelling charge.*	1.3G	UN0019	II	1.3G, 8, 6.1.			62	None	Forbidden	Forbidden	E	17E, 20E
	Ammunition, tear-producing *with burster, expelling charge or propelling charge.*	1.4G	UN0301	II	1.4G, 8, 6.1.			62	None	Forbidden	75 kg	E	17E, 20E
	Ammunition, toxic, non-explosive, *without burster or expelling charge, non-fuzed.*	6.1	UN2016	II	6.1 ...		None	212	None	Forbidden	100 kg	E	13, 40
	Ammunition, toxic (water-activated contrivances), with burster, expelling charge or propelling charge, see Contrivances, water-activated, etc.												

§ 172.101 Hazardous Materials Table—Continued

Symbols (1)	Hazardous materials descriptions and proper shipping names (2)	Hazard class or Division (3)	Identification Numbers (4)	PG (5)	Label Codes (6)	Special provisions (7)	Packaging (§173.***) Exceptions (8A)	Packaging Non-bulk (8B)	Packaging Bulk (8C)	Quantity limitations Passenger aircraft/rail (9A)	Quantity limitations Cargo aircraft only (9B)	Vessel stowage Location (10A)	Vessel stowage Other (10B)
G	Ammunition, toxic *with burster, expelling charge, or propelling charge*.	1.2K	UN0020	II	1.2K, 6.1.			62	None	Forbidden	Forbidden	E	2E, 8E, 11E, 17E
G	Ammunition, toxic *with burster, expelling charge, or propelling charge*.	1.3K	UN0021	II	1.3K, 6.1.			62	None	Forbidden	Forbidden	E	2E, 8E, 11E, 17E
	Amyl acetates	3	UN1104	III	3	B1, T1	150	203	242	60 L	220 L	A	
	Amyl acid phosphate	8	UN2819	III	8	T7	154	203	241	5 L	60 L	A	
	Amyl butyrates	3	UN2620	III	3	B1, T1	150	203	242	60 L	220 L	A	
	Amyl chlorides	3	UN1107	II	3	T1	150	202	242	5 L	60 L	B	
	Amyl formates	3	UN1109	III	3	B1, T1	150	203	242	60 L	220 L	A	
	Amyl mercaptans	3	UN1111	II	3	A3, T8	None	202	242	5 L	60 L	B	95, 102
	n-Amyl methyl ketone	3	UN1110	III	3	B1, T1	150	203	242	60 L	220 L	A	
	Amyl nitrate	3	UN1112	III	3	B1, T1	150	203	242	60 L	220 L	A	40
	Amyl nitrites	3	UN1113	II	3	T8	150	202	242	5 L	60 L	E	40
	Amylamines	3	UN1106	II	3, 8	T1	None	202	243	1 L	5 L	B	
				III	3, 8		150	203	242	5 L	60 L	A	
	Amyltrichlorosilane	8	UN1728	II	8	A7, B2, B6, N34, T8, T26	None	202	242	Forbidden	30 L	C	40
	Anhydrous ammonia see Ammonia, anhydrous.												
	Anhydrous hydrofluoric acid, see Hydrogen fluoride, anhydrous.												
+	Aniline	6.1	UN1547	II	6.1	T8	None	202	243	5 L	60 L	A	40
	Aniline hydrochloride	6.1	UN1548	III	6.1		153	213	240	100 kg	200 kg	A	
	Aniline oil, see Aniline												
	Anisidines	6.1	UN2431	III	6.1	T1	153	203	241	60 L	220 L	A	
	Anisole	3	UN2222	III	3	B1, T1	150	203	242	60 L	220 L	A	
	Anisoyl chloride	8	UN1729	II	8	B2, T8	154	202	242	1 L	30 L	C	40
	Anti-freeze, liquid, see Flammable liquids, n.o.s.												
	Antimonous chloride, see Antimony trichloride.												
	Antimony compounds, inorganic, liquid, n.o.s.	6.1	UN3141	III	6.1	35, T7	153	203	241	60 L	220 L	A	
	Antimony compounds, inorganic, solid, n.o.s.	6.1	UN1549	III	6.1	35	153	213	240	100 kg	200 kg	A	
	Antimony lactate	6.1	UN1550	III	6.1		153	213	240	100 kg	200 kg	A	
	Antimony pentachloride, liquid	8	UN1730	II	8	B2, T8, T26	None	202	242	1 L	30 L	C	40
	Antimony pentachloride, solutions.	8	UN1731	II	8	B2, T8, T27	154	202	242	1 L	30 L	C	40
				III	8	T7, T26	154	203	241	5 L	60 L	C	40
	Antimony pentafluoride	8	UN1732	II	8, 6.1	A3, A6, A7, A10, N3, T12, T26	None	202	243	Forbidden	30 L	D	40
	Antimony potassium tartrate	6.1	UN1551	III	6.1		153	213	240	100 kg	200 kg	A	
	Antimony powder	6.1	UN2871	III	6.1		153	213	240	100 kg	200 kg	A	
	Antimony sulfide and a chlorate, mixtures of.	Forbidden											
	Antimony sulfide, solid, see Antimony compounds, inorganic, n.o.s.												
D	Antimony tribromide, solid	8	NA1549	II	8		154	212	240	25 kg	100 kg	A	13
D	Antimony tribromide, solution	8	NA1549	II	8	B2	154	202	242	1 L	30 L	C	13
	Antimony trichloride, liquid	8	UN1733	II	8	B2	154	202	242	1 L	30 L	C	40
	Antimony trichloride, solid	8	UN1733	II	8	B106	154	212	240	15 kg	50 kg	A	40
D	Antimony trifluoride, solid	8	NA1549	II	8		154	212	240	25 kg	25 kg	A	13
D	Antimony trifluoride, solution	8	NA1549	II	8	B2	154	202	242	1 L	30 L	C	13
	Aqua ammonia, see Ammonia solution, etc.												
	Argon, compressed	2.2	UN1006		2.2		306	302	314, 315.	75 kg	150 kg	A	
	Argon, refrigerated liquid *(cryogenic liquid)*.	2.2	UN1951		2.2		320	316	318	50 kg	500 kg	B	
	Arsenic	6.1	UN1558	II	6.1		None	212	242	25 kg	100 kg	A	
	Arsenic acid, liquid	6.1	UN1553	I	6.1	T18, T27	None	201	243	1 L	30 L	B	46
	Arsenic acid, solid	6.1	UN1554	II	6.1		None	212	242	25 kg	100 kg	A	
	Arsenic bromide	6.1	UN1555	II	6.1		None	212	242	25 kg	100 kg	A	12, 40
	Arsenic chloride, see Arsenic trichloride.												
	Arsenic compounds, liquid, n.o.s. *inorganic, including arsenates, n.o.s.; arsenites, n.o.s.; arsenic sulfides, n.o.s.; and organic compounds of arsenic, n.o.s.*.	6.1	UN1556	I	6.1		None	201	243	1 L	30 L	B	40
				II	6.1		None	202	243	5 L	60 L	B	40
				III	6.1		153	203	241	60 L	220 L	B	40
	Arsenic compounds, solid, n.o.s. *inorganic, including arsenates, n.o.s.; arsenites, n.o.s.; arsenic sulfides, n.o.s.; and organic compounds of arsenic, n.o.s.*.	6.1	UN1557	I	6.1		None	211	242	5 kg	50 kg	A	
				II	6.1		None	212	242	25 kg	100 kg	A	
				III	6.1		153	213	240	100 kg	200 kg	A	
	Arsenic pentoxide	6.1	UN1559	II	6.1		None	212	242	25 kg	100 kg	A	
D	Arsenic sulfide	6.1	NA1557	II	6.1		None	212	242	25 kg	100 kg	A	
	Arsenic sulfide and a chlorate, mixtures of.	Forbidden											

Appendix A

§ 172.101 HAZARDOUS MATERIALS TABLE—Continued

Symbols	Hazardous materials descriptions and proper shipping names	Hazard class or Division	Identification Numbers	PG	Label Codes	Special provisions	(8) Packaging (§173.***)			(9) Quantity limitations		(10) Vessel stowage	
							Exceptions	Non-bulk	Bulk	Passenger aircraft/rail	Cargo aircraft only	Location	Other
(1)	(2)	(3)	(4)	(5)	(6)	(7)	(8A)	(8B)	(8C)	(9A)	(9B)	(10A)	(10B)
	Arsenic trichloride	6.1	UN1560	I	6.1	2, B9, B14, B32, B74, T38, T43, T45	None	227	244	Forbidden	Forbidden	B	40
	Arsenic trioxide	6.1	UN1561	II	6.1		None	212	242	25 kg	100 kg	A	
D	Arsenic trisulfide	6.1	NA1557	II	6.1		None	212	242	25 kg	100 kg	A	
	Arsenic, white, solid, see Arsenic trioxide.												
	Arsenical dust	6.1	UN1562	II	6.1		None	212	242	25 kg	100 kg	A	
	Arsenical pesticides, liquid, flammable, toxic, flash point less than 23 degrees C.	3	UN2760	I	3, 6.1		None	201	243	Forbidden	30 L	B	40
				II	3, 6.1		None	202	243	1 L	60 L	B	40
	Arsenical pesticides, liquid, toxic	6.1	UN2994	I	6.1	T42	None	201	243	1 L	30 L	B	40
				II	6.1	T14	None	202	243	5 L	60 L	B	40
				III	6.1	T14	153	203	241	60 L	220 L	A	40
	Arsenical pesticides, liquid, toxic, flammable flashpoint not less than 23 degrees C.	6.1	UN2993	I	6.1, 3	T42	None	201	243	1 L	30 L	B	40
				II	6.1, 3	T14	None	202	243	5 L	60 L	B	40
				III	6.1, 3	B1, T14	153	203	242	60 L	220 L	A	40
	Arsenical pesticides, solid, toxic	6.1	UN2759	I	6.1		None	211	242	5 kg	50 kg	A	40
				II	6.1		None	212	242	25 kg	100 kg	A	40
				III	6.1		153	213	240	100 kg	200 kg	A	40
	Arsenious acid, solid, see Arsenic trioxide.												
	Arsenious and mercuric iodide solution, see Arsenic compounds, liquid, n.o.s.												
	Arsine	2.3	UN2188		2.3, 2.1	1	None	192	245	Forbidden	Forbidden	D	40
	Articles, explosive, extremely insensitive or Articles, EEI.	1.6N	UN0486	II	1.6N	101	None	62	None	Forbidden	Forbidden	B	
G	Articles, explosive, n.o.s	1.4S	UN0349	II	1.4S	101	None	62	None	25 kg	100 kg	A	
G	Articles, explosive, n.o.s	1.4B	UN0350	II	1.4B	101	None	62	None	Forbidden	Forbidden	A	24E
G	Articles, explosive, n.o.s	1.4C	UN0351	II	1.4C	101	None	62	None	Forbidden	75 kg	A	24E
G	Articles, explosive, n.o.s	1.4D	UN0352	II	1.4D	101	None	62	None	Forbidden	75 kg	A	24E
G	Articles, explosive, n.o.s	1.4G	UN0353	II	1.4G	101	None	62	None	Forbidden	75 kg	A	24E
G	Articles, explosive, n.o.s	1.1L	UN0354	II	1.1L	101	None	62	None	Forbidden	Forbidden	E	2E, 8E, 11E, 17E
G	Articles, explosive, n.o.s	1.2L	UN0355	II	1.2L	101	None	62	None	Forbidden	Forbidden	E	2E, 8E, 11E, 17E
G	Articles, explosive, n.o.s	1.3L	UN0356	II	1.3L	101	None	62	None	Forbidden	Forbidden	E	2E, 8E, 11E, 17E
G	Articles, explosive, n.o.s	1.1C	UN0462	II	1.1C	101	None	62	None	Forbidden	Forbidden	B	
G	Articles, explosive, n.o.s	1.1D	UN0463	II	1.1D	101	None	62	None	Forbidden	Forbidden	B	
G	Articles, explosive, n.o.s	1.1E	UN0464	II	1.1E	101	None	62	None	Forbidden	Forbidden	B	
G	Articles, explosive, n.o.s	1.1F	UN0465	II	1.1F	101	None	62	None	Forbidden	Forbidden	E	
G	Articles, explosive, n.o.s	1.2C	UN0466	II	1.2C	101	None	62	None	Forbidden	Forbidden	B	
G	Articles, explosive, n.o.s	1.2D	UN0467	II	1.2D	101	None	62	None	Forbidden	Forbidden	B	
G	Articles, explosive, n.o.s	1.2E	UN0468	II	1.2E	101	None	62	None	Forbidden	Forbidden	B	
G	Articles, explosive, n.o.s	1.2F	UN0469	II	1.2F	101	None	62	None	Forbidden	Forbidden	E	
G	Articles, explosive, n.o.s	1.3C	UN0470	II	1.3C	101	None	62	None	Forbidden	Forbidden	B	
G	Articles, explosive, n.o.s	1.4E	UN0471	II	1.4E	101	None	62	None	Forbidden	75 kg	A	24E
G	Articles, explosive, n.o.s	1.4F	UN0472	II	1.4F	101	None	62	None	Forbidden	Forbidden	E	
	Articles, pressurized pneumatic or hydraulic containing non-flammable gas.	2.2	UN3164		2.2		306	302, 304.	None	No limit	No limit	A	
	Articles, pyrophoric	1.2L	UN0380	II	1.2L		None	62	None	Forbidden	Forbidden	E	2E, 8E, 11E, 17E
	Articles, pyrotechnic for technical purposes.	1.1G	UN0428	II	1.1G		None	62	None	Forbidden	Forbidden	B	
	Articles, pyrotechnic for technical purposes.	1.2G	UN0429	II	1.2G		None	62	None	Forbidden	Forbidden	B	
	Articles, pyrotechnic for technical purposes.	1.3G	UN0430	II	1.3G		None	62	None	Forbidden	Forbidden	B	
	Articles, pyrotechnic for technical purposes.	1.4G	UN0431	II	1.4G		None	62	None	Forbidden	75 kg	A	24E
	Articles, pyrotechnic for technical purposes.	1.4S	UN0432	II	1.4S		None	62	None	25 kg	100 kg	A	
D	Asbestos	9	NA2212	III	9		155	216	240	200 kg	200 kg	A	34, 40
	Ascaridole (organic peroxide)	Forbidden											
D	Asphalt, at or above its flashpoint.	3	NA1999	III	3		150	203	247	Forbidden	Forbidden	D	
D	Asphalt, cut back, see Tars, liquid, etc.												
	Automobile, motorcycle, tractor, other self-propelled vehicle, engine, or other mechanical apparatus, see Vehicles or Battery etc.												
A G	Aviation regulated liquid, n.o.s	9	UN3334		9	A35	155	204		No limit	No limit	A	
A G	Aviation regulated solid, n.o.s	9	UN3335		9	A35	155	204		No limit	No limit	A	
	Azaurolic acid (salt of) (dry)	Forbidden											
	Azido guanidine picrate (dry)	Forbidden											
	5-Azido-1-hydroxy tetrazole	Forbidden											

§172.101 HAZARDOUS MATERIALS TABLE—Continued

Symbols (1)	Hazardous materials descriptions and proper shipping names (2)	Hazard class or Division (3)	Identification Numbers (4)	PG (5)	Label Codes (6)	Special provisions (7)	(8) Packaging (§173.***) Exceptions (8A)	Non-bulk (8B)	Bulk (8C)	(9) Quantity limitations Passenger aircraft/rail (9A)	Cargo aircraft only (9B)	(10) Vessel stowage Location (10A)	Other (10B)
	Azido hydroxy tetrazole (mercury and silver salts).	Forbidden											
	3-Azido-1,2-Propylene glycol dinitrate.	Forbidden											
	Azidodithiocarbonic acid	Forbidden											
	Azidoethyl nitrate	Forbidden											
	1-Aziridinylphosphine oxide-(tris), see Tris-(1-aziridinyl) phosphine oxide, solution.												
	Azodicarbonamide	4.1	UN3242	II	4.1	38	151	212	240	Forbidden	Forbidden	D	12, 61, 74
	Azotetrazole (dry)	Forbidden											
	Barium	4.3	UN1400	II	4.3	A19, B100, B106	151	212	241	15 kg	50 kg	E	
	Barium alloys, pyrophoric	4.2	UN1854	I	4.2		None	181	None	Forbidden	Forbidden	D	
	Barium azide, dry or wetted with less than 50 percent water, by mass.	1.1A	UN0224	II	1.1A, 6.1	111, 117	None	62	None	Forbidden	Forbidden	E	2E, 6E
	Barium azide, wetted with not less than 50 percent water, by mass.	4.1	UN1571	I	4.1, 6.1	A2	None	182	None	Forbidden	0.5 kg	D	28
	Barium bromate	5.1	UN2719	II	5.1, 6.1		None	212	242	5 kg	25 kg	A	56, 58, 106
	Barium chlorate	5.1	UN1445	II	5.1, 6.1	A9, N34, T8	None	212	242	5 kg	25 kg	A	56, 58, 106
	Barium compounds, n.o.s	6.1	UN1564	II	6.1		None	212	242	25 kg	100 kg	A	
				III	6.1		153	213	240	100 kg	200 kg	A	
	Barium cyanide	6.1	UN1565	I	6.1	N74, N75	None	211	242	5 kg	50 kg	A	26, 40
	Barium hypochlorite with more than 22 percent available chlorine.	5.1	UN2741	II	5.1, 6.1	A7, A9, N34	152	212	None	5 kg	25 kg	B	56, 58, 106
	Barium nitrate	5.1	UN1446	II	5.1, 6.1		None	212	242	5 kg	25 kg	A	
	Barium oxide	6.1	UN1884	III	6.1		153	213	240	100 kg	200 kg	A	
	Barium perchlorate	5.1	UN1447	II	5.1, 6.1	T8	None	212	242	5 kg	25 kg	A	56, 58, 106
	Barium permanganate	5.1	UN1448	II	5.1, 6.1		None	212	242	5 kg	25 kg	D	56, 58, 69, 106, 107
	Barium peroxide	5.1	UN1449	II	5.1, 6.1		None	212	242	5 kg	25 kg	A	13, 75, 106
	Barium selenate, see Selenates or Selenites.												
	Barium selenite, see Selenates or Selenites.												
D	Barium styphnate	1.1A	NA0473	II	1.1A	111, 117	None	62	None	Forbidden	Forbidden	E	2E, 6E
	Batteries, containing sodium	4.3	UN3292	II	4.3		189	189	189	Forbidden	No limit	A	
	Batteries, dry, containing potassium hydroxide solid, electric, storage.	8	UN3028	III	8		None	213	None	25 kg gross	230 kg gross	A	
	Batteries, wet, filled with acid, electric storage.	8	UN2794	III	8		159	159	159	30 kg gross	No limit	A	
	Batteries, wet, filled with alkali, electric storage.	8	UN2795	III	8		159	159	159	30 kg gross	No limit	A	
	Batteries, wet, filled with alkali, electric storage.	8	UN2795	III	8		159	159	159	25 kg gross	No limit	A	
	Batteries, wet, non-spillable, electric storage.	8	UN2800	III	8		159	159	159	No Limit	No Limit	A	
	Batteries, dry, not subject to the requirements of this subchapter.					130							
	Battery fluid, acid	8	UN2796	II	8	A3, A7, B2, B15, N6, N34, T9, T27	154	202	242	1 L	30 L	B	
	Battery fluid, alkali	8	UN2797	II	8	B2, N6, T8	154	202	242	1 L	30 L	A	
	Battery lithium type, see Lithium batteries etc.												
	Battery-powered vehicle or Battery-powered equipment.	9	UN3171		9	134	220	220	None	No limit	No limit		
	Battery, wet, filled with acid or alkali with automobile (or named self-propelled vehicle or mechanical equipment containing internal combustion engine) see Vehicles, self-propelled etc.												
+	Benzaldehyde	9	UN1990	III	9	T1	155	203	241	100 L	220 L	A	
	Benzene	3	UN1114	II	3	B101, T8	150	202	242	5 L	60 L	B	40
	Benzene diazonium chloride (dry).	Forbidden											
	Benzene diazonium nitrate (dry)	Forbidden											
	Benzene phosphorus dichloride, see Phenyl phosphorus dichloride.												
	Benzene phosphorus thiodichloride, see Phenyl phosphorus thiodichloride.												
	Benzene sulfonyl chloride	8	UN2225	III	8	T8	154	203	241	5 L	60 L	A	40
	Benzene triozonide	Forbidden											

Appendix A

§ 172.101 Hazardous Materials Table—Continued

Symbols	Hazardous materials descriptions and proper shipping names	Hazard class or Division	Identification Numbers	PG	Label Codes	Special provisions	(8) Packaging (§173.***)			(9) Quantity limitations		(10) Vessel stowage	
							Exceptions	Non-bulk	Bulk	Passenger aircraft/rail	Cargo aircraft only	Location	Other
(1)	(2)	(3)	(4)	(5)	(6)	(7)	(8A)	(8B)	(8C)	(9A)	(9B)	(10A)	(10B)
	Benzenethiol, see Phenyl mercaptan.												
	Benzidine	6.1	UN1885	II	6.1		None	212	242	25 kg	100 kg	A	
	Benzol, see Benzene												
	Benzonitrile	6.1	UN2224	II	6.1	T14	None	202	243	5 L	60 L	A	26, 40
	Benzoquinone	6.1	UN2587	II	6.1		None	212	242	25 kg	100 kg	A	
	Benzotrichloride	8	UN2226	II	8	B2, B101, T15	154	202	242	1 L	30L	A	40
	Benzotrifluoride	3	UN2338	II	3	T2	150	202	242	5 L	60 L	B	40
	Benzoxidiazoles (dry)	Forbidden											
	Benzoyl azide	Forbidden											
	Benzoyl chloride	8	UN1736	II	8	B2, T9, T26	154	202	242	1 L	30 L	C	40
	Benzyl bromide	6.1	UN1737	II	6.1, 8	A3, A7, N33, N34, T12, T26	None	202	243	1 L	30 L	D	13, 40
	Benzyl chloride	6.1	UN1738	II	6.1, 8	A3, A7, B70, N33, N42, T12, T26	None	202	243	1 L	30 L	D	13, 40
	Benzyl chloride *unstabilized*	6.1	UN1738	II	6.1, 8	A3, A7, B8, B11, N33, N34, N43, T12, T26	None	202	243	1 L	30L	D	13, 40
	Benzyl chloroformate	8	UN1739	I	8	A3, A6, B4, N41, T18, T26	None	201	243	Forbidden	2.5 L	D	40
	Benzyl iodide	6.1	UN2653	II	6.1	T8	None	202	243	5 L	60 L	B	12, 40
	Benzyldimethylamine	8	UN2619	II	8, 3	B2, T1	154	202	243	1 L	30 L	A	40, 48
	Benzylidene chloride	6.1	UN1886	II	6.1	T8	None	202	243	5 L	60 L	D	40
	Beryllium compounds, n.o.s	6.1	UN1566	II	6.1		None	212	242	25 kg	100 kg	A	
				III	6.1		153	213	240	100 kg	200 kg	A	
	Beryllium nitrate	5.1	UN2464	II	5.1, 6.1		None	212	242	5 kg	25 kg	A	
	Beryllium, powder	6.1	UN1567	II	6.1, 4.1		None	212	242	15 kg	50 kg	A	
	Bicyclo [2,2,1] hepta-2, 5-diene, inhibited or 2,5-Norbornadiene, inhibited	3	UN2251	II	3		150	202	242	5 L	60 L	D	
	Biphenyl triozonide	Forbidden											
	Bipyridilium pesticides, liquid, flammable, toxic, *flash point less than 23 degrees C.*	3	UN2782	I	3, 6.1		None	201	243	Forbidden	30 L	E	
				II	3, 6.1		None	202	243	1 L	60 L	B	40
	Bipyridilium pesticides, liquid, toxic.	6.1	UN3016	I	6.1	T42	None	201	243	1 L	30 L	B	40
				II	6.1	T14	None	202	243	5 L	60 L	B	40
				III	6.1	T14	153	203	241	60 L	220 L	A	40
	Bipyridilium pesticides, liquid, toxic, flammable, *flashpoint not less than 23 degrees C.*	6.1	UN3015	I	6.1, 3	T42	None	201	243	1 L	30 L	B	21, 40
				II	6.1, 3	T14	None	202	243	5 L	60 L	B	21, 40
				III	6.1, 3	B1, T14	153	203	242	60 L	220 L	A	21, 40
	Bipyridilium pesticides, solid, toxic.	6.1	UN2781	I	6.1		None	211	242	5 kg	50 kg	A	40
				II	6.1		None	212	242	25 kg	100 kg	A	40
				III	6.1		153	213	240	100 kg	200 kg	A	40
	Bis (Aminopropyl) piperazine, see Corrosive liquid, n.o.s.												
	Bisulfate, aqueous solution	8	UN2837	II	8	A7, B2, N34, T8, T26	154	202	242	1 L	30 L	A	
				III	8	A7, N34, T7, T26	154	203	241	5 L	60 L	A	
	Bisulfites, aqueous solutions, n.o.s.	8	UN2693	III	8	T8	154	203	241	1 L	30 L	A	26, 40
	Black powder, compressed *or* Gunpowder, compressed *or* Black powder, in pellets *or* Gunpowder, in pellets.	1.1D	UN0028	II	1.1D		None	62	None	Forbidden	Forbidden	B	1E, 5E
	Black powder *or* Gunpowder, granular or as a meal.	1.1D	UN0027	II	1.1D		None	62	None	Forbidden	Forbidden	B	10E, 26E
D	Black powder for small arms	4.1	NA0027	I	4.1	70	None	170	None	Forbidden	Forbidden	E	
	Blasting agent, n.o.s., see Explosives, blasting etc.												
	Blasting cap assemblies, see Detonator assemblies, non-electric, for blasting.												
	Blasting caps, electric, see Detonators, electric for blasting.												
	Blasting caps, non-electric, see Detonators, non-electric, for blasting.												
	Bleaching powder, see Calcium hypochlorite mixtures, etc.												
I	Blue asbestos *(Crocidolite)* or Brown asbestos *(amosite, mysorite).*	9	UN2212	II	9		155	216	240	Forbidden	Forbidden	A	34, 40
	Bombs, photo-flash	1.1F	UN0037	II	1.1F			62	None	Forbidden	Forbidden	E	
	Bombs, photo-flash	1.1D	UN0038	II	1.1D			62	None	Forbidden	Forbidden	B	
	Bombs, photo-flash	1.2G	UN0039	II	1.2G			62	None	Forbidden	Forbidden	B	
	Bombs, photo-flash	1.3G	UN0299	II	1.3G			62	None	Forbidden	Forbidden	B	
	Bombs, smoke, non-explosive, with *corrosive liquid, without initiating device.*	8	UN2028	II	8		None	160	None	Forbidden	50 kg	E	40
	Bombs, *with bursting charge*	1.1F	UN0033	II	1.1F			62	None	Forbidden	Forbidden	E	
	Bombs, *with bursting charge*	1.1D	UN0034	II	1.1D			62	None	Forbidden	Forbidden	B	3E, 7E
	Bombs, *with bursting charge*	1.2D	UN0035	II	1.2D			62	None	Forbidden	Forbidden	B	3E, 7E
	Bombs, *with bursting charge*	1.2F	UN0291	II	1.2F			62	None	Forbidden	Forbidden	E	

§ 172.101 Hazardous Materials Table—Continued

Symbols	Hazardous materials descriptions and proper shipping names	Hazard class or Division	Identification Numbers	PG	Label Codes	Special provisions	(8) Packaging (§173.***)			(9) Quantity limitations		(10) Vessel stowage	
							Exceptions	Non-bulk	Bulk	Passenger aircraft/rail	Cargo aircraft only	Location	Other
(1)	(2)	(3)	(4)	(5)	(6)	(7)	(8A)	(8B)	(8C)	(9A)	(9B)	(10A)	(10B)
	Bombs with flammable liquid, with bursting charge.	1.1J	UN0399	II	1.1J			62	None	Forbidden	Forbidden	E	7E, 16E, 23E
	Bombs with flammable liquid, with bursting charge.	1.2J	UN0400	II	1.2J			62	None	Forbidden	Forbidden	E	7E, 16E, 23E
	Boosters with detonator	1.1B	UN0225	II	1.1B		None	62	None	Forbidden	Forbidden	B	2E, 6E
	Boosters with detonator	1.2B	UN0268	II	1.2B		None	62	None	Forbidden	Forbidden	E	1E, 7E
	Boosters, without detonator	1.1D	UN0042	II	1.1D		None	62	None	Forbidden	Forbidden	B	
	Boosters, without detonator	1.2D	UN0283	II	1.2D		None	62	None	Forbidden	Forbidden	B	
	Borate and chlorate mixtures, see Chlorate and borate mixtures.												
	Borneol	4.1	UN1312	III	4.1	A1	None	213	240	25 kg	100 kg	A	
+	Boron tribromide	8	UN2692	I	8, 6.1	2, A3, A7, B9, B14, B32, B74, N34, T38, T43, T45	None	227	244	Forbidden	2.5 L	C	12
	Boron trichloride	2.3	UN1741		2.3, 8	3, B9, B14	None	304	314	Forbidden	Forbidden	D	25, 40
	Boron trifluoride, compressed.	2.3	UN1008		2.3	2, B9, B14	None	302	314, 315.	Forbidden	Forbidden	D	40
	Boron trifluoride acetic acid complex.	8	UN1742	II	8	B2, B6, T9, T27	154	202	242	1 L	30 L	A	
	Boron trifluoride diethyl etherate	8	UN2604	I	8, 3	A19, T8, T26	None	201	243	0.5 L	2.5 L	D	40
	Boron trifluoride dihydrate	8	UN2851	II	8	T9, T27	154	212	240	15 kg	50 kg	B	12, 40
	Boron trifluoride dimethyl etherate.	4.3	UN2965	I	4.3, 8, 3.	A19, T12, T26	None	201	243	Forbidden	1 L	D	21, 28, 40, 49, 100
	Boron trifluoride propionic acid complex.	8	UN1743	II	8	B2, T9, T27	154	202	242	1 L	30 L	A	
	Box toe gum, see Nitrocellulose etc.												
	Bromates, inorganic, aqueous solution, n.o.s.	5.1	UN3213	II	5.1	T8	152	202	242	1L	5 L	B	56, 58, 106
	Bromates, inorganic, n.o.s	5.1	UN1450	II	5.1		152	212	242	5 kg	25 kg	A	56, 58, 106
	Bromine azide	Forbidden											
+	Bromine or Bromine solutions	8	UN1744	I	8, 6.1	1, A3, A6, B9, B64, B85, N34, N43, T18, T41	None	226	249	Forbidden	Forbidden		12, 40, 66, 74, 89, 90
	Bromine chloride	2.3	UN2901		2.3, 8, 5.1.	2, B9, B14	None	304	314, 315.	Forbidden	Forbidden	D	40, 89, 90
+	Bromine pentafluoride	5.1	UN1745	I	5.1, 6.1, 8.	1, B9, B14, B30, B72, T38, T43, T44	None	228	244	Forbidden	Forbidden	D	25, 40, 66, 90
+	Bromine trifluoride	5.1	UN1746	I	5.1, 6.1, 8.	2, B9, B14, B32, B74, T38, T43, T45	None	228	244	Forbidden	Forbidden	D	25, 40, 66, 90
	4-Bromo-1,2-dinitrobenzene	Forbidden											
	4-Bromo-1,2-dinitrobenzene (unstable at 59 degrees C.).	Forbidden											
	1-Bromo-3-methylbutane	3	UN2341	III	3	B1, T7, T30	150	203	242	60 L	220 L	A	
	1-Bromo-3-nitrobenzene (unstable at 56 degrees C).	Forbidden											
	2-Bromo-2-nitropropane-1,3-diol	4.1	UN3241	III	4.1	46	151	213	None	25 kg	50 kg	C	12, 25, 40
	Bromoacetic acid, solid	8	UN1938	II	8	A7, N34, T9	154	212	240	15 kg	50 kg	A	
	Bromoacetic acid, solution	8	UN1938	II	8	B2, T9	154	202	242	1 L	30 L	A	40
+	Bromoacetone	6.1	UN1569	II	6.1, 3	2	None	193	245	Forbidden	Forbidden	D	40
	Bromoacetyl bromide	8	UN2513	II	8	B2, T9, T26	154	202	242	1 L	30 L	C	40
	Bromobenzene	3	UN2514	III	3	B1, T1	150	203	242	60 L	220 L	A	
	Bromobenzyl cyanides, liquid	6.1	UN1694	I	6.1	T18	None	201	243	Forbidden	30 L	D	12, 40
	Bromobenzyl cyanides, solid	6.1	UN1694	I	6.1	T18	None	211	242	Forbidden	50 kg	D	12, 40
	1-Bromobutane	3	UN1126	II	3	T1	150	202	242	5L	60L	B	40
	2-Bromobutane	3	UN2339	II	3	B1, T1	150	202	242	5 L	60 L	B	40
	Bromochloromethane	6.1	UN1887	III	6.1	T7	153	203	241	60 L	220 L	A	
	2-Bromoethyl ethyl ether	3	UN2340	II	3	T7	150	202	242	5 L	60 L	B	40
	Bromoform	6.1	UN2515	III	6.1	T7	153	203	241	60 L	220 L	A	12, 40
	Bromomethylpropanes	3	UN2342	II	3	T7, T30	150	202	242	5 L	60 L	B	
	2-Bromopentane	3	UN2343	II	3	T1	150	202	242	5 L	60 L	B	
	Bromopropanes	3	UN2344	II	3	T7	150	202	242	5 L	60 L	B	40
	3-Bromopropyne	3	UN2345	II	3	T8	150	202	242	5 L	60 L	D	40
	Bromosilane	Forbidden											
	Bromotoluene-alpha, see Benzyl bromide.												
	Bromotrifluoroethylene	2.1	UN2419		2.1		None	304	314, 315.	Forbidden	150 kg	B	40
	Bromotrifluoromethane or Refrigerant gas, R 13B1..	2.2	UN1009		2.2		306	304	314, 315.	75 kg	150 kg	A	
	Brucine	6.1	UN1570	I	6.1		None	211	242	5 kg	50 kg	A	
	Bursters, explosive	1.1D	UN0043	II	1.1D		None	62	None	Forbidden	Forbidden	B	
	Butadienes, inhibited	2.1	UN1010		2.1		306	304	314, 315.	Forbidden	150 kg	B	40
	Butane see also Petroleum gases, liquefied.	2.1	UN1011		2.1	19	306	304	314, 315.	Forbidden	150 kg	E	40

Appendix A

§ 172.101 Hazardous Materials Table—Continued

Symbols	Hazardous materials descriptions and proper shipping names	Hazard class or Division	Identification Numbers	PG	Label Codes	Special provisions	(8) Packaging (§173.***)			(9) Quantity limitations		(10) Vessel stowage	
							Exceptions	Non-bulk	Bulk	Passenger aircraft/rail	Cargo aircraft only	Location	Other
(1)	(2)	(3)	(4)	(5)	(6)	(7)	(8A)	(8B)	(8C)	(9A)	(9B)	(10A)	(10B)
	Butane, butane mixtures and mixtures having similar properties in cartridges each not exceeding 500 grams, see Receptacles, etc.												
	Butanedione	3	UN2346	II	3	T1	150	202	242	5 L	60 L	B	
	1,2,4-Butanetriol trinitrate	Forbidden											
	Butanols	3	UN1120	II	3	T1	150	202	242	5 L	60 L	B	
				III	3	B1, T1	150	203	242	60 L	220 L	A	
	tert-Butoxycarbonyl azide	Forbidden											
	Butyl acetates	3	UN1123	II	3	T1	150	202	242	5 L	60 L	B	
				III	3	B1, T1	150	203	242	60 L	220 L	A	
	Butyl acid phosphate	8	UN1718	III	8	T7	154	203	241	5 L	60 L	A	
	Butyl acrylates, inhibited	3	UN2348	III	3	B1, T8, T31	150	203	242	60 L	220 L	A	
	Butyl alcohols, see Butanols												
	Butyl benzenes	3	UN2709	III	3	B1, T1	150	203	242	60 L	220 L	A	
	n-Butyl bromide, see 1-Bromobutane.												
	n-Butyl chloride, see Chlorobutanes.												
D	sec-Butyl chloroformate	6.1	NA2742	I	6.1, 3, 8.	2, B9, B14, B32, B74, T38, T43, T45	None	227	244	1 L	30 L	A	12, 13, 22, 25, 40, 48, 100
	n-Butyl chloroformate	6.1	UN2743	I	6.1, 8, 3.	2, B9, B14, B32, B74, T38, T43, T45	None	227	244	1 L	30 L	A	12, 13, 21, 25, 40, 100
	Butyl ethers, see Dibutyl ethers												
	Butyl ethyl ether, see Ethyl butyl ether.												
	n-Butyl formate	3	UN1128	II	3	T1	150	202	242	5 L	60 L	B	
	tert-Butyl hydroperoxide, with more than 90 percent with water.	Forbidden											
	tert-Butyl hypochlorite	4.2	UN3255	I	4.2, 8		None	211	243	Forbidden	Forbidden	D	
	N-n-Butyl imidazole	6.1	UN2690	II	6.1	T8	None	202	243	5 L	60 L	A	
	tert-Butyl isocyanate	6.1	UN2484	I	6.1, 3	1, A7, B9, B14, B30, B72, T38, T43, T44	None	226	244	Forbidden	Forbidden	D	40
	n-Butyl isocyanate	6.1	UN2485	I	6.1, 3	2, A7, B9, B14, B32, B74, B77, T38, T43, T45	None	227	244	Forbidden	30 L	D	40
	Butyl mercaptans	3	UN2347	II	3	A3, T8	150	202	242	5 L	60 L	D	26, 95
	n-Butyl methacrylate, inhibited.	3	UN2227	III	3	B1, T1	150	203	242	60 L	220 L	A	
	Butyl methyl ether	3	UN2350	II	3	T8	150	202	242	5 L	60 L	B	
	Butyl nitrites	3	UN2351	I	3	T8	150	201	243	1 L	30 L	E	40
				II	3	T8	150	202	242	5 L	60 L	B	40
				III	3	B1, T8	150	203	242	60 L	220 L	A	40
	tert-Butyl peroxyacetate, with more than 76 percent in solution.	Forbidden											
	n-Butyl peroxydicarbonate, with more than 52 percent in solution.	Forbidden											
	tert-Butyl peroxyisobutyrate, with more than 77 percent in solution.	Forbidden											
	Butyl phosphoric acid, see Butyl acid phosphate.												
	Butyl propionates	3	UN1914	III	3	B1,T1	150	203	242	60 L	220 L	A	
	5-tert-Butyl-2,4,6-trinitro-m-xylene or Musk xylene.	4.1	UN2956	III	4.1		None	214	None	Forbidden	Forbidden	D	12
	Butyl vinyl ether, inhibited	3	UN2352	II	3	B101, T7	150	202	242	5 L	60 L	B	40
	n-Butylamine	3	UN1125	II	3, 8	B101, T8	None	202	242	1 L	5 L	B	40
	N-Butylaniline	6.1	UN2738	II	6.1	T8	None	202	243	5 L	60 L	A	
	tert-Butylcyclohexylchloroformate	6.1	UN2747	III	6.1	T8	153	203	241	60 L	220 L	A	12, 13, 25
	Butylene see also Petroleum gases, liquefied.	2.1	UN1012		2.1	19	None	304	314, 315.	Forbidden	150 kg	E	40
	1,2-Butylene oxide, stabilized	3	UN3022	II	3	T8	150	202	242	5 L	60 L	B	49
	Butyltoluenes	6.1	UN2667	III	6.1	T2	153	203	241	60 L	220 L	A	
	Butyltrichlorosilane	8	UN1747	II	8, 3	A7, B2, B6, N34, T8, T26	None	202	243	Forbidden	30 L	C	40
	1,4-Butynediol	6.1	UN2716	III	6.1	A1	None	213	240	100 kg	200 kg	A	61, 70
	Butyraldehyde	3	UN1129	II	3	T8	150	202	242	5 L	60 L	B	
	Butyraldoxime	3	UN2840	III	3	B1, T1	150	203	242	60 L	220 L	A	
	Butyric acid	8	UN2820	III	8	T1	154	203	241	5 L	60 L	A	12
	Butyric anhydride	8	UN2739	III	8	T2	154	203	241	5 L	60 L	A	
	Butyronitrile	3	UN2411	II	3, 6.1	T14	None	202	243	1 L	60 L	E	40
	Butyryl chloride	3	UN2353	II	3, 8	B100, T9, T26	None	202	243	1 L	5 L	C	40
	Cacodylic acid	6.1	UN1572	II	6.1		None	212	242	25 kg	100 kg	E	26
	Cadmium compounds	6.1	UN2570	I	6.1		None	211	242	5 kg	50 kg	A	
				II	6.1		None	212	242	25 kg	100 kg	A	
				III	6.1		153	213	240	100 kg	200 kg	A	
	Caesium hydroxide	8	UN2682	II	8		154	212	240	15 kg	50 kg	A	
	Caesium hydroxide solution	8	UN2681	II	8	B2, T8	154	202	242	1 L	30 L	A	
				III	8	T7	154	203	241	5 L	60 L	A	
	Calcium	4.3	UN1401	II	4.3	B101, B106	151	212	241	15 kg	50 kg	E	
	Calcium arsenate	6.1	UN1573	II	6.1		None	212	242	25 kg	100 kg	A	

§172.101 HAZARDOUS MATERIALS TABLE—Continued

Symbols	Hazardous materials descriptions and proper shipping names	Hazard class or Division	Identification Numbers	PG	Label Codes	Special provisions	(8) Packaging (§173.***)			(9) Quantity limitations		(10) Vessel stowage	
							Exceptions	Non-bulk	Bulk	Passenger aircraft/rail	Cargo aircraft only	Location	Other
(1)	(2)	(3)	(4)	(5)	(6)	(7)	(8A)	(8B)	(8C)	(9A)	(9B)	(10A)	(10B)
	Calcium arsenate and calcium arsenite, mixtures, solid.	6.1	UN1574	II	6.1		None	212	242	25 kg	100 kg	A	
D	Calcium arsenite, solid	6.1	NA1574	II	6.1		None	212	242	25 kg	100 kg	A	
	Calcium bisulfite solution, see Bisulfites, inorganic, aqueous solutions, n.o.s.												
	Calcium carbide	4.3	UN1402	I	4.3	A1, A8, B55, B59, B101, B106, N34	None	211	242	Forbidden	15 kg	B	
				II	4.3	A1, A8, B55, B59, B101, B106, N34	151	212	241	15 kg	50 kg	B	
	Calcium chlorate	5.1	UN1452	II	5.1	N34	152	212	242	5 kg	25 kg	A	56, 58, 106
	Calcium chlorate aqueous solution.	5.1	UN2429	II	5.1	A2, N41, T8	152	202	242	1 L	5 L	B	56, 58, 106
				III	5.1	A2, N41, T8	152	203	241	2.5 L	30 L	B	56, 58, 106
	Calcium chlorite	5.1	UN1453	II	5.1	A9, N34	152	212	242	5 kg	25 kg	A	56, 58, 106
	Calcium cyanamide with more than 0.1 percent of calcium carbide.	4.3	UN1403	III	4.3	A1, A19, B105	151	213	241	25 kg	100 kg	A	
	Calcium cyanide	6.1	UN1575	I	6.1	N79, N80	None	211	242	5 kg	50 kg	A	26, 40
	Calcium dithionite or Calcium hydrosulfite.	4.2	UN1923	II	4.2	A19, A20	None	212	241	15 kg	50 kg	E	13
	Calcium hydride	4.3	UN1404	I	4.3	A19, B100, N40	None	211	242	Forbidden	15 kg	E	
	Calcium hydrosulfite, see Calcium dithionite.												
	Calcium hypochlorite, dry or Calcium hypochlorite mixtures dry with more than 39 percent available chlorine (8.8 percent available oxygen).	5.1	UN1748	II	5.1	A7, A9, N34	152	212	None	5 kg	25 kg	D	48, 56, 58, 69, 106, 118
	Calcium hypochlorite, hydrated or Calcium hypochlorite, hydrated mixtures, with not less than 5.5 percent but not more than 10 percent water.	5.1	UN2880	II	5.1		152	212	240	5 kg	25 kg	A	50, 56, 58, 69, 106
	Calcium hypochlorite mixtures, dry, with more than 10 percent but not more than 39 percent available chlorine.	5.1	UN2208	III	5.1	A1, A29, B103, N34	152	213	240	25 kg	100 kg	A	56, 58, 69, 106
	Calcium manganese silicon	4.3	UN2844	III	4.3	A1, A19, B105, B106	151	213	241	25 kg	100 kg	A	85, 103
	Calcium nitrate	5.1	UN1454	III	5.1	34	152	213	240	25 kg	100 kg	A	
A	Calcium oxide	8	UN1910	III	8		154	213	240	25 kg	100 kg	A	
	Calcium perchlorate	5.1	UN1455	II	5.1		152	212	242	5 kg	25 kg	A	56, 58, 106
	Calcium permanganate	5.1	UN1456	II	5.1		152	212	242	5 kg	25 kg	D	56, 58, 69, 106, 107
	Calcium peroxide	5.1	UN1457	II	5.1		152	212	242	5 kg	25 kg	A	13, 75, 106
	Calcium phosphide	4.3	UN1360	I	4.3, 6.1	A8, A19, B100, N40	None	211	242	Forbidden	15 kg	E	40, 85
	Calcium, pyrophoric or Calcium alloys, pyrophoric.	4.2	UN1855	I	4.2		None	187	None	Forbidden	Forbidden	D	
	Calcium resinate	4.1	UN1313	III	4.1	A1, A19	None	213	240	25 kg	100 kg	A	
	Calcium resinate, fused	4.1	UN1314	III	4.1	A1, A19	None	213	240	25 kg	100 kg	A	
	Calcium selenate, see Selenates or Selenites.												
	Calcium silicide	4.3	UN1405	II	4.3	A19, B105, B106	151	212	241	15 kg	50 kg	B	85, 103
				III	4.3	A1, A19, B106, B108	151	213	241	25 kg	100 kg	B	85, 103
	Camphor oil	3	UN1130	III	3	B1, T1	150	203	242	60 L	220 L	A	
	Camphor, synthetic	4.1	UN2717	III	4.1	A1	None	213	240	25 kg	100 kg	A	
	Cannon primers, see Primers, tubular.												
	Caproic acid	8	UN2829	III	8	T1	154	203	241	5 L	60 L	A	
	Caps, blasting, see Detonators, etc.												
	Carbamate pesticides, liquid, flammable, toxic, flash point less than 23 degrees C.	3	UN2758	I	3, 6.1		None	201	243	Forbidden	30 L	B	40
				II	3, 6.1		None	202	243	1 L	60 L	B	40
	Carbamate pesticides, liquid, toxic.	6.1	UN2992	I	6.1	T42	None	201	243	1 L	30 L	B	40
				II	6.1	T14	None	202	243	5 L	60 L	B	40
				III	6.1	T14	153	203	241	60 L	220 L	A	40
	Carbamate pesticides, liquid, toxic, flammable, flash point not less than 23 degrees C.	6.1	UN2991	I	6.1, 3	T42	None	201	243	1 L	30 L	B	40
				II	6.1, 3	T14	None	202	243	5 L	60 L	B	40
				III	6.1, 3	B1, T14	153	203	241	60 L	220 L	A	40
	Carbamate pesticides, solid, toxic.	6.1	UN2757	I	6.1		None	211	242	5 kg	50 kg	A	40
				II	6.1		None	212	242	25 kg	100 kg	A	40

Appendix A

§ 172.101 Hazardous Materials Table—Continued

Symbols	Hazardous materials descriptions and proper shipping names	Hazard class or Division	Identification Numbers	PG	Label Codes	Special provisions	(8) Packaging (§173.***)			(9) Quantity limitations		(10) Vessel stowage	
							Exceptions	Non-bulk	Bulk	Passenger aircraft/rail	Cargo aircraft only	Location	Other
(1)	(2)	(3)	(4)	(5)	(6)	(7)	(8A)	(8B)	(8C)	(9A)	(9B)	(10A)	(10B)
	Carbolic acid, see Phenol, solid or Phenol, molten.			III	6.1		153	213	240	100 kg	200 kg	A	40
	Carbolic acid solutions, see Phenol solutions.												
I	Carbon, activated	4.2	UN1362	III	4.2		None	213	241	0.5 kg	0.5 kg	A	12
I	Carbon, animal or vegetable origin.	4.2	UN1361	II	4.2		None	212	242	Forbidden	Forbidden	A	12
				III	4.2		None	213	241	Forbidden	Forbidden	A	12
	Carbon bisulfide, see Carbon disulfide.												
	Carbon dioxide	2.2	UN1013		2.2		306	302, 304.	302, 314, 315.	75 kg	150 kg	A	
	Carbon dioxide and nitrous oxide mixtures.	2.2	UN1015		2.2		306	None	314, 315.	75 kg	150 kg	A	
	Carbon dioxide and oxygen mixtures, compressed.	2.2	UN1014		2.2, 5.1	77	306	304	314, 315.	75 kg	150 kg	A	
	Carbon dioxide, refrigerated liquid.	2.2	UN2187		2.2		306	304	314, 315.	50 kg	500 kg	B	
AW	Carbon dioxide, solid or Dry ice	9	UN1845	III	None		217	217	240	200 kg	200 kg	C	40
	Carbon disulfide	3	UN1131	I	3, 6.1	B16, T18, T26, T29	None	201	243	Forbidden	Forbidden	D	18, 40, 115
	Carbon monoxide, compressed.	2.3	UN1016		2.3, 2.1	4	None	302	314, 315.	Forbidden	25 kg	D	40
	Carbon monoxide and hydrogen mixture, compressed.	2.3	UN2600		2.3, 2.1	6	None	302	302	Forbidden	Forbidden	D	40
D	Carbon monoxide, refrigerated liquid (cryogenic liquid).	2.3	NA9202		2.3, 2.1	4	None	316	318	Forbidden	Forbidden	D	
	Carbon tetrabromide	6.1	UN2516	III	6.1		153	213	240	100 kg	200 kg	A	25
	Carbon tetrachloride	6.1	UN1846	II	6.1	N36, T8	None	202	243	5 L	60 L	A	40
	Carbonyl chloride, see Phosgene.												
	Carbonyl fluoride, compressed	2.3	UN2417		2.3, 8	2	None	302	None	Forbidden	Forbidden	D	40
	Carbonyl sulfide	2.3	UN2204		2.3, 2.1	3, B14	None	304	314, 315.	Forbidden	25 kg	D	40
	Cartridge cases, empty primed, see Cases, cartridge, empty, with primer.												
	Cartridges, actuating, for aircraft ejector seat catapult, fire extinguisher, canopy removal or apparatus, see Cartridges, power device.												
	Cartridges, explosive, see Charges, demolition.												
	Cartridges, flash	1.1G	UN0049	II	1.1G		None	62	None	Forbidden	Forbidden	B	
	Cartridges, flash	1.3G	UN0050	II	1.3G		None	62	None	Forbidden	75 kg	B	
	Cartridges for weapons, blank	1.1C	UN0326	II	1.1C		None	62	None	Forbidden	Forbidden	B	
	Cartridges for weapons, blank	1.2C	UN0413	II	1.2C		None	62	None	Forbidden	Forbidden	B	
	Cartridges for weapons, blank or Cartridges, small arms, blank.	1.4S	UN0014	II	None		63	62	None	25 kg	100 kg	A	
	Cartridges for weapons, blank or Cartridges, small arms, blank.	1.3C	UN0327	II	1.3C		None	62	None	Forbidden	Forbidden	B	
	Cartridges for weapons, blank or Cartridges, small arms, blank.	1.4C	UN0338	II	1.4C		None	62	None	Forbidden	75 kg	A	24E
	Cartridges for weapons, inert projectile.	1.2C	UN0328	II	1.2C		None	62	None	Forbidden	Forbidden	B	
	Cartridges for weapons, inert projectile or Cartridges, small arms.	1.4S	UN0012	II	None		63	62	None	25 kg	100 kg	A	
	Cartridges for weapons, inert projectile or Cartridges, small arms.	1.4C	UN0339	II	1.4C		None	62	None	Forbidden	75 kg	B	
	Cartridges for weapons, inert projectile or Cartridges, small arms.	1.3C	UN0417	II	1.3C		None	62	None	Forbidden	Forbidden	B	
	Cartridges for weapons, with bursting charge.	1.1F	UN0005	II	1.1F		None	62	None	Forbidden	Forbidden	E	
	Cartridges for weapons, with bursting charge.	1.1E	UN0006	II	1.1E		None	62	None	Forbidden	Forbidden	B	
	Cartridges for weapons, with bursting charge.	1.2F	UN0007	II	1.2F		None	62	None	Forbidden	Forbidden	E	
	Cartridges for weapons, with bursting charge.	1.2E	UN0321	II	1.2E		None	62	None	Forbidden	Forbidden	B	
	Cartridges for weapons, with bursting charge.	1.4F	UN0348	II	1.4F		None	62	None	Forbidden	Forbidden	E	
	Cartridges for weapons, with bursting charge.	1.4E	UN0412	II	1.4E		None	62	None	Forbidden	75 kg	A	24E
	Cartridges, oil well	1.3C	UN0277	II	1.3C		None	62	None	Forbidden	Forbidden	B	
	Cartridges, oil well	1.4C	UN0278	II	1.4C		None	62	None	Forbidden	75 kg	A	24E
	Cartridges, power device	1.3C	UN0275	II	1.3C		None	62	None	Forbidden	75 kg	B	
	Cartridges, power device	1.4C	UN0276	II	1.4C	110	None	62	None	Forbidden	75 kg	A	24E
	Cartridges, power device	1.4S	UN0323	II	1.4S	110,	63	62	None	25 kg	100 kg	A	
	Cartridges, power device	1.2C	UN0381	II	1.2C		None	62	None	Forbidden	Forbidden	B	
	Cartridges, safety, blank, see Cartridges for weapons, blank (UN 0014).												

§172.101 Hazardous Materials Table—Continued

Symbols	Hazardous materials descriptions and proper shipping names	Hazard class or Division	Identification Numbers	PG	Label Codes	Special provisions	Packaging (§173.***)			Quantity limitations		Vessel stowage	
							Exceptions	Non-bulk	Bulk	Passenger aircraft/rail	Cargo aircraft only	Location	Other
(1)	(2)	(3)	(4)	(5)	(6)	(7)	(8A)	(8B)	(8C)	(9A)	(9B)	(10A)	(10B)
	Cartridges, safety, see Cartridges for weapons, other than blank or *Cartridges, power device (UN 0323).*												
	Cartridges, signal	1.3G	UN0054	II	1.3G		None	62	None	Forbidden	75 kg	B	
	Cartridges, signal	1.4G	UN0312	II	1.4G		None	62	None	Forbidden	75 kg	A	24E
	Cartridges, signal	1.4S	UN0405	II	1.4S		None	62	None	25 kg	100 kg	A	
D	Cartridges, small arms	ORM-D			None		63	None	None	30 kg gross	30 kg gross	A	
	Cartridges, sporting, see Cartridges for weapons, other than blank.												
	Cartridges, starter, jet engine, see Cartridges, power device.												
	Cases, cartridge, empty with primer.	1.4S	UN0055	II	1.4S	50	None	62	None	25 kg	100 kg	A	
	Cases, cartridges, empty with primer.	1.4C	UN0379	II	1.4C	50	None	62	None	Forbidden	75 kg	A	24E
	Cases, combustible, empty, without primer.	1.4C	UN0446	II	1.4C		None	62	None	Forbidden	75 kg	A	24E
	Cases, combustible, empty, without primer.	1.3C	UN0447	II	1.3C		None	62	None	Forbidden	Forbidden	B	
	Casinghead gasoline see Gasoline.												
AW	Castor beans or Castor meal or Castor pomace or Castor flake.	9	UN2969	II	None		155	204	240	No limit	No limit	E	34, 40
G	Caustic alkali liquids, n.o.s	8	UN1719	II	8	B2, T14	154	202	242	1 L	30 L	A	
				III	8	T7	154	203	241	5 L	60 L	A	
	Caustic potash, see Potassium hydroxide etc.												
	Caustic soda, (etc.) see Sodium hydroxide etc.												
	Cells, containing sodium	4.3	UN3292	II	4.3		189	189	189	25 kg	No limit	A	
	Celluloid, in block, rods, rolls, sheets, tubes, etc., except scrap.	4.1	UN2000	III	4.1		None	213	240	25 kg	100 kg	A	
	Celluloid, scrap	4.2	UN2002	III	4.2		None	213	241	Forbidden	Forbidden	D	
	Cement, see Adhesives containing flammable liquid.												
	Cerium, slabs, ingots, or rods	4.1	UN1333	II	4.1	N34	None	212	240	15 kg	50 kg	A	74, 91
	Cerium, turnings or gritty powder	4.3	UN3078	II	4.3	A1, B106, B109	151	212	242	15 kg	50 kg	E	
	Cesium or Caesium	4.3	UN1407	I	4.3	A19, B100, N34, N40	None	211	242	Forbidden	15 kg	D	
	Cesium nitrate or Caesium nitrate.	5.1	UN1451	III	5.1	A1, A29	152	213	240	25 kg	100 kg	A	
D	Charcoal briquettes, shell, screenings, wood, etc.	4.2	NA1361	III	4.2		151	213	240	25 kg	100 kg	A	12
	Charges, bursting, plastics bonded.	1.1D	UN0457	II	1.1D		None	62	None	Forbidden	Forbidden	B	
	Charges, bursting, plastics bonded.	1.2D	UN0458	II	1.2D		None	62	None	Forbidden	Forbidden	B	
	Charges, bursting, plastics bonded.	1.4D	UN0459	II	1.4D		None	62	None	Forbidden	75 kg	A	24E
	Charges, bursting, plastics bonded.	1.4S	UN0460	II	1.4S		None	62	None	25 kg	100 kg	A	
	Charges, demolition	1.1D	UN0048	II	1.1D		None	62	None	Forbidden	Forbidden	B	
	Charges, depth	1.1D	UN0056	II	1.1D		None	62	None	Forbidden	Forbidden	B	3E, 7E
	Charges, expelling, explosive, for fire extinguishers, see Cartridges, power device.												
	Charges, explosive, commercial without detonator.	1.1D	UN0442	II	1.1D		None	62	None	Forbidden	Forbidden	B	
	Charges, explosive, commercial without detonator.	1.2D	UN0443	II	1.2D		None	62	None	Forbidden	Forbidden	B	
	Charges, explosive, commercial without detonator.	1.4D	UN0444	II	1.4D		None	62	None	Forbidden	75 kg	A	24E
	Charges, explosive, commercial without detonator.	1.4S	UN0445	II	1.4S		None	62	None	25 kg	100 kg	A	
	Charges, propelling	1.1C	UN0271	II	1.1C		None	62	None	Forbidden	Forbidden	B	
	Charges, propelling	1.3C	UN0272	II	1.3C		None	62	None	Forbidden	Forbidden	B	
	Charges, propelling	1.2C	UN0415	II	1.2C		None	62	None	Forbidden	Forbidden	B	
	Charges, propelling	1.4C	UN0491	II	1.4C		None	62	None	Forbidden	75 kg	A	1E, 5E
	Charges, propelling, for cannon	1.3C	UN0242	II	1.3C		None	62	None	Forbidden	Forbidden	B	1E, 5E
	Charges, propelling, for cannon	1.1C	UN0279	II	1.1C		None	62	None	Forbidden	Forbidden	B	1E, 5E
	Charges, propelling, for cannon	1.2C	UN0414	II	1.2C		None	62	None	Forbidden	Forbidden	B	1E, 5E
	Charges, shaped, flexible, linear	1.4D	UN0237	II	1.4D		None	62	None	Forbidden	75 kg	A	24E
	Charges, shaped, flexible, linear	1.1D	UN0288	II	1.1D	101	None	62	None	Forbidden	Forbidden	B	
	Charges, shaped, without detonator.	1.1D	UN0059	II	1.1D		None	62	None	Forbidden	Forbidden	B	
	Charges, shaped, without detonator.	1.2D	UN0439	II	1.2D		None	62	None	Forbidden	Forbidden	B	
	Charges, shaped, without detonator.	1.4D	UN0440	II	1.4D		None	62	None	Forbidden	75 kg	A	24E
	Charges, shaped, without detonator.	1.4S	UN0441	II	1.4S		None	62	None	25 kg	100 kg	A	
	Charges, supplementary explosive.	1.1D	UN0060	II	1.1D		None	62	None	Forbidden	Forbidden	B	1E, 5E
D	Chemical kit	8	NA1760	II	8		154	161	None	1 L	30 L	B	40

Appendix A

§ 172.101 HAZARDOUS MATERIALS TABLE—Continued

Symbols	Hazardous materials descriptions and proper shipping names	Hazard class or Division	Identification Numbers	PG	Label Codes	Special provisions	Packaging (§173.***)			Quantity limitations		Vessel stowage	
							Exceptions	Non-bulk	Bulk	Passenger aircraft/rail	Cargo aircraft only	Location	Other
(1)	(2)	(3)	(4)	(5)	(6)	(7)	(8A)	(8B)	(8C)	(9A)	(9B)	(10A)	(10B)
	Chemical kits or First aid kits (containing hazardous materials).	9	UN3316		9	15	None	None	None	10 kg	10 kg	A	
	Chloral, anhydrous, inhibited	6.1	UN2075	II	6.1	B101, T14	None	202	243	5 L	60 L	D	40
	Chlorate and borate mixtures	5.1	UN1458	II	5.1	A9, N34	152	212	240	5 kg	25 kg	A	56, 58, 106
				III	5.1	A9, N34	152	213	240	25 kg	100 kg	A	56, 58, 106
	Chlorate and magnesium chloride mixtures.	5.1	UN1459	II	5.1	A9, N34, T8	152	212	240	5 kg	25 kg	A	56, 58, 106
				III	5.1	A9, N34, T8	152	213	240	25 kg	100 kg	A	56, 58, 106
	Chlorate of potash, see Potassium chlorate.												
	Chlorate of soda, see Sodium chlorate.												
	Chlorates, inorganic, aqueous solution, n.o.s.	5.1	UN3210	II	5.1	T8	152	202	242	1 L	5 L	B	56, 58, 106
	Chlorates, inorganic, n.o.s	5.1	UN1461	II	5.1	A9, N34	152	212	242	5 kg	25 kg	A	56, 58, 106
	Chloric acid aqueous solution, with not more than 10 percent chloric acid.	5.1	UN2626	II	5.1	T25	None	229	None	Forbidden	Forbidden	D	56, 58, 106
	Chloride of phosphorus, see Phosphorus trichloride.												
	Chloride of sulfur, see Sulfur chloride.												
	Chlorinated lime, see Calcium hypochlorite mixtures, etc.												
	Chlorine	2.3	UN1017		2.3, 8	2, B9, B14	None	304	314, 315.	Forbidden	Forbidden	D	40, 51, 55, 62, 68, 89, 90
	Chlorine azide	Forbidden											
D	Chlorine dioxide, hydrate, frozen	5.1	NA9191	II	5.1, 6.1		None	229	None	Forbidden	Forbidden	E	
	Chlorine dioxide (not hydrate)	Forbidden											
	Chlorine pentafluoride	2.3	UN2548		2.3, 5.1, 8	1, B7, B9, B14	None	304	314	Forbidden	Forbidden	D	40, 89, 90
	Chlorite solution	8	UN1908	II	8	A3, A6, A7, B2, N34, T8	154	202	242	1 L	30 L	B	26
				III	8	A3, A6, A7, B2, N34, T8	154	203	241	5 L	60 L	B	26
	Chlorine trifluoride	2.3	UN1749		2.3, 5.1, 8	2, B7, B9, B14	None	304	314	Forbidden	Forbidden	D	40, 89, 90
	Chlorites, inorganic, n.o.s.	5.1	UN1462	II	5.1	A7, N34	152	212	242	5 kg	25 kg	A	56, 58, 106
	1-Chloro-3-bromopropane	6.1	UN2688	III	6.1	T2	153	203	241	60 L	220 L	A	
	1-Chloro-1,1-difluoroethane, see Chlorodifluoroethanes.												
	1-Chloro-1,1-difluoroethane or Refrigerant gas R 142b.	2.1	UN2517		2.1		306	304	314, 315.	Forbidden	150 kg	B	40
	3-Chloro-4-methylphenyl isocyanate.	6.1	UN2236	II	6.1		None	202	243	5 L	60 L	B	40
	1-Chloro-1,2,2,2-tetrafluoroethane or Refrigerant gas R 124.	2.2	UN1021		2.2		306	304	314, 315.	75 kg	150 kg	A	
	4-Chloro-o-toluidine hydrochloride.	6.1	UN1579	III	6.1		153	213	240	100 kg	200 kg	A	
	1-Chloro-2,2,2-trifluoroethane or Refrigerant gas R 133a.	2.2	UN1983		2.2		306	304	314, 315.	75 kg	150 kg	A	
	Chloroacetic acid, molten	6.1	UN3250	II	6.1, 8	T9	None	202	243	Forbidden	Forbidden	C	40
	Chloroacetic acid, solid	6.1	UN1751	II	6.1, 8	A3, A7, N34	None	212	242	15 kg	50 kg	A	40
	Chloroacetic acid, solution	6.1	UN1750	II	6.1, 8	A7, N34, T8, T27	None	202	243	1 L	30 L	C	40
	Chloroacetone, stabilized	6.1	UN1695	I	6.1, 3, 8	2, B9, B14, B32, B74, N12, N32, N34, T38, T43, T45	None	227	244	Forbidden	Forbidden	D	20, 40, 95
	Chloroacetone (unstabilized)	Forbidden											
+	Chloroacetonitrile	6.1	UN2668	II	6.1, 3	2, B9, B14, B32, B74, T38, T43, T45	None	227	244	Forbidden	60 L	A	12, 26, 40
	Chloroacetophenone (CN), liquid	6.1	UN1697	II	6.1	A3, N12, N32, N33	None	202	243	Forbidden	60 L	D	12, 40
	Chloroacetophenone (CN), solid	6.1	UN1697	II	6.1	A3, N12, N32, N33, N34	None	212	None	Forbidden	100 kg	D	12, 40
	Chloroacetyl chloride	6.1	UN1752	I	6.1, 8	2, A3, A6, A7, B3, B8, B9, B14, B32, B74, B77, N34, N43, T38, T43, T45	None	227	244	Forbidden	Forbidden	D	40
	Chloroanilines, liquid	6.1	UN2019	II	6.1	T14	None	202	243	5 L	60 L	A	
	Chloroanilines, solid	6.1	UN2018	II	6.1	T14, T38	None	212	242	25 kg	100 kg	A	
	Chloroanisidines	6.1	UN2233	III	6.1		153	213	240	100 kg	200 kg	A	
	Chlorobenzene	3	UN1134	III	3	B1, T1	150	203	242	60 L	220 L	A	
	Chlorobenzol, see Chlorobenzene.												
	Chlorobenzotrifluorides	3	UN2234	III	3	B1, T1	150	203	242	60 L	220 L	A	40

§172.101 HAZARDOUS MATERIALS TABLE—Continued

Symbols	Hazardous materials descriptions and proper shipping names	Hazard class or Division	Identification Numbers	PG	Label Codes	Special provisions	(8) Packaging (§173.***)			(9) Quantity limitations		(10) Vessel stowage	
							Exceptions	Non-bulk	Bulk	Passenger aircraft/rail	Cargo aircraft only	Location	Other
(1)	(2)	(3)	(4)	(5)	(6)	(7)	(8A)	(8B)	(8C)	(9A)	(9B)	(10A)	(10B)
	Chlorobenzyl chlorides	6.1	UN2235	III	6.1	T8	153	203	241	60 L	220 L	A	
	Chlorobutanes	3	UN1127	II	3	B101, T8	150	202	242	5 L	60 L	B	
	Chlorocresols, liquid	6.1	UN2669	II	6.1	T8	None	202	243	5 L	60 L	A	12
	Chlorocresols, solid	6.1	UN2669	II	6.1		None	212	242	25 kg	100 kg	A	12
	Chlorodifluorobromomethane or Refrigerant gas R 12B1.	2.2	UN1974		2.2		306	304	314, 315.	75 kg	150 kg	A	
	Chlorodifluoromethane and chloropentafluoroethane mixture or Refrigerant gas R 502 with fixed boiling point, with approximately 49 percent chlorodifluoromethane.	2.2	UN1973		2.2		306	304	314, 315.	75 kg	150 kg	A	
	Chlorodifluoromethane or Refrigerant gas R 22.	2.2	UN1018		2.2		306	304	314, 315.	75 kg	150 kg	A	
+	Chlorodinitrobenzenes	6.1	UN1577	II	6.1	T14	None	212	242	25 kg	100 kg	A	91
	2-Chloroethanal	6.1	UN2232	I	6.1	2, B9, B14, B32, B74, T38, T43, T45	None	227	244	Forbidden	Forbidden	D	40
	Chloroform	6.1	UN1888	III	6.1	N36, T14	153	203	241	5 L	60 L	A	40
G	Chloroformates, toxic, corrosive, flammable, n.o.s.	6.1	UN2742	II	6.1, 8, 3.	5	None	202	243	1 L	30 L	A	12, 13, 21, 25, 40, 100
G	Chloroformates, toxic, corrosive, n.o.s.	6.1	UN3277	II	6.1, 8	T12, T26	None	202	243	1 L	30 L	A	12, 13, 25, 40
	Chloromethyl chloroformate	6.1	UN2745	II	6.1, 8	T18	None	202	243	1 L	30 L	A	12, 13, 21, 25, 40, 100
	Chloromethyl ethyl ether	3	UN2354	II	3, 6.1	T8	None	202	243	1 L	60 L	E	40
	Chloronitroanilines	6.1	UN2237	III	6.1		153	213	240	100 kg	200 kg	A	
+	Chloronitrobenzene, ortho, liquid	6.1	UN1578	II	6.1	T14	None	202	243	5 L	60 L	A	
+	Chloronitrobenzenes meta or para, solid.	6.1	UN1578	II	6.1	T14	None	212	242	25 kg	100 kg	A	
	Chloronitrotoluenes liquid	6.1	UN2433	III	6.1		153	203	241	60 L	220 L	A	
	Chloronitrotoluenes, solid	6.1	UN2433	III	6.1		153	213	240	100 kg	200 kg	A	
	Chloropentafluoroethane or Refrigerant gas R 115.	2.2	UN1020		2.2		306	304	314, 315.	75 kg	150 kg	A	
	Chlorophenolates, liquid or Phenolates, liquid.	8	UN2904	III	8		154	203	241	5 L	60 L	A	
	Chlorophenolates, solid or Phenolates, solid.	8	UN2905	III	8		154	213	240	25 kg	100 kg	A	
	Chlorophenols, liquid	6.1	UN2021	III	6.1	T7	153	203	241	60 L	220 L	A	
	Chlorophenols, solid	6.1	UN2020	III	6.1	T7	153	213	240	100 kg	200 kg	A	
	Chlorophenyltrichlorosilane	8	UN1753	II	8	A7, B2, B6, N34, T8, T26	None	202	242	Forbidden	30 L	C	40
+	Chloropicrin	6.1	UN1580	I	6.1	2, B7, B9, B14, B32, B46, B74, T38, T43, T45	None	227	244	Forbidden	Forbidden	D	40
	Chloropicrin and methyl bromide mixtures.	2.3	UN1581		2.3	2, B9, B14	None	193	314, 315.	Forbidden	Forbidden	D	25, 40
	Chloropicrin and methyl chloride mixtures.	2.3	UN1582		2.3	2	None	193	245	Forbidden	Forbidden	D	25, 40
	Chloropicrin mixture, flammable (pressure not exceeding 14.7 psia at 115 degrees F flash point below 100 degrees F) see Toxic liquids, flammable, etc.												
	Chloropicrin mixtures, n.o.s	6.1	UN1583	I	6.1	5	None	201	243	Forbidden	Forbidden	C	40
				II	6.1		None	202	243	Forbidden	Forbidden	C	40
				III	6.1		153	203	241	Forbidden	Forbidden	C	40
D	Chloropivaloyl chloride	6.1	NA9263	I	6.1, 8	2, B9, B14, B32, B74, T38, T43, T45	None	227	244	Forbidden	Forbidden	B	40
	Chloroplatinic acid, solid	8	UN2507	III	8		154	213	240	25 kg	100 kg	A	
	Chloroprene, inhibited	3	UN1991	I	3, 6.1	B57, T15	None	201	243	Forbidden	30 L	D	40
	Chloroprene, uninhibited	Forbidden											
	2-Chloropropane	3	UN2356	I	3	N36, T14	150	201	243	1 L	30 L	E	
	3-Chloropropanol-1	6.1	UN2849	III	6.1	T8	153	203	241	60 L	220 L	A	
	2-Chloropropene	3	UN2456	I	3	A3, N36, T20	150	201	243	1 L	30 L	E	
	2-Chloropropionic acid	8	UN2511	III	8	T8	154	203	241	5 L	60 L	A	8
	2-Chloropyridine	6.1	UN2822	II	6.1	T14	None	202	243	5 L	60 L	A	40
	Chlorosilanes, corrosive, flammable, n.o.s.	8	UN2986	II	8, 3	B100, T18, T26	None	202	243	1 L	30 L	C	40
	Chlorosilanes, corrosive, n.o.s	8	UN2987	II	8	B2, T14, T26	154	202	242	1 L	30 L	C	40
	Chlorosilanes, flammable, corrosive, n.o.s.	3	UN2985	II	3, 8	B100, T17, T26	None	201	243	1 L	5 L	B	40
	Chlorosilanes, water-reactive, flammable, corrosive, n.o.s.	4.3	UN2988	I	4.3, 3, 8	A2, T18, T26	None	201	244	Forbidden	1 L	D	21, 28, 40, 49, 100
+	Chlorosulfonic acid (with or without sulfur trioxide).	8	UN1754	I	8, 6.1	2, A3, A6, A10, B9, B10, B14, B32, B74, T38, T43, T45	None	227	244	Forbidden	Forbidden	C	40
	Chlorotoluenes	3	UN2238	III	3	B1, T1	150	203	242	60 L	220 L	A	
	Chlorotoluidines liquid	6.1	UN2239	III	6.1	T7	153	203	241	60 L	220 L	A	
	Chlorotoluidines solid	6.1	UN2239	III	6.1		153	213	240	100 kg	200 kg	A	

Appendix A

§172.101 Hazardous Materials Table—Continued

Symbols	Hazardous materials descriptions and proper shipping names	Hazard class or Division	Identification Numbers	PG	Label Codes	Special provisions	Packaging (§173.***)			Quantity limitations		Vessel stowage	
							Exceptions	Non-bulk	Bulk	Passenger aircraft/rail	Cargo aircraft only	Location	Other
(1)	(2)	(3)	(4)	(5)	(6)	(7)	(8A)	(8B)	(8C)	(9A)	(9B)	(10A)	(10B)
	Chlorotrifluoromethane and trifluoromethane azeotropic mixture *or* Refrigerant gas R 503 with approximately 60 percent chlorotrifluoromethane.	2.2	UN2599		2.2		306	304	314, 315.	75 kg	150 kg	A	
	Chlorotrifluoromethane *or* Refrigerant gas R 13.	2.2	UN1022		2.2		306	304	314, 315.	75 kg	150 kg	A	
D	Chromic acid, solid	5.1	NA1463	II	5.1, 8		None	212	242	5 kg	25 kg	A	
	Chromic acid solution	8	UN1755	II	8	B2, T9, T27	154	202	242	1 L	30 L	C	40
				III	8	T8, T26	154	203	241	5 L	60 L	C	40
	Chromic anhydride, see Chromium trioxide, anhydrous.												
	Chromic fluoride, solid	8	UN1756	II	8		154	212	240	15 kg	50 kg	A	26
	Chromic fluoride, solution	8	UN1757	II	8	B2, T8	154	202	242	1 L	30 L	A	
				III	8	T7	154	203	241	5 L	60 L	A	
	Chromium nitrate	5.1	UN2720	III	5.1	A1, A29	152	213	240	25 kg	100 kg	A	
	Chromium oxychloride	8	UN1758	I	8	A3, A6, A7, B10, N34, T12, T26	None	201	243	0.5 L	2.5 L	C	40, 66, 74, 89, 90
	Chromium trioxide, anhydrous	5.1	UN1463	II	5.1, 8	B106	None	212	242	5 kg	25 kg	A	
	Chromosulfuric acid	8	UN2240	I	8	A3, A6, A7, B4, B6, N34, T12, T27	None	201	243	0.5 L	2.5 L	B	40, 66, 74, 89, 90
	Chromyl chloride, see Chromium oxychloride.												
	Cigar and cigarette lighters, charged with fuel, see Lighters for cigars, cigarettes, etc.												
	Coal briquettes, hot	Forbidden											
	Coal gas, compressed	2.3	UN1023		2.3, 2.1	3	None	302	314, 315.	Forbidden	25 kg	D	40
	Coal tar distillates, flammable	3	UN1136	II	3	T8, T31	150	202	242	5 L	60 L	B	
				III	3	B1, T7, T30	150	203	242	60 L	220 L	A	
	Coal tar dye, corrosive, liquid, n.o.s., see Dyes, liquid or solid, n.o.s. or Dye intermediates, liquid or solid, n.o.s., corrosive.												
	Coating solution (*includes surface treatments or coatings used for industrial or other purposes such as vehicle undercoating, drum or barrel lining*).	3	UN1139	I	3	T42	150	201	243	1 L	30 L	E	
				II	3	T7, T30	150	202	242	5 L	60 L	B	
				III	3	B1, T7, T30	150	203	242	60 L	220 L	A	
	Cobalt naphthenates, powder	4.1	UN2001	III	4.1	A19	151	213	240	25 kg	100 kg	A	
	Cobalt resinate, precipitated	4.1	UN1318	III	4.1	A1, A19	151	213	240	25 kg	100 kg	A	
	Coke, hot	Forbidden											
	Collodion, see Nitrocellulose etc												
DG	Combustible liquid, n.o.s	Combustible liquid	NA1993	III	None	T1	150	203	241	60 L	220 L	A	
G	Components, explosive train, n.o.s.	1.2B	UN0382	II	1.2B	101	None	62	None	Forbidden	Forbidden	B	1E, 6E
G	Components, explosive train, n.o.s.	1.4B	UN0383	II	1.4B	101	None	62	None	Forbidden	75 kg	A	24E
G	Components, explosive train, n.o.s.	1.4S	UN0384	II	1.4S	101	None	62	None	25 kg	100 kg	A	
G	Components, explosive train, n.o.s.	1.1B	UN0461	II	1.1B	101	None	62	None	Forbidden	Forbidden	B	1E, 6E
	Composition B, see Hexolite, etc												
DG	Compounds, cleaning liquid	8	NA1760	I	8	A7, B10, T42	None	201	243	0.5 L	2.5 L	B	40
				II	8	B2, N37, T14	154	202	242	1 L	30 L	B	40
				III	8	N37, T7	154	203	241	5 L	60 L	A	40
DG	Compounds, cleaning liquid	3	NA1993	I	3	T42	150	201	243	1 L	30 L	E	
				II	3	T8, T31	150	202	242	5 L	60 L	B	
				III	3	B1, B52, T7, T30	150	203	242	60 L	220 L	A	
D G	Compounds, tree killing, liquid *or* Compounds, weed killing, liquid.	8	NA1760	I	8	A7, B10, T42	None	201	243	0.5 L	2.5 L	B	40
				II	8	B2, N37, T14	154	202	242	1 L	30 L	B	40
				III	8	N37, T7	154	203	241	5 L	60 L	A	40
D G	Compounds, tree killing, liquid *or* Compounds, weed killing, liquid.	3	NA1993	I	3	T42	150	201	243	1 L	30 L	E	
				II	3	T8, T31	150	202	242	5 L	60 L	B	
				III	3	B1, B52, T7, T30	150	203	242	60 L	220 L	A	
D G	Compounds, tree killing, liquid *or* Compounds, weed killing, liquid.	6.1	NA2810	I	6.1		None	201	243	1 L	30 L	B	40
				II	6.1		None	202	243	5 L	60 L	B	40
				III	6.1		153	203	241	60 L	220 L	A	40
G	Compressed gas, flammable, n.o.s.	2.1	UN1954		2.1		306	302, 305.	314, 315.	Forbidden	150 kg	D	40
G	Compressed gas, n.o.s	2.2	UN1956		2.2		306, 307.	302, 305.	314, 315.	75 kg	150 kg	A	
G	Compressed gas, oxidizing, n.o.s.	2.2	UN3156		2.2, 5.1		306	302	314, 315.	75 kg	150 kg	D	

§ 172.101 HAZARDOUS MATERIALS TABLE—Continued

Symbols	Hazardous materials descriptions and proper shipping names	Hazard class or Division	Identification Numbers	PG	Label Codes	Special provisions	(8) Packaging (§173.***)			(9) Quantity limitations		(10) Vessel stowage	
							Exceptions	Non-bulk	Bulk	Passenger aircraft/rail	Cargo aircraft only	Location	Other
(1)	(2)	(3)	(4)	(5)	(6)	(7)	(8A)	(8B)	(8C)	(9A)	(9B)	(10A)	(10B)
GI	Compressed gas, toxic, corrosive, n.o.s. *Inhalation Hazard Zone A.*	2.3	UN3304		2.3, 8	1	None	192	245	Forbidden	Forbidden	D	40
GI	Compressed gas, toxic, corrosive, n.o.s. *Inhalation Hazard Zone B.*	2.3	UN3304		2.3, 8	2	None	302, 305.	314, 315.	Forbidden	Forbidden	D	40
GI	Compressed gas, toxic, corrosive, n.o.s. *Inhalation Hazard Zone C.*	2.3	UN3304		2.3, 8	3	None	302, 305.	314, 315.	Forbidden	Forbidden	D	40
GI	Compressed gas, toxic, corrosive, n.o.s. *Inhalation Hazard Zone D.*	2.3	UN3304		2.3, 8	4	None	302, 305.	314, 315.	Forbidden	Forbidden	D	40
GI	Compressed gas, toxic, flammable, corrosive, n.o.s. *Inhalation Hazard Zone A.*	2.3	UN3305		2.3, 2.1, 8.	1	None	192	245	Forbidden	Forbidden	D	17, 40
GI	Compressed gas, toxic, flammable, corrosive, n.o.s. *Inhalation Hazard Zone B.*	2.3	UN3305		2.3, 2.1, 8.	2	None	302, 305.	314, 315.	Forbidden	Forbidden	D	17, 40
GI	Compressed gas, toxic, flammable, corrosive, n.o.s. *Inhalation Hazard Zone C.*	2.3	UN3305		2.3, 2.1, 8.	3	None	302, 305.	314, 315.	Forbidden	Forbidden	D	17, 40
GI	Compressed gas, toxic, flammable, corrosive, n.o.s. *Inhalation Hazard Zone D.*	2.3	UN3305		2.3, 2.1, 8.	4	None	302, 305.	314, 315.	Forbidden	Forbidden	D	17, 40
G	Compressed gas, toxic, flammable, n.o.s. *Inhalation hazard Zone A.*	2.3	UN1953		2.3, 2.1	1	None	192	245	Forbidden	Forbidden	D	40, 95
G	Compressed gas, toxic, flammable, n.o.s. *Inhalation hazard Zone B.*	2.3	UN1953		2.3, 2.1	2, B9, B14	None	302, 305.	314, 315.	Forbidden	Forbidden	D	40
G	Compressed gas, toxic, flammable, n.o.s. *Inhalation Hazard Zone C.*	2.3	UN1953		2.3, 2.1	3, B14	None	302, 305.	314, 315.	Forbidden	Forbidden	D	40
G	Compressed gas, toxic, flammable, n.o.s. *Inhalation Hazard Zone D.*	2.3	UN1953		2.3, 2.1	4	None	302, 305.	314, 315.	Forbidden	Forbidden	D	40
G	Compressed gas, toxic, n.o.s. *Inhalation Hazard Zone A*	2.3	UN1955		2.3	1	None	192	245	Forbidden	Forbidden	D	40
G	Compressed gas, toxic, n.o.s. *Inhalation Hazard Zone B.*	2.3	UN1955		2.3	2, B9, B14	None	302, 305.	314, 315.	Forbidden	Forbidden	D	40
G	Compressed gas, toxic, n.o.s. *Inhalation Hazard Zone C.*	2.3	UN1955		2.3	3, B14	None	302, 305.	314, 315.	Forbidden	Forbidden	D	40
G	Compressed gas, toxic, n.o.s. *Inhalation Hazard Zone D.*	2.3	UN1955		2.3	4	None	302, 305.	314, 315.	Forbidden	Forbidden	D	40
GI	Compressed gas, toxic, oxidizing, corrosive, n.o.s. *Inhalation Hazard Zone A.*	2.3	UN3306		2.3, 5.1, 8.	1	None	192	244	Forbidden	Forbidden	D	40, 89, 90
GI	Compressed gas, toxic, oxidizing, corrosive, n.o.s. *Inhalation Hazard Zone B.*	2.3	UN3306		2.3, 5.1, 8.	2	None	302, 305.	314, 315.	Forbidden	Forbidden	D	40, 89, 90
GI	Compressed gas, toxic, oxidizing, corrosive, n.o.s. *Inhalation Hazard Zone C.*	2.3	UN3306		2.3, 5.1, 8.	3	None	302, 305.	314, 315.	Forbidden	Forbidden	D	40, 89, 90
GI	Compressed gas, toxic, oxidizing, corrosive, n.o.s. *Inhalation Hazard Zone D.*	2.3	UN3306		2.3, 5.1, 8.	4	None	302, 305.	314, 315.	Forbidden	Forbidden	D	40, 89, 90
G	Compressed gas, toxic, oxidizing, n.o.s. *Inhalation Hazard Zone A.*	2.3	UN3303		2.3, 5.1	1	None	192	245	Forbidden	Forbidden	D	40
G	Compressed gas, toxic, oxidizing, n.o.s. *Inhalation Hazard Zone B.*	2.3	UN3303		2.3, 5.1	2	None	302, 305.	314, 315.	Forbidden	Forbidden	D	40
G	Compressed gas, toxic, oxidizing, n.o.s. *Inhalation Hazard Zone C.*	2.3	UN3303		2.3, 5.1	3	None	302, 305.	314, 315.	Forbidden	Forbidden	D	40
G	Compressed gas, toxic, oxidizing, n.o.s. *Inhalation Hazard Zone D.*	2.3	UN3303		2.3, 5.1	4	None	302, 305.	314, 315.	Forbidden	Forbidden	D	40
D	Consumer commodity	ORM-D			None		156, 306.	156, 306.	None	30 kg gross	30 kg gross	A	
	Contrivances, water-activated, with burster, expelling charge or propelling charge.	1.2L	UN0248	II	1.2L	101	None	62	None	Forbidden	Forbidden	E	2E, 8E, 11E, 17E
	Contrivances, water-activated, with burster, expelling charge or propelling charge.	1.3L	UN0249	II	1.3L	101	None	62	None	Forbidden	Forbidden	E	2E, 8E, 11E, 17E
	Copper acetoarsenite	6.1	UN1585	II	6.1		None	212	242	25 kg	100 kg	A	
	Copper acetylide	Forbidden											
	Copper amine azide	Forbidden											
	Copper arsenite	6.1	UN1586	II	6.1		None	212	242	25 kg	100 kg	A	
	Copper based pesticides, liquid, flammable, toxic, *flash point less than 23 degrees C.*	3	UN2776	I	3, 6.1		None	201	243	Forbidden	30 L	B	40
				II	3, 6.1		None	202	243	1 L	60 L	B	40
	Copper based pesticides, liquid, toxic.	6.1	UN3010	I	6.1	T42	None	201	243	1 L	30 L	B	40
				II	6.1	T14	None	202	243	5 L	60 L	B	40
				III	6.1	T14	153	203	241	60 L	220 L	A	40

Appendix A

§172.101 HAZARDOUS MATERIALS TABLE—Continued

Symbols	Hazardous materials descriptions and proper shipping names	Hazard class or Division	Identification Numbers	PG	Label Codes	Special provisions	(8) Packaging (§173.***)			(9) Quantity limitations		(10) Vessel stowage	
							Exceptions	Non-bulk	Bulk	Passenger aircraft/rail	Cargo aircraft only	Location	Other
(1)	(2)	(3)	(4)	(5)	(6)	(7)	(8A)	(8B)	(8C)	(9A)	(9B)	(10A)	(10B)
	Copper based pesticides, liquid, toxic, flammable *flashpoint not less than 23 degrees C.*	6.1	UN3009	I	6.1, 3	T42	None	201	243	1 L	30 L	B	40
				II	6.1, 3	T14	None	202	243	5 L	60 L	B	40
				III	6.1, 3	B1, T14	153	203	242	60 L	220 L	A	40
	Copper based pesticides, solid, toxic.	6.1	UN2775	I	6.1		None	211	242	5 kg	50 kg	A	40
				II	6.1		None	212	242	25 kg	100 kg	A	40
				III	6.1		153	213	240	100 kg	200 kg	A	40
	Copper chlorate	5.1	UN2721	II	5.1	A1	152	212	242	5 kg	25 kg	A	56, 58, 106
	Copper chloride	8	UN2802	III	8		154	213	240	25 kg	100 kg	A	
	Copper cyanide	6.1	UN1587	II	6.1		None	204	242	25 kg	100 kg	A	26
	Copper selenate, see Selenates or Selenites.												
	Copper selenite, see Selenates or Selenites.												
	Copper tetramine nitrate	Forbidden											
AW	Copra	4.2	UN1363	III	4.2		None	213	241	Forbidden	Forbidden	A	13, 19, 48, 119
	Cord, detonating, *flexible*	1.1D	UN0065	II	1.1D	102	63(a)	62	None	Forbidden	Forbidden	B	
	Cord, detonating, *flexible*	1.4D	UN0289	II	1.4D		None	62	None	Forbidden	75 kg	A	24E
	Cord detonating *or* Fuse detonating *metal clad.*	1.2D	UN0102	II	1.2D		None	62	None	Forbidden	Forbidden	B	
	Cord, detonating *or* Fuse, detonating *metal clad.*	1.1D	UN0290	II	1.1D		None	62	None	Forbidden	Forbidden	B	
	Cord, detonating, mild effect *or* Fuse, detonating, mild effect *metal clad.*	1.4D	UN0104	II	1.4D		None	62	None	Forbidden	75 kg	A	24E
	Cord, igniter	1.4G	UN0066	II	1.4G		None	62	None	Forbidden	75 kg	A	24E
	Cordeau detonant fuse, see Cord, detonating, *etc*; Cord, detonating, *flexible.*												
	Cordite, see Powder, smokeless												
G	Corrosive liquid, acidic, inorganic, n.o.s.	8	UN3264	I	8	B10	None	201	243	0.5 L	2.5 L	B	40
				II	8	B2, T14	154	202	242	1 L	30 L	B	40
				III	8	T7	154	203	241	5 L	60 L	A	40
G	Corrosive liquid, acidic, organic, n.o.s.	8	UN3265	I	8	B10	None	201	243	0.5 L	2.5 L	B	40
				II	8	B2, T14	154	202	242	1 L	30 L	B	40
				III	8	T7	154	203	241	5 L	60 L	A	40
G	Corrosive liquid, basic, inorganic, n.o.s.	8	UN3266	I	8	B10	None	201	243	0.5 L	2.5 L	B	40
				II	8	B2, T14	154	202	242	1 L	30 L	B	40
				III	8	T7	154	203	241	5 L	60 L	A	40
G	Corrosive liquid, basic, organic, n.o.s.	8	UN3267	I	8	B10	None	201	243	0.5 L	2.5 L	B	40
				II	8	B2, T14	154	202	242	1 L	30 L	B	40
				III	8	T7	154	203	241	5 L	60 L	A	40
G	Corrosive liquid, self-heating, n.o.s.	8	UN3301	I	8, 4.2	B10	None	201	243	0.5 L	2.5 L	D	
				II	8, 4.2	B2	154	202	242	1 L	30 L	D	
G	Corrosive liquids, flammable, n.o.s.	8	UN2920	I	8, 3	B10, T42	None	201	243	0.5 L	2.5 L	C	25, 40
				II	8, 3	B2, T15, T26	None	202	243	1 L	30 L	C	25, 40
G	Corrosive liquids, n.o.s	8	UN1760	I	8	A7, B10, T42	None	201	243	0.5 L	2.5 L	B	40
				II	8	B2, T14	154	202	242	1 L	30 L	B	40
				III	8	T7	154	203	241	5 L	60 L	A	40
G	Corrosive liquids, oxidizing, n.o.s	8	UN3093	I	8, 5.1		None	201	243	Forbidden	2.5 L	C	89
				II	8, 5.1		None	202	243	1 L	30 L	C	89
G	Corrosive liquids, toxic, n.o.s	8	UN2922	I	8, 6.1	A7, B10, T18, T27	None	201	243	0.5 L	2.5 L	B	40
				II	8, 6.1	B3, T18, T26	None	202	243	1 L	30 L	B	40
				III	8, 6.1	T8	154	203	241	5 L	60 L	B	40
G	Corrosive liquids, water-reactive, n.o.s.	8	UN3094	I	8, 4.3		None	201	243	Forbidden	1 L	E	
				II	8, 4.3		None	202	243	1 L	5 L	E	
G	Corrosive solid, acidic, inorganic, n.o.s.	8	UN3260	I	8		None	211	242	1 kg	25 kg	B	
				II	8		154	212	240	15 kg	50 kg	B	
				III	8		154	213	240	25 kg	100 kg	A	
G	Corrosive solid, acidic, organic, n.o.s.	8	UN3261	I	8		None	211	242	1 kg	25 kg	B	
				II	8		154	212	240	15 kg	50 kg	B	
				III	8		154	213	240	25 kg	100 kg	A	
G	Corrosive solid, basic, inorganic, n.o.s.	8	UN3262	I	8		None	211	242	1 kg	25 kg	B	
				II	8		154	212	240	15 kg	50 kg	B	
				III	8		154	213	240	25 kg	100 kg	A	
G	Corrosive solid, basic, organic, n.o.s.	8	UN3263	I	8		None	211	242	1 kg	25 kg	B	
				II	8		154	212	240	15 kg	50 kg	B	
				III	8		154	213	240	25 kg	100 kg	A	
G	Corrosive solids, flammable, n.o.s.	8	UN2921	I	8, 4.1	B106	None	211	242	1 kg	25 kg	B	12, 25
				II	8, 4.1		None	212	240	15 kg	50 kg	B	12, 25
G	Corrosive solids, n.o.s	8	UN1759	I	8		None	211	242	1 kg	25 kg	B	
				II	8	128	154	212	240	15 kg	50 kg	A	

§172.101 HAZARDOUS MATERIALS TABLE—Continued

Symbols (1)	Hazardous materials descriptions and proper shipping names (2)	Hazard class or Division (3)	Identification Numbers (4)	PG (5)	Label Codes (6)	Special provisions (7)	Packaging (§173.***) (8) Exceptions (8A)	Packaging (§173.***) (8) Non-bulk (8B)	Packaging (§173.***) (8) Bulk (8C)	Quantity limitations (9) Passenger aircraft/rail (9A)	Quantity limitations (9) Cargo aircraft only (9B)	Vessel stowage (10) Location (10A)	Vessel stowage (10) Other (10B)
G	Corrosive solids, oxidizing, n.o.s	8	UN3084	III	8	128	154	213	240	25 kg	100 kg	A	
				I	8, 5.1	B100	None	211	242	1 kg	25 kg	C	
				II	8, 5.1	B100	None	212	242	15 kg	50 kg	C	
G	Corrosive solids, self-heating, n.o.s.	8	UN3095	I	8, 4.2	B100	None	211	243	1 kg	25 kg	C	
				II	8, 4.2		None	212	242	15 kg	50 kg	C	
G	Corrosive solids, toxic, n.o.s	8	UN2923	I	8, 6.1		None	211	242	1 kg	25 kg	B	40
				II	8, 6.1		None	212	240	15 kg	50 kg	B	40
				III	8, 6.1		154	213	240	25 kg	100 kg	B	40, 95
G	Corrosive solids, water-reactive, n.o.s.	8	UN3096	I	8, 4.3	B105	None	211	243	1 kg	25 kg	D	
				II	8, 4.3	B105	None	212	242	15 kg	50 kg	D	
D, W	Cotton	9	NA1365		9	137, W41	None	None	None	No limit	No limit	A	
A, W	Cotton waste, oily	4.2	UN1364	III	4.2		None	213	None	Forbidden	Forbidden	A	54
A,W	Cotton, wet	4.2	UN1365	III	4.2		None	204	241	Forbidden	Forbidden	A	
	Coumarin derivative pesticides, liquid, flammable, toxic, *flashpoint less than 23 degrees C.*	3	UN3024	I	3, 6.1		None	201	243	Forbidden	30 L	B	40
				II	3, 6.1		None	202	243	1 L	60 L	B	40
	Coumarin derivative pesticides, liquid, toxic.	6.1	UN3026	I	6.1		None	201	243	1 L	30 L	B	40
				II	6.1		None	202	243	5 L	60 L	B	40
				III	6.1		153	203	241	60 L	220 L	A	40
	Coumarin derivative pesticides, liquid, toxic, flammable *flashpoint not less than 23 degrees C.*	6.1	UN3025	I	6.1, 3		None	201	243	1 L	30 L	B	40
				II	6.1, 3		None	202	243	5 L	60 L	B	40
				III	6.1, 3	B1	153	203	242	60 L	220 L	A	40
	Coumarin derivative pesticides, solid, toxic.	6.1	UN3027	I	6.1		None	211	242	5 kg	50 kg	A	40
				II	6.1		None	212	242	25 kg	100 kg	A	40
				III	6.1		153	213	240	100 kg	200 kg	A	40
	Cresols	6.1	UN2076	II	6.1, 8	B110, T8	None	202	243	1 L	30 L	B	
	Cresylic acid	6.1	UN2022	II	6.1, 8	B110, T8	None	202	243	1 L	30 L	B	
	Crotonaldehyde, stabilized	6.1	UN1143	I	6.1, 3	2, B9, B14, B32, B74, B77, T38, T43, T45	None	227	244	Forbidden	30 L	B	40
	Crotonic acid *liquid*	8	UN2823	III	8		154	203	241	5 L	60 L	A	12
	Crotonic acid, *solid*	8	UN2823	III	8		154	213	240	25 kg	100 kg	A	12
	Crotonylene	3	UN1144	I	3	T20	150	201	243	1 L	30 L	E	
	Cupriethylenediamine solution	8	UN1761	II	8, 6.1	T8, T26	None	202	243	1 L	30 L	A	95
				III	8, 6.1	T7	154	203	242	5 L	60 L	A	95
	Cutters, cable, explosive	1.4S	UN0070	II	1.4S		None	62	None	25 kg	100 kg	A	
	Cyanide or cyanide mixtures, dry, see Cyanides, inorganic, solid, n.o.s.												
	Cyanide solutions, n.o.s	6.1	UN1935	I	6.1	B37, T18, T26	None	201	243	1 L	30 L	B	40, 52
				II	6.1	T18, T26	None	202	243	5 L	60 L	A	40, 52
				III	6.1	T18, T26	153	203	241	60 L	220 L	A	40, 52
	Cyanides, inorganic, solid, n.o.s	6.1	UN1588	I	6.1	N74, N75	None	211	242	5 kg	50 kg	A	52
				II	6.1	N74, N75	None	212	242	25 kg	100 kg	A	52
				III	6.1	N74, N75	153	213	240	100 kg	200 kg	A	52
	Cyanogen bromide	6.1	UN1889	I	6.1, 8	A6, A8	None	211	242	1 kg	15 kg	D	40
	Cyanogen chloride, inhibited	2.3	UN1589		2.3, 8	1	None	192	245	Forbidden	Forbidden	D	40
	Cyanogen	2.3	UN1026		2.3, 2.1	2	None	192	245	Forbidden	Forbidden	D	40
	Cyanuric chloride	8	UN2670	II	8		None	212	240	15 kg	50 kg	A	12, 40
	Cyanuric triazide	Forbidden											
	Cyclobutane	2.1	UN2601		2.1		306	304	314, 315.	Forbidden	150 kg	B	40
	Cyclobutyl chloroformate	6.1	UN2744	II	6.1, 8, 3.	T18	None	202	243	1 L	30 L	A	12, 13, 21, 25, 40, 100
	1,5,9-Cyclododecatriene	6.1	UN2518	III	6.1	T7	153	203	241	60 L	220 L	A	40
	Cycloheptane	3	UN2241	II	3	T1	150	202	242	5 L	60 L	B	40
	Cycloheptatriene	3	UN2603	II	3, 6.1	T14	None	202	243	1 L	60 L	E	40
	Cycloheptene	3	UN2242	II	3	B1, T7	150	202	242	5 L	60 L	B	
	Cyclohexane	3	UN1145	II	3	B101, T8	150	202	242	5 L	60 L	E	
	Cyclohexanone	3	UN1915	III	3	B1, T1	150	203	242	60 L	220 L	A	
	Cyclohexene	3	UN2256	II	3	B101, T7	150	202	242	5 L	60 L	E	
	Cyclohexenyltrichlorosilane	8	UN1762	II	8	A7, B2, N34, T8, T26	None	202	242	Forbidden	30 L	C	40
	Cyclohexyl acetate	3	UN2243	III	3	B1, T1	150	203	242	60 L	220 L	A	
	Cyclohexyl isocyanate	6.1	UN2488	I	6.1, 3	2,B9, B14,B32, B74,B77, T38, T43, T45	None	227	244	Forbidden	Forbidden	D	20, 40, 95
	Cyclohexyl mercaptan	3	UN3054	III	3	B1, T1	150	203	242	60 L	220 L	A	40, 95
	Cyclohexylamine	8	UN2357	II	8, 3	B101, T8, T26	None	202	243	1 L	30 L	A	40
	Cyclohexyltrichlorosilane	8	UN1763	II	8	A7, B2, N34, T8, T26	None	202	242	Forbidden	30 L	C	40
	Cyclonite and cyclotetramethylenetetranitramine mixtures, wetted *or* desensitized *see* RDX and HMX mixtures, wetted *or* desensitized *etc.*												

Appendix A

§ 172.101 Hazardous Materials Table—Continued

Symbols	Hazardous materials descriptions and proper shipping names	Hazard class or Division	Identification Numbers	PG	Label Codes	Special provisions	(8) Packaging (§173.***)			(9) Quantity limitations		(10) Vessel stowage	
							Exceptions	Non-bulk	Bulk	Passenger aircraft/rail	Cargo aircraft only	Location	Other
(1)	(2)	(3)	(4)	(5)	(6)	(7)	(8A)	(8B)	(8C)	(9A)	(9B)	(10A)	(10B)
	Cyclonite and HMX mixtures, wetted or desensitized see RDX and HMX mixtures, wetted or desensitized etc.												
	Cyclonite and octogen mixtures, wetted or desensitized see RDX and HMX mixtures, wetted or desensitized etc.												
	Cyclonite, see Cyclotrimethylenetrinitramine, etc.												
	Cyclooctadiene phosphines, see 9-Phosphabicyclononanes.												
	Cyclooctadienes	3	UN2520	III	3	B1, T1	150	203	242	60 L	220 L	A	
	Cyclooctatetraene	3	UN2358	II	3	T8	150	202	242	5 L	60 L	B	
	Cyclopentane	3	UN1146	II	3	B101, T14	150	202	242	5 L	60 L	E	
	Cyclopentane, methyl, see Methylcyclopentane.												
	Cyclopentanol	3	UN2244	III	3	B1, T1	150	203	242	60 L	220 L	A	
	Cyclopentanone	3	UN2245	III	3	B1, T1	150	203	242	60 L	220 L	A	
	Cyclopentene	3	UN2246	II	3	B101, T13	150	202	242	5 L	60 L	E	
	Cyclopropane	2.1	UN1027		2.1		306	304	314, 315.	Forbidden	150 kg	E	40
	Cyclotetramethylene tetranitramine (dry or unphlegmatized) (HMX).	Forbidden											
	Cyclotetramethylenetetranitramine, desensitized or Octogen, desensitized or HMX, desensitized.	1.1D	UN0484	II	1.1D		None	62	None	Forbidden	Forbidden	B	1E, 5E
	Cyclotetramethylenetetranitramine, wetted or HMX, wetted or Octogen, wetted with not less than 15 percent water, by mass.	1.1D	UN0226	II	1.1D		None	62	None	Forbidden	Forbidden	B	1E, 5E
	Cyclotrimethylenenitramine and octogen, mixtures, wetted or desensitized see RDX and HMX mixtures, wetted or desensitized etc.												
	Cyclotrimethylenetrinitramine and cyclotetramethylenetetranitramine mixtures, wetted or desensitized see RDX and HMX mixtures, wetted or desensitized etc.												
	Cyclotrimethylenetrinitramine and HMX mixtures, wetted or desensitized see RDX and HMX mixtures, wetted or desensitized etc.												
	Cyclotrimethylenetrinitramine, desensitized or Cyclonite, desensitized or Hexogen, desensitized or RDX, desensitized.	1.1D	UN0483	II	1.1D		None	62	None	Forbidden	Forbidden	B	1E, 5E
	Cyclotrimethylenetrinitramine, wetted or Cyclonite, wetted or Hexogen, wetted or RDX, wetted with not less than 15 percent water by mass.	1.1D	UN0072	II	1.1D		None	62	None	Forbidden	Forbidden	B	1E, 5E
	Cymenes	3	UN2046	III	3	B1, T1	150	203	242	60 L	220 L	A	
D	Dangerous Goods in Machinery or Dangerous Goods in Apparatus.		NA8001			136	None	222	None	No limit	No limit	A	
	Decaborane	4.1	UN1868	II	4.1, 6.1	A19, A20	None	212	None	Forbidden	50 kg	A	
	Decahydronaphthalene	3	UN1147	III	3	B1, T1	150	203	242	60 L	220 L	A	
	n-Decane	3	UN2247	III	3	B1, T1	150	203	242	60 L	220 L	A	
	Deflagrating metal salts of aromatic nitroderivatives, n.o.s.	1.3C	UN0132	II	1.3C		None	62	None	Forbidden	Forbidden	B	1E, 5E
	Delay electric igniter, see Igniters.												
D	Denatured alcohol	3	NA1986	I	3, 6.1	T8, T31	None	201	243	Forbidden	30 L	E	40
				II	3, 6.1	T8, T31	None	202	243	1 L	60 L	E	40
				III	3, 6.1	B1, T8, T31	150	203	242	60 L	220 L	E	40
D	Denatured alcohol	3	NA1987	II	3	T8, T31	None	202	242	5 L	60 L	B	
				III	3	B1, T7, T30	150	203	242	60 L	220 L	A	
	Depth charges, see Charges, depth.												
	Detonating relays, see Detonators, etc.												
	Detonator assemblies, non-electric for blasting.	1.1B	UN0360	II	1.1B		None	62	None	Forbidden	Forbidden	B	2E, 6E
	Detonator assemblies, non-electric, for blasting.	1.4B	UN0361	II	1.4B	103	63(f), 63(g).	62	None	Forbidden	75 kg	A	24E

§ 172.101 Hazardous Materials Table—Continued

Symbols	Hazardous materials descriptions and proper shipping names	Hazard class or Division	Identification Numbers	PG	Label Codes	Special provisions	Packaging (§173.***) Exceptions (8A)	Packaging Non-bulk (8B)	Packaging Bulk (8C)	Quantity limitations Passenger aircraft/rail (9A)	Quantity limitations Cargo aircraft only (9B)	Vessel stowage Location (10A)	Vessel stowage Other (10B)
(1)	(2)	(3)	(4)	(5)	(6)	(7)	(8A)	(8B)	(8C)	(9A)	(9B)	(10A)	(10B)
	Detonator, assemblies, non-electric for blasting.	1.4S	UN0500	II	1.4S		63(f), 63(g).	62	None	25 kg	100 kg	A	
	Detonators, electric, for blasting	1.1B	UN0030	II	1.1B		63(f), 63(g).	62	None	Forbidden	Forbidden	B	2E, 6E
	Detonators, electric, for blasting	1.4B	UN0255	II	1.4B	103	63(f), 63(g).	62	None	Forbidden	75 kg	A	24E
	Detonators, electric for blasting	1.4S	UN0456	II	1.4S		63(f), 63(g).	62	None	25 kg	100 kg	A	
	Detonators for ammunition	1.1B	UN0073	II	1.1B		None	62	None	Forbidden	Forbidden	B	2E, 6E
	Detonators for ammunition	1.2B	UN0364	II	1.2B		None	62	None	Forbidden	Forbidden	B	2E, 6E
	Detonators for ammunition	1.4B	UN0365	II	1.4B	103	None	62	None	Forbidden	75 kg	A	24E
	Detonators for ammunition	1.4S	UN0366	II	1.4S		None	62	None	25 kg	100 kg	A	
	Detonators, non-electric, for blasting	1.1B	UN0029	II	1.1B		None	62	None	Forbidden	Forbidden	B	2E, 6E
	Detonators, non-electric, for blasting	1.4B	UN0267	II	1.4B	103	63(f), 63(g).	62	None	Forbidden	75 kg	A	24E
	Detonators, non-electric for blasting	1.4S	UN0455	II	1.4S	104	63(f), 63(g).	62	None	25 kg	100 kg	A	
	Deuterium, compressed	2.1	UN1957		2.1		306	302	None	Forbidden	150 kg	E	40
	Devices, small, hydrocarbon gas powered or Hydrocarbon gas refills for small devices with release device.	2.1	UN3150		2.1		306	304	None	Forbidden	150 kg	B	40
	Di-n-amylamine	3	UN2841	III	3, 6.1	B1, T8	150	203	242	60 L	220 L	A	
	Di-n-butyl peroxydicarbonate, with more than 52 percent in solution.	Forbidden											
	Di-n-butylamine	8	UN2248	II	8, 3	T8	None	202	243	1 L	30 L	A	
	2,2-Di-(tert-butylperoxy) butane, with more than 55 percent in solution.	Forbidden											
	Di-(tert-butylperoxy) phthalate, with more than 55 percent in solution.	Forbidden											
	2,2-Di-(4,4-di-tert-butylperoxycyclohexyl) propane, with more than 42 percent with inert solid.	Forbidden											
	Di-2,4-dichlorobenzoyl peroxide, with more than 75 percent with water.	Forbidden											
	1,2-Di-(dimethylamino)ethane	3	UN2372	II	3	T8	150	202	242	5 L	60 L	B	
	Di-2-ethylhexyl phosphoric acid, see Diisooctyl acid phosphate.												
	Di-(1-hydroxytetrazole) (dry)	Forbidden											
	Di-(1-naphthoyl) peroxide	Forbidden											
	a,a'-Di-(nitroxy) methylether	Forbidden											
	Di-(beta-nitroxyethyl) ammonium nitrate.	Forbidden											
	Diacetone alcohol	3	UN1148	II	3	T1	150	202	242	5 L	60 L	B	
				III	3	B1, T1	150	203	242	60 L	220 L	A	
	Diacetone alcohol peroxides, with more than 57 percent in solution with more than 9 percent hydrogen peroxide, less than 26 percent diacetone alcohol and less than 9 percent water; total active oxygen content more than 9 percent by mass.	Forbidden											
	Diacetyl, see Butanedione												
	Diacetyl peroxide, solid, or with more than 25 percent in solution.	Forbidden											
	Diallylamine	3	UN2359	II	3, 6.1, 8.	T8	None	202	243	1 L	5 L	B	21, 40, 100
	Diallylether	3	UN2360	II	3, 6.1	N12, T8	None	202	243	1 L	60 L	E	40
	4,4'-Diaminodiphenyl methane	6.1	UN2651	III	6.1		153	213	240	100 kg	200 kg	A	
	p-Diazidobenzene	Forbidden											
	1,2-Diazidoethane	Forbidden											
	1,1'-Diazoaminonaphthalene	Forbidden											
	Diazoaminotetrazole (dry)	Forbidden											
	Diazodinitrophenol (dry)	Forbidden											
	Diazodinitrophenol, wetted with not less than 40 percent water or mixture of alcohol and water, by mass.	1.1A	UN0074	II	1.1A	111, 117	None	62	None	Forbidden	Forbidden	E	2E, 6E
	Diazodiphenylmethane	Forbidden											
	Diazonium nitrates (dry)	Forbidden											
	Diazonium perchlorates (dry)	Forbidden											
	1,3-Diazopropane	Forbidden											
	Dibenzyl peroxydicarbonate, with more than 87 percent with water.	Forbidden											
	Dibenzyldichlorosilane	8	UN2434	II	8	B2, T8, T26	154	202	242	1 L	30 L	C	40
	Diborane, compressed	2.3	UN1911		2.3, 2.1	1	None	302	None	Forbidden	Forbidden	D	40, 57
	Diborane mixtures	2.1	NA1911		2.1	5	None	302	245	Forbidden	Forbidden	D	40, 57
	Dibromoacetylene	Forbidden											

Appendix A

§ 172.101 Hazardous Materials Table—Continued

Symbols	Hazardous materials descriptions and proper shipping names	Hazard class or Division	Identification Numbers	PG	Label Codes	Special provisions	Packaging (§173.***) Exceptions	Packaging (§173.***) Non-bulk	Packaging (§173.***) Bulk	Quantity limitations Passenger aircraft/rail	Quantity limitations Cargo aircraft only	Vessel stowage Location	Vessel stowage Other
(1)	(2)	(3)	(4)	(5)	(6)	(7)	(8A)	(8B)	(8C)	(9A)	(9B)	(10A)	(10B)
	1,2-Dibromobutan-3-one	6.1	UN2648	II	6.1		None	202	243	5 L	60 L	B	40
	Dibromochloropropane	6.1	UN2872	III	6.1	T7	153	203	241	60 L	220 L	A	
A	Dibromodifluoromethane, R12B2	9	UN1941	III	None	T22	155	203	241	100 L	220 L	A	25
	1,2-Dibromoethane, see Ethylene dibromide.												
	Dibromomethane	6.1	UN2664	III	6.1	T7	153	203	241	60 L	220 L	A	
	Dibutyl ethers	3	UN1149	III	3	B1, T1	150	203	242	60 L	220 L	A	
	Dibutylaminoethanol	6.1	UN2873	III	6.1	T1	153	203	241	60 L	220 L	A	
	N,N'-Dichlorazodicarbonamidine (salts of) (dry).	Forbidden											
	1,1-Dichloro-1-nitroethane	6.1	UN2650	II	6.1	T8	None	202	243	5 L	60 L	A	12, 40
D	3,5-Dichloro-2,4,6-trifluoropyridine.	6.1	NA9264	I	6.1	2, B9, B14, B32, B74, T38, T43, T45	None	227	244	Forbidden	Forbidden	A	40, 95
	Dichloroacetic acid	8	UN1764	II	8	A3, A6, A7, B2, N34, T9, T27	154	202	242	1 L	30 L	A	
	1,3-Dichloroacetone	6.1	UN2649	II	6.1		None	212	242	25 kg	100 kg	B	12, 40
	Dichloroacetyl chloride	8	UN1765	II	8	A3, A6, A7, B2, B6, N34, T8, T26	154	202	242	1 L	30 L	D	40
	Dichloroacetylene	Forbidden											
+	Dichloroanilines, liquid	6.1	UN1590	II	6.1	T14	None	202	243	5 L	60 L	A	40
+	Dichloroanilines, solid	6.1	UN1590	II	6.1	T14	None	212	242	25 kg	100 kg	A	40
+	o-Dichlorobenzene	6.1	UN1591	III	6.1	T7	153	203	241	60 L	220 L	A	
D	Dichlorobutene	8	NA2920	I	8, 3		None	201	243	0.5 L	2.5 L	C	12, 21, 25, 40, 48
	2,2'-Dichlorodiethyl ether	6.1	UN1916	II	6.1, 3	N33, N34, T8	None	202	243	5 L	60 L	A	
	Dichlorodifluoromethane and difluoroethane azeotropic mixture or Refrigerant gas R 500 with approximately 74 percent dichlorodifluoromethane.	2.2	UN2602		2.2		306	304	314, 315.	75 kg	150 kg	A	
	Dichlorodifluoromethane or Refrigerant gas R 12.	2.2	UN1028		2.2		306	304	314, 315.	75 kg	150 kg	A	
	Dichlorodimethyl ether, symmetrical.	6.1	UN2249	I	6.1	T25	None	201	243	Forbidden	Forbidden	D	40
	1,1-Dichloroethane	3	UN2362	II	3	B101, T7	150	202	242	5 L	60 L	B	40
	1,2-Dichloroethane, see Ethylene dichloride.												
	Dichloroethyl sulfide	Forbidden											
	1,2-Dichloroethylene	3	UN1150	II	3	T14	150	202	242	5 L	60 L	B	
	Dichlorofluoromethane or refrigerant gas R21.	2.2	UN1029		2.2		306	304	314, 315.	75 kg	150 kg	A	
	Dichloroisocyanuric acid, dry or Dichloroisocyanuric acid salts.	5.1	UN2465	II	5.1	28	152	212	240	5 kg	25 kg	A	13
	Dichloroisopropyl ether	6.1	UN2490	II	6.1	T8	None	202	243	5 L	60L	B	
	Dichloromethane	6.1	UN1593	III	6.1	N36, T13	153	203	241	60 L	220 L	A	
	Dichloropentanes	3	UN1152	III	3	B1, T1	150	203	242	60 L	220 L	A	
	Dichlorophenyl isocyanates	6.1	UN2250	II	6.1		None	212	242	25 kg	100 kg	B	25, 40, 48
	Dichlorophenyltrichlorosilane	8	UN1766	II	8	A7, B2, B6, N34, T8, T26	None	202	242	Forbidden	30 L	C	40
	1,2-Dichloropropane	3	UN1279	II	3	N36,T1	150	202	242	5 L	60 L	B	
	1,3-Dichloropropanol-2	6.1	UN2750	II	6.1	T8	None	202	243	5 L	60 L	A	12, 40
	Dichloropropene and propylene dichloride mixture, see Propylene dichloride.												
	Dichloropropenes	3	UN2047	II	3	T8	150	202	242	5 L	60 L	B	
		3	UN2047	III	3	B1, T8	150	203	242	60 L	220 L	A	
	Dichlorosilane	2.3	UN2189		2.3, 2.1, 8.	2, B9, B14	None	304	314, 315.	Forbidden	Forbidden	D	17, 40
	1,2-Dichloro-1,1,2,2-Tetrafluoroethane or Refrigerant gas R 114.	2.2	UN1958		2.2		306	304	314, 315.	75 kg	150 kg	A	
	Dichlorovinylchloroarsine	Forbidden											
	Dicycloheptadiene, see 2,5-Norbornadiene.												
	Dicyclohexylamine	8	UN2565	III	8	T8	154	203	241	5 L	60 L	A	
	Dicyclohexylammonium nitrite	4.1	UN2687	III	4.1		151	213	240	25 kg	100 kg	A	48
	Dicyclopentadiene	3	UN2048	III	3	B1, T1	150	203	242	60 L	220 L	A	
	Didymium nitrate	5.1	UN1465	III	5.1	A1	152	213	240	25 kg	100 kg	A	
D	Dieldrin	6.1	NA2761	II	6.1		None	212	242	0.5 kg	5 kg	A	40
D	Diesel fuel	3	NA1993	III	None	B1	150	203	242	60 L	220 L	A	
	Diethanol nitrosamine dinitrate (dry).	Forbidden											
	Diethoxymethane	3	UN2373	II	3	T8	150	202	242	5 L	60 L	E	
	3,3-Diethoxypropene	3	UN2374	II	3	T1	150	202	242	5 L	60 L	B	
	Diethyl carbonate	3	UN2366	III	3	B1, T1	150	203	242	60 L	220 L	A	
	Diethyl cellosolve, see Ethylene glycol diethyl ether.												
	Diethyl ether or Ethyl ether	3	UN1155	I	3	T21	150	201	243	1 L	30 L	E	40
	Diethyl ketone	3	UN1156	II	3	T1	150	202	242	5 L	60 L	B	
	Diethyl peroxydicarbonate, with more than 27 percent in solution.	Forbidden											
	Diethyl sulfate	6.1	UN1594	II	6.1	B101, T14	None	202	243	5 L	60 L	C	
	Diethyl sulfide	3	UN2375	II	3	B101, T14	None	202	243	1 L	60 L	E	
	Diethylamine	3	UN1154	II	3, 8	B101, N34, T8	None	202	243	1 L	5 L	E	40

§172.101 HAZARDOUS MATERIALS TABLE—Continued

Symbols (1)	Hazardous materials descriptions and proper shipping names (2)	Hazard class or Division (3)	Identification Numbers (4)	PG (5)	Label Codes (6)	Special provisions (7)	Packaging (§173.***) (8)			Quantity limitations (9)		Vessel stowage (10)	
							Exceptions (8A)	Non-bulk (8B)	Bulk (8C)	Passenger aircraft/rail (9A)	Cargo aircraft only (9B)	Location (10A)	Other (10B)
	2-Diethylaminoethanol	8	UN2686	II	8, 3	B2, T15, T26	None	202	243	1 L	30 L	A	
	Diethylaminopropylamine	3	UN2684	III	3, 8	B1, T8	150	203	242	5 L	60 L	A	
+	N,N-Diethylaniline	6.1	UN2432	III	6.1	T2	153	203	241	60 L	220 L	A	
	Diethylbenzene	3	UN2049	III	3	B1, T1	150	203	242	60 L	220 L	A	
	Diethyldichlorosilane	8	UN1767	II	8, 3	A7, B6, B100, N34, T8, T26	None	202	243	Forbidden	30 L	C	40
	Diethylene glycol dinitrate	Forbidden											
	Diethyleneglycol dinitrate, desensitized with not less than 25 percent non-volatile water-insoluble phlegmatizer, by mass.	1.1D	UN0075	II	1.1D		None	62	None	Forbidden	Forbidden	B	1E, 4E, 21E
	Diethylenetriamine	8	UN2079	II	8	B2, T8	154	202	242	1 L	30 L	A	40
	N,N-Diethylethylenediamine	8	UN2685	II	8, 3	T8	None	202	243	1 L	30 L	A	
	Diethylgold bromide	Forbidden											
	Diethylthiophosphoryl chloride	8	UN2751	II	8	B2, T8	None	212	240	15 kg	50 kg	C	40
	Diethylzinc	4.2	UN1366	I	4.2, 4.3	B11, T28, T40	None	181	244	Forbidden	Forbidden	D	18
	Difluorochloroethanes, see 1-Chloro-1,1-difluoroethanes.												
	1,1-Difluoroethane or Refrigerant gas R 152a.	2.1	UN1030		2.1		306	304	314, 315.	Forbidden	150 kg	B	40
	1,1-Difluoroethylene or Refrigerant gas R 1132a.	2.1	UN1959		2.1		306	304	None	Forbidden	150 kg	E	40
	Difluoromethane or Refrigerant gas R 32.	2.1	UN3252		2.1		306	302	314, 315.	Forbidden	150 kg	D	40
	Difluorophosphoric acid, anhydrous.	8	UN1768	II	8	A6, A7, B2, N5, N34, T9, T27	None	202	242	1 L	30 L	A	40
	2,3-Dihydropyran	3	UN2376	II	3	T7	150	202	242	5 L	60 L	B	
	1,8-Dihydroxy-2,4,5,7-tetranitroanthraquinone (chrysamminic acid).	Forbidden											
	Diiodoacetylene	Forbidden											
	Diisobutyl ketone	3	UN1157	III	3	B1, T1	150	203	242	60 L	220 L	A	
	Diisobutylamine	3	UN2361	III	3, 8	B1, T1	150	203	242	5 L	60 L	A	
	Diisobutylene, isomeric compounds.	3	UN2050	II	3	T1	150	202	242	5 L	60 L	B	
	Diisooctyl acid phosphate	8	UN1902	III	8	T7	154	203	241	5 L	60 L	A	
	Diisopropyl ether	3	UN1159	II	3	B101, T8	150	202	242	5 L	60 L	E	40
	Diisopropylamine	3	UN1158	II	3, 8	B101, T8	None	202	243	1 L	5 L	B	
	Diisopropylbenzene hydroperoxide, with more than 72 percent in solution.	Forbidden											
	Diketene, inhibited	6.1	UN2521	I	6.1, 3	2, B9, B14, B32, B74, T38, T43, T45	None	227	244	Forbidden	Forbidden	D	40, 49
	1,2-Dimethoxyethane	3	UN2252	II	3	T1	150	202	242	5 L	60 L	B	
	1,1-Dimethoxyethane	3	UN2377	II	3	T13	150	202	242	5 L	60 L	B	
	Dimethyl carbonate	3	UN1161	II	3	T8	150	202	242	5 L	60 L	B	
	Dimethyl chlorothiophosphate, see Dimethyl thiophosphoryl chloride.												
	2,5-Dimethyl-2,5-dihydroperoxy hexane, with more than 82 percent with water.	Forbidden											
	Dimethyl disulfide	3	UN2381	II	3	T8	150	202	242	5 L	60 L	B	40
	Dimethyl ether	2.1	UN1033		2.1		306	304	314, 315.	Forbidden	150 kg	B	40
	Dimethyl-N-propylamine	3	UN2266	II	3, 8	T14, T26	None	202	243	1 L	5 L	B	40
	Dimethyl sulfate	6.1	UN1595	I	6.1, 8	2, B9, B14, B32, B74, B77, T38, T43, T45	None	227	244	Forbidden	Forbidden	D	40
	Dimethyl sulfide	3	UN1164	II	3	B100, T14	None	202	242	5 L	60 L	E	40
	Dimethyl thiophosphoryl chloride	6.1	UN2267	II	6.1, 8	T7	None	202	243	1 L	30 L	B	25
	Dimethylamine, anhydrous	2.1	UN1032		2.1		None	304	314, 315.	Forbidden	150 kg	D	40
	Dimethylamine solution	3	UN1160	II	3, 8	T8, T34	None	202	243	1 L	5 L	B	
	2-Dimethylaminoacetonitrile	3	UN2378	II	3, 6.1	T8	None	202	243	1 L	60 L	A	26, 40
	2-Dimethylaminoethanol	8	UN2051	II	8, 3	B2, T8	154	202	243	1 L	30 L	A	
	2-Dimethylaminoethyl acrylate	6.1	UN3302	II	6.1	T8	None	202	243	5 L	60 L	D	25
	2-Dimethylaminoethyl methacrylate.	6.1	UN2522	II	6.1	T8	None	202	243	5 L	60 L	B	40
	N,N-Dimethylaniline	6.1	UN2253	II	6.1	T8	None	202	243	5 L	60 L	A	
	2,3-Dimethylbutane	3	UN2457	II	3	T13	150	202	242	5 L	60L	E	
	1,3-Dimethylbutylamine	3	UN2379	II	3, 8	T8	None	202	243	1 L	5 L	B	
	Dimethylcarbamoyl chloride	8	UN2262	II	8	B2, T8	154	202	243	1 L	30 L	A	40
	Dimethylcyclohexanes	3	UN2263	II	3	T1	150	202	242	5 L	60 L	B	
	Dimethylcyclohexylamine	8	UN2264	II	8, 3	B2, T8	154	202	243	1 L	30 L	A	40
	Dimethyldichlorosilane	3	UN1162	II	3, 8	B77, T15, T26	None	202	243	Forbidden	Forbidden	B	40
	Dimethyldiethoxysilane	3	UN2380	II	3	T8	150	202	242	5 L	60 L	B	
	Dimethyldioxanes	3	UN2707	II	3	T8, T31	150	202	242	5 L	60 L	B	
				III	3	B1, T7, T30	150	203	242	60 L	220 L	A	
	N,N-Dimethylformamide	3	UN2265	III	3	B1, T1	150	203	242	60 L	220 L	A	
	Dimethylhexane dihydroperoxide (dry).	Forbidden											
	Dimethylhydrazine, symmetrical	6.1	UN2382	I	6.1, 3	2, A7, B9, B14, B32, B74, B77, T38, T43, T45	None	227	244	Forbidden	Forbidden	D	40

Appendix A

§ 172.101 Hazardous Materials Table—Continued

Symbols	Hazardous materials descriptions and proper shipping names	Hazard class or Division	Identification Numbers	PG	Label Codes	Special provisions	Packaging (§173.***) Exceptions	Packaging (§173.***) Non-bulk	Packaging (§173.***) Bulk	Quantity limitations Passenger aircraft/rail	Quantity limitations Cargo aircraft only	Vessel stowage Location	Vessel stowage Other
(1)	(2)	(3)	(4)	(5)	(6)	(7)	(8A)	(8B)	(8C)	(9A)	(9B)	(10A)	(10B)
	Dimethylhydrazine, unsymmetrical.	6.1	UN1163	I	6.1, 3, 8.	2, B7, B9, B14, B32, B74, T38, T43, T45	None	227	244	Forbidden	Forbidden	D	21, 38, 40, 100
	2,2-Dimethylpropane	2.1	UN2044		2.1		306	304	314, 315.	Forbidden	150 kg	E	40
	Dimethylzinc	4.2	UN1370	I	4.2, 4.3	B11, B16, T28, T29, T40	None	181	244	Forbidden	Forbidden	D	18
	Dinitro-o-cresol, *solid*	6.1	UN1598	II	6.1	T14	None	212	242	25 kg	100 kg	A	
	Dinitro-o-cresol, *solution*	6.1	UN1598	II	6.1	T14	None	202	243	5 L	60 L	A	
	1,3-Dinitro-5,5-dimethyl hydantoin.	Forbidden											
	Dinitro-7,8-dimethylglycoluril (dry).	Forbidden											
	1,3-Dinitro-4,5-dinitrosobenzene	Forbidden											
	1,4-Dinitro-1,1,4,4-tetramethylolbutanetetranitrate (dry).	Forbidden											
	2,4-Dinitro-1,3,5-trimethylbenzene.	Forbidden											
	Dinitroanilines	6.1	UN1596	II	6.1	T14	None	212	242	25 kg	100 kg	A	91
	Dinitrobenzenes, *liquid*	6.1	UN1597	II	6.1	11, T14	None	202	243	5 L	60 L	A	91
	Dinitrobenzenes, *solid*	6.1	UN1597	II	6.1	11	None	212	242	25 kg	100 kg	A	91
	Dinitrochlorobenzene, see Chlorodinitrobenzene.												
	1,2-Dinitroethane	Forbidden											
	1,1-Dinitroethane (dry)	Forbidden											
	Dinitrogen tetroxide	2.3	UN1067		2.3, 5.1, 8.	1, B7, B14, B45, B46, B61, B66, B67, B77	None	336	314	Forbidden	Forbidden	D	40, 89, 90
	Dinitroglycoluril or Dingu	1.1D	UN0489	II	1.1D		None	62	None	Forbidden	Forbidden	B	1E, 5E
	Dinitromethane	Forbidden											
	Dinitrophenol, dry or wetted with less than 15 percent water, by mass.	1.1D	UN0076	II	1.1D, 6.1.		None	62	None	Forbidden	Forbidden	B	1E, 5E
	Dinitrophenol solutions	6.1	UN1599	II	6.1	T8	None	202	243	5 L	60 L	A	36
				III	6.1	T7	153	203	241	60 L	220 L	A	36
	Dinitrophenol, wetted *with not less than 15 percent water, by mass.*	4.1	UN1320	I	4.1, 6.1	23, A8, A19, A20, N41	None	211	None	1 kg	15 kg	E	28, 36
	Dinitrophenolates alkali metals, dry or wetted with less than 15 percent water, by mass.	1.3C	UN0077	II	1.3C, 6.1.		None	62	None	Forbidden	Forbidden	B	1E, 5E
	Dinitrophenolates, wetted *with not less than 15 percent water, by mass.*	4.1	UN1321	I	4.1, 6.1	23, A8, A19, A20, N41	None	211	None	1 kg	15 kg	E	28, 36
	Dinitropropylene glycol	Forbidden											
	Dinitroresorcinol, dry or wetted with less than 15 percent water, by mass.	1.1D	UN0078	II	1.1D		None	62	None	Forbidden	Forbidden	B	1E, 5E
	2,4-Dinitroresorcinol (heavy metal salts of) (dry).	Forbidden											
	4,6-Dinitroresorcinol (heavy metal salts of) (dry).	Forbidden											
	Dinitroresorcinol, wetted *with not less than 15 percent water, by mass.*	4.1	UN1322	I	4.1	23, A8, A19, A20, N41	None	211	None	1 kg	15 kg	E	28, 36
	3,5-Dinitrosalicylic acid (lead salt) (dry).	Forbidden											
	Dinitrosobenzene	1.3C	UN0406	II	1.3C		None	62	None	Forbidden	Forbidden	B	1E, 5E
	Dinitrosobenzylamidine and salts of (dry).	Forbidden											
	2,2-Dinitrostilbene	Forbidden											
	Dinitrotoluenes, *liquid*	6.1	UN2038	II	6.1	T8	None	202	243	5 L	60 L	A	
	Dinitrotoluenes, *molten*	6.1	UN1600	II	6.1	B100, T14	None	202	243	Forbidden	Forbidden	C	
	Dinitrotoluenes, *solid*	6.1	UN2038	II	6.1	T8	None	212	242	25 kg	100kg	A	
	1,9-Dinitroxy pentamethylene-2,4,6,8-tetramine (dry).	Forbidden											
	Dioxane	3	UN1165	II	3	T8	150	202	242	5 L	60 L	B	
	Dioxolane	3	UN1166	II	3	T8	150	202	242	5 L	60 L	B	40
	Dipentene	3	UN2052	III	3	B1, T1	150	203	242	60 L	220 L	A	40
	Diphenylamine chloroarsine	6.1	UN1698	I	6.1		None	201	None	Forbidden	Forbidden	D	40
	Diphenylchloroarsine, liquid	6.1	UN1699	I	6.1	A8, B14, B32, N33, N34	None	201	243	Forbidden	30 L	D	40
	Diphenylchloroarsine, solid	6.1	UN1699	I	6.1	A8, B14, B32, N33, N34	None	211	242	Forbidden	15 kg	D	40
	Diphenyldichlorosilane	8	UN1769	II	8	A7, B2, N34, T8, T26	None	202	242	Forbidden	30 L	C	40
	Diphenylmethyl bromide	8	UN1770	II	8		154	212	240	15 kg	50 kg	D	40
	Dipicryl sulfide, dry or wetted with less than 10 percent water, by mass.	1.1D	UN0401	II	1.1D		None	62	None	Forbidden	Forbidden	B	1E, 5E
	Dipicryl sulfide, wetted *with not less than 10 percent water, by mass.*	4.1	UN2852	I	4.1	A2, N41	None	211	None	Forbidden	0.5 kg	D	28
	Dipicrylamine, see Hexanitrodiphenylamine.												

§ 172.101 HAZARDOUS MATERIALS TABLE—Continued

Symbols	Hazardous materials descriptions and proper shipping names	Hazard class or Division	Identification Numbers	PG	Label Codes	Special provisions	(8) Packaging (§173.***)			(9) Quantity limitations		(10) Vessel stowage	
							Exceptions	Non-bulk	Bulk	Passenger aircraft/rail	Cargo aircraft only	Location	Other
(1)	(2)	(3)	(4)	(5)	(6)	(7)	(8A)	(8B)	(8C)	(9A)	(9B)	(10A)	(10B)
	Dipropionyl peroxide, with more than 28 percent in solution.	Forbidden											
	Di-n-propyl ether	3	UN2384	II	3	T1	150	202	242	5 L	60 L	B	
	Dipropyl ketone	3	UN2710	III	3	B1, T1	150	203	242	60 L	220 L	A	
	Dipropylamine	3	UN2383	II	3, 8	T8	None	202	243	1 L	5 L	B	
G	Disinfectant, liquid, corrosive, n.o.s.	8	UN1903	I	8	A7, B10, T42	None	201	243	0.5 L	2.5 L	B	
G	Disinfectants, liquid, corrosive n.o.s.	8	UN1903	II	8	B2	154	202	242	1 L	30 L	B	
				III	8		154	203	241	5 L	60 L	A	
G	Disinfectants, liquid, toxic, n.o.s	6.1	UN3142	I	6.1	A4, T42	None	201	243	1 L	30 L	A	40
				II	6.1	T14	None	202	243	5 L	60 L	A	40
				III	6.1	T7	153	203	241	60 L	220 L	A	40
G	Disinfectants, solid, toxic, n.o.s	6.1	UN1601	II	6.1		None	212	242	25 kg	100 kg	A	40
				III	6.1		153	213	240	100 kg	200 kg	A	40
	Disodium trioxosilicate	8	UN3253	III	8		154	213	240	25 kg	100 kg	A	
G	Dispersant gases, n.o.s. see Refrigerant gases, n.o.s.												
	Divinyl ether, inhibited	3	UN1167	I	3	T14	None	201	243	1 L	60 L	E	40
D	Dodecylbenzenesulfonic acid	8	NA2584	II	8	B2	154	202	242	1 L	30 L	B	9
	Dodecyltrichlorosilane	8	UN1771	II	8	A7, B2, B6, N34, T8, T26	None	202	242	Forbidden	30 L	C	40
	Dry ice, see Carbon dioxide, solid.												
G	Dyes, liquid, corrosive, n.o.s. or Dye intermediates, liquid, corrosive, n.o.s..	8	UN2801	I	8	11, B10	None	201	243	0.5 L	2.5 L	A	
				II	8	11, B2, T14	154	202	242	1 L	30 L	A	
				III	8	11, T7	154	203	241	5 L	60 L	A	
G	Dyes, liquid, toxic, n.o.s or Dye intermediates, liquid, toxic, n.o.s.	6.1	UN1602	II	6.1		None	202	243	5 L	60 L	A	
				III	6.1		153	203	241	60 L	220 L	A	
G	Dyes, solid, corrosive, n.o.s. or Dye intermediates, solid, corrosive, n.o.s.	8	UN3147	I	8		None	211	242	1 kg	25 kg	A	
G	Dyes, solid, corrosive, n.o.s. or Dye intermediates, solid, corrosive, n.o.s.	8	UN3147	II	8		154	212	240	15 kg	50 kg	A	
				III	8		154	213	240	25 kg	100 kg	A	
G	Dyes, solid, toxic, n.o.s. or Dye intermediates, solid, toxic, n.o.s.	6.1	UN3143	I	6.1	A5	None	211	242	5 kg	50 kg	A	
				II	6.1		None	212	242	25 kg	100 kg	A	
				III	6.1		153	213	240	100 kg	200 kg	A	
	Dynamite, see Explosive, blasting, type A.												
	Electrolyte (acid or alkali) for batteries, see Battery fluid, acid or Battery fluid, alkali.												
	Elevated temperature liquid, flammable, n.o.s., with flash point above 37.8 C, at or above its flash point.	3	UN3256	III	3	T1	None	None	247	Forbidden	Forbidden	A	
	Elevated temperature liquid, n.o.s., at or above 100 C and below its flash point (including molten metals, molten salts, etc.).	9	UN3257	III	9	T1	None	None	247	Forbidden	Forbidden	A	85
	Elevated temperature solid, n.o.s., at or above 240 C, see section 173.247(h)(4).	9	UN3258	III	9		247(h)(4).	None	247	Forbidden	Forbidden	A	85
	Engines, internal combustion, flammable gas powered.	9	UN3166		9	135	220	220	220	Forbidden	No limit	A	
	Engines, internal combustion, flammable liquid powered.	9	UN3166		9	135	220	220	220	No limit	No limit	A	
G	Environmentally hazardous substances, liquid, n.o.s.	9	UN3082	III	9	8, T1	155	203	241	No limit	No limit	A	
G	Environmentally hazardous substances, solid, n.o.s.	9	UN3077	III	9	8, B54, N20	155	213	240	No limit	No limit	A	s
	Epibromohydrin	6.1	UN2558	I	6.1, 3	T18, T26	None	201	243	Forbidden	Forbidden	D	40
+	Epichlorohydrin	6.1	UN2023	II	6.1, 3	T14	None	202	243	5 L	60 L	A	40
	1,2-Epoxy-3-ethoxypropane	3	UN2752	III	3	B1, T1	150	203	242	60 L	220 L	A	
	Esters, n.o.s	3	UN3272	II	3	T8	150	202	242	5 L	60 L	B	
				III	3	B1, T7	150	203	242	60 L	220 L	A	
	Etching acid, liquid, n.o.s., see Hydrofluoric acid, solution etc.												
	Ethane	2.1	UN1035		2.1		306	304	302	Forbidden	150 kg	E	40
D	Ethane-Propane mixture, refrigerated liquid.	2.1	NA1961		2.1		None	316	314, 315.	Forbidden	Forbidden	D	40
	Ethane, refrigerated liquid	2.1	UN1961		2.1		None	None	315	Forbidden	Forbidden	D	40
	Ethanol amine dinitrate	Forbidden											
	Ethanol or Ethyl alcohol or Ethanol solutions or Ethyl alcohol solutions.	3	UN1170	II	3	24, T1	150	202	242	5 L	60 L	A	
				III	3	24, B1, T1	150	203	242	60 L	220 L	A	
	Ethanolamine or Ethanolamine solutions.	8	UN2491	III	8	T7	154	203	241	5 L	60 L	A	
	Ether, see Diethyl ether												

Appendix A

§172.101 Hazardous Materials Table—Continued

Symbols	Hazardous materials descriptions and proper shipping names	Hazard class or Division	Identification Numbers	PG	Label Codes	Special provisions	(8) Packaging (§173.***)			(9) Quantity limitations		(10) Vessel stowage	
							Exceptions	Non-bulk	Bulk	Passenger aircraft/rail	Cargo aircraft only	Location	Other
(1)	(2)	(3)	(4)	(5)	(6)	(7)	(8A)	(8B)	(8C)	(9A)	(9B)	(10A)	(10B)
	Ethers, n.o.s	3	UN3271	II	3	T8	150	202	242	5 L	60 L	B	
				III	3	B1, T7	150	203	242	60 L	220 L	A	
	Ethyl acetate	3	UN1173	II	3	T2	150	202	242	5 L	60 L	B	
	Ethyl acrylate, inhibited	3	UN1917	II	3	T8	150	202	242	5 L	60 L	B	40
	Ethyl alcohol, see Ethanol												
	Ethyl aldehyde, see Acetaldehyde.												
	Ethyl amyl ketone	3	UN2271	III	3	B1, T1	150	203	242	60 L	220 L	A	
	N-Ethyl-N-benzylaniline	6.1	UN2274	III	6.1	T2	153	203	241	60 L	220 L	A	
	Ethyl borate	3	UN1176	II	3	T8	150	202	242	5 L	60 L	B	
	Ethyl bromide	6.1	UN1891	II	6.1	B100, T17	None	202	243	5 L	60 L	B	40, 85
	Ethyl bromoacetate	6.1	UN1603	II	6.1, 3	T14	None	202	243	Forbidden	Forbidden	D	40
	Ethyl butyl ether	3	UN1179	II	3	B1, B101, T1	150	202	242	5 L	60 L	B	
	Ethyl butyrate	3	UN1180	III	3	B1, T1	150	203	242	60 L	220 L	A	
	Ethyl chloride	2.1	UN1037		2.1	B43, B77	None	322	314, 315.	Forbidden	150 kg	B	40
	Ethyl chloroacetate	6.1	UN1181	II	6.1, 3	T14	None	202	243	5 L	60 L	A	
	Ethyl chloroformate	6.1	UN1182	I	6.1, 3, 8.	2, A3, A6, A7, B9, B14, B32, B74, N34, T38, T43, T45	None	227	244	Forbidden	Forbidden	D	21, 40, 100
	Ethyl 2-chloropropionate	3	UN2935	III	3	B1, T1	150	203	242	60 L	220 L	A	
+	Ethyl chlorothioformate	8	UN2826	II	8, 6.1, 3.	2, B9, B14, B32, B74, T38, T43, T45	None	227	244	Forbidden	Forbidden	A	40
	Ethyl crotonate	3	UN1862	II	3	T1	150	202	242	5 L	60 L	B	
	Ethyl ether, see Diethyl ether												
	Ethyl fluoride or Refrigerant gas R161.	2.1	UN2453		2.1		306	304	314, 315.	Forbidden	150 kg	E	40
	Ethyl formate	3	UN1190	II	3	T8	150	202	242	5 L	60 L	E	
	Ethyl hydroperoxide	Forbidden											
	Ethyl isobutyrate	3	UN2385	II	3	T1	150	202	242	5 L	60 L	B	
+	Ethyl isocyanate	3	UN2481	I	3, 6.1	1, A7, B9, B14, B30, B72, T38, T43, T44	None	226	244	Forbidden	Forbidden	D	40
	Ethyl lactate	3	UN1192	III	3	B1, T1	150	203	242	60 L	220 L	A	
	Ethyl mercaptan	3	UN2363	I	3	T21	None	201	243	Forbidden	30 L	E	95, 102
	Ethyl methacrylate	3	UN2277	II	3	T1	150	202	242	5 L	60 L	B	
	Ethyl methyl ether	2.1	UN1039		2.1	B43	None	201	314, 315.	Forbidden	150 kg	B	40
	Ethyl methyl ketone or Methyl ethyl ketone.	3	UN1193	II	3	T8	150	202	242	5 L	60 L	B	
	Ethyl nitrite solutions	3	UN1194	I	3, 6.1		None	201	None	Forbidden	Forbidden	E	40, 105
	Ethyl orthoformate	3	UN2524	III	3	B1, T7	150	203	242	60 L	220 L	A	
	Ethyl oxalate	6.1	UN2525	III	6.1	T1	153	203	241	60 L	220 L	A	
	Ethyl perchlorate	Forbidden											
D	Ethyl phosphonothioic dichloride, anhydrous.	6.1	NA2927	I	6.1, 8	2, B9, B14, B32, B74, T38, T43, T45	None	227	244	Forbidden	Forbidden	D	20, 40, 95
D	Ethyl phosphonous dichloride, anhydrous pyrophoric liquid.	6.1	NA2845	I	6.1, 4.2	2, B9, B14, B32, B74, T38, T43, T45	None	227	244	Forbidden	Forbidden	D	18
D	Ethyl phosphorodichloridate	6.1	NA2927	I	6.1, 8	2, B9, B14, B32, B74, T38, T43, T45	None	227	244	Forbidden	Forbidden	D	20, 40, 95
	Ethyl propionate	3	UN1195	II	3	T1	150	202	242	5 L	60 L	B	
	Ethyl propyl ether	3	UN2615	II	3	B101, T8	150	202	242	5 L	60 L	E	
	Ethyl silicate, see Tetraethyl silicate.												
	Ethylacetylene, inhibited	2.1	UN2452		2.1		None	304	314, 315.	Forbidden	150 kg	B	40
	Ethylamine	2.1	UN1036		2.1	B77	None	321	314, 315.	Forbidden	150 kg	D	40
	Ethylamine, aqueous solution with not less than 50 percent but not more than 70 percent ethylamine.	3	UN2270	II	3, 8	T14	None	202	243	1 L	5 L	B	40
	N-Ethylaniline	6.1	UN2272	III	6.1	T2	153	203	241	60 L	220 L	A	
	2-Ethylaniline	6.1	UN2273	III	6.1	T2	153	203	241	60 L	220 L	A	
	Ethylbenzene	3	UN1175	II	3	T1	150	202	242	5 L	60 L	B	
	N-Ethylbenzyltoluidines liquid	6.1	UN2753	III	6.1	T14	153	203	241	60 L	220 L	A	
	N-Ethylbenzyltoluidines solid	6.1	UN2753	III	6.1		153	213	240	100 kg	200 kg	A	
	2-Ethylbutanol	3	UN2275	III	3	B1, T1	150	203	242	60 L	220 L	A	
	Ethylbutyl acetate	3	UN1177	III	3	B1, T1	150	203	242	60 L	220 L	A	
	2-Ethylbutyraldehyde	3	UN1178	II	3	B1, T1	150	202	242	5 L	60 L	B	
	Ethyldichloroarsine	6.1	UN1892	I	6.1	2, B9, B14, B32, B74, T38, T43, T45	None	227	244	Forbidden	Forbidden	D	40
	Ethyldichlorosilane	4.3	UN1183	I	4.3, 8, 3.	A2, A3, A7, N34, T18, T26	None	201	244	Forbidden	1 L	D	21, 28, 40, 49, 100
	Ethylene, acetylene and propylene in mixture, refrigerated liquid with at least 71.5 percent ethylene with not more than 22.5 percent acetylene and not more than 6 percent propylene.	2.1	UN3138		2.1		None	304	314, 315.	Forbidden	Forbidden	D	40

§172.101 HAZARDOUS MATERIALS TABLE—Continued

Symbols	Hazardous materials descriptions and proper shipping names	Hazard class or Division	Identification Numbers	PG	Label Codes	Special provisions	(8) Packaging (§173.***)			(9) Quantity limitations		(10) Vessel stowage	
							Exceptions	Non-bulk	Bulk	Passenger aircraft/rail	Cargo aircraft only	Location	Other
(1)	(2)	(3)	(4)	(5)	(6)	(7)	(8A)	(8B)	(8C)	(9A)	(9B)	(10A)	(10B)
	Ethylene chlorohydrin	6.1	UN1135	I	6.1, 3	2, B9, B14, B32, B74, T38, T43, T45	None	227	244	Forbidden	Forbidden	D	40
	Ethylene, compressed	2.1	UN1962		2.1		306	304	302	Forbidden	150 kg	E	40
	Ethylene diamine diperchlorate	Forbidden											
	Ethylene dibromide	6.1	UN1605	I	6.1	2, B9, B14, B32, B74, B77, T38, T43, T45	None	227	244	Forbidden	Forbidden	D	40
	Ethylene dibromide and methyl bromide liquid mixtures, see Methyl bromide and ethylene dibromide, liquid mixtures.												
	Ethylene dichloride	3	UN1184	II	3, 6.1	T14	None	202	243	1 L	60 L	B	40
	Ethylene glycol diethyl ether	3	UN1153	III	3	B1, T1	150	203	242	60 L	220 L	A	
	Ethylene glycol dinitrate	Forbidden											
	Ethylene glycol monoethyl ether	3	UN1171	III	3	B1, T1	150	203	242	60 L	220 L	A	
	Ethylene glycol monoethyl ether acetate.	3	UN1172	III	3	B1, T1	150	203	242	60 L	220 L	A	
	Ethylene glycol monomethyl ether.	3	UN1188	III	3	B1, T1	150	203	242	60 L	220 L	A	
	Ethylene glycol monomethyl ether acetate.	3	UN1189	III	3	B1, T1	150	203	242	60 L	220 L	A	
	Ethylene oxide and carbon dioxide mixture *with more than 87 percent ethylene oxide.*	2.3	UN3300		2.3, 2.1	4	None	304	314, 315.	Forbidden	Forbidden	D	40
	Ethylene oxide and carbon dioxide mixtures *with more than 9 percent but not more than 87 percent ethylene oxide.*	2.1	UN1041		2.1		306	304	314, 315.	Forbidden	25 kg	B	40
	Ethylene oxide and carbon dioxide mixtures *with not more than 9 percent ethylene oxide.*	2.2	UN1952		2.2		306	304	314, 315.	75 kg	150 kg	A	
	Ethylene oxide and chlorotetrafluoroethane mixture *with not more than 8.8 percent ethylene oxide.*	2.2	UN3297		2.2		306	304	314, 315.	75 kg	150 kg	A	
	Ethylene oxide and dichlorodifluoromethane mixture, *with not more than 12.5 percent ethylene oxide.*	2.2	UN3070		2.2		306	304	314, 315.	75 kg	150 kg	A	
	Ethylene oxide and pentafluoroethane mixture *with not more than 7.9 percent ethylene oxide.*	2.2	UN3298		2.2		306	304	314, 315.	75 kg	150 kg	A	
	Ethylene oxide and propylene oxide mixtures, *with not more than 30 percent ethylene oxide.*	3	UN2983	I	3, 6.1	5, A11, N4, N34, T24, T29	None	201	243	Forbidden	30 L	E	40
	Ethylene oxide and tetrafluoroethane mixture *with not more than 5.6 percent ethylene oxide.*	2.2	UN3299		2.2		306	304	314, 315.	75 kg	150 kg	A	
	Ethylene oxide or Ethylene oxide with nitrogen *up to a total pressure of 1MPa (10 bar) at 50 degrees C.*	2.3	UN1040		2.3, 2.1	4	None	323	323	Forbidden	25 kg	D	40
	Ethylene, refrigerated liquid (cryogenic liquid).	2.1	UN1038		2.1		None	316	318, 319.	Forbidden	Forbidden	D	40
	Ethylenediamine	8	UN1604	II	8, 3	T14	154	202	243	1 L	30 L	A	40
	Ethyleneimine, inhibited	6.1	UN1185	I	6.1, 3	1, B9, B14, B30, B72, B77, N25, N32, T38, T43, T44	None	226	244	Forbidden	Forbidden	D	40
	Ethylhexaldehyde, see Octyl aldehydes etc.												
	2-Ethylhexyl chloroformate	6.1	UN2748	II	6.1, 8	T12	None	202	243	1 L	30 L	A	12, 13, 21, 25, 40, 100
	2-Ethylhexylamine	3	UN2276	III	3, 8	B1, T2	150	203	242	5 L	60 L	A	40
	Ethylphenyldichlorosilane	8	UN2435	II	8	A7, B2, N34, T8, T26	None	202	242	Forbidden	30 L	C	
	1-Ethylpiperidine	3	UN2386	II	3, 8	T8	None	202	243	1 L	5 L	B	
	N-Ethyltoluidines	6.1	UN2754	II	6.1	T14	None	202	243	5 L	60 L	A	
	Ethyltrichlorosilane	3	UN1196	II	3, 8	A7, B100, N34, T15, T26	None	202	243	1 L	5 L	B	40
	Etiologic agent, see Infectious substances, etc).												
	Explosive articles, see Articles, explosive, n.o.s. etc.												
	Explosive, blasting, type A	1.1D	UN0081	II	1.1D		None	62	None	Forbidden	Forbidden	B	1E, 5E, 21E
	Explosive, blasting, type B	1.1D	UN0082	II	1.1D		None	62	None	Forbidden	Forbidden	B	1E, 5E
	Explosive, blasting, type B *or* Agent blasting, Type B.	1.5D	UN0331	II	1.5D	105, 106	None	62	None	Forbidden	Forbidden	B	1E, 5E
	Explosive, blasting, type C	1.1D	UN0083	II	1.1D	123	None	62	None	Forbidden	Forbidden	B	1E, 5E
	Explosive, blasting, type D	1.1D	UN0084	II	1.1D		None	62	None	Forbidden	Forbidden	B	1E, 5E

Appendix A

§ 172.101 HAZARDOUS MATERIALS TABLE—Continued

Symbols	Hazardous materials descriptions and proper shipping names	Hazard class or Division	Identification Numbers	PG	Label Codes	Special provisions	Packaging (§173.***)			Quantity limitations		Vessel stowage	
							Exceptions	Non-bulk	Bulk	Passenger aircraft/rail	Cargo aircraft only	Location	Other
(1)	(2)	(3)	(4)	(5)	(6)	(7)	(8A)	(8B)	(8C)	(9A)	(9B)	(10A)	(10B)
	Explosive, blasting, type E	1.1D	UN0241	II	1.1D		None	62	None	Forbidden	Forbidden	B	1E, 5E, 19E
	Explosive, blasting, type E or Agent blasting, Type E.	1.5D	UN0332	II	1.5D	105, 106	None	62	None	Forbidden	Forbidden	B	1E, 5E
	Explosive, forbidden. See Sec. 173.54.	Forbidden											
	Explosive substances, see Substances, explosive, n.o.s. etc.												
	Explosives, slurry, see Explosive, blasting, type E.												
	Explosives, water gels, see Explosive, blasting, type E.												
	Extracts, aromatic, liquid	3	UN1169	II	3	T7, T30	150	202	242	5 L	60 L	B	
				III	3	B1, T7, T30	150	203	242	60 L	220 L	A	
	Extracts, flavoring, liquid	3	UN1197	II	3	T7, T30	150	202	242	5 L	60 L	B	
				III	3	B1, T7, T30	150	203	242	60 L	220 L	A	
	Fabric with animal or vegetable oil, see Fibers or fabrics, etc.												
	Ferric arsenate	6.1	UN1606	II	6.1		None	212	242	25 kg	100 kg	A	
	Ferric arsenite	6.1	UN1607	II	6.1		None	212	242	25 kg	100kg	A	
	Ferric chloride, anhydrous	8	UN1773	III	8		154	213	240	25 kg	100 kg	A	
	Ferric chloride, solution	8	UN2582	III	8	B15, T8	154	203	241	5 L	60 L	A	
	Ferric nitrate	5.1	UN1466	III	5.1	A1, A29	152	213	240	25 kg	100 kg	A	
	Ferrocerium	4.1	UN1323	II	4.1	59, A19	151	212	240	15 kg	50 kg	A	
	Ferrosilicon, with 30 percent or more but less than 90 percent silicon.	4.3	UN1408	III	4.3, 6.1	A1, A19	151	213	240	25 kg	100 kg	A	13, 40, 85, 103
	Ferrous arsenate	6.1	UN1608	II	6.1		None	212	242	25 kg	100 kg	A	
D	Ferrous chloride, solid	8	NA1759	II	8		154	212	240	15 kg	50 kg	A	
D	Ferrous chloride, solution	8	NA1760	II	8	B3	154	202	242	1 L	30 L	B	40
	Ferrous metal borings or Ferrous metal shavings or Ferrous metal turnings or Ferrous metal cuttings in a form liable to self-heating.	4.2	UN2793	III	4.2	A1, A19, B101	None	213	241	25 kg	100 kg	A	
	Fertilizer ammoniating solution with free ammonia.	2.2	UN1043		2.2		306	304	314, 315.	Forbidden	150 kg	E	40
A, W	Fibers or Fabrics, animal or vegetable or Synthetic, n.o.s. with animal or vegetable oil.	4.2	UN1373	III	4.2	137	None	213	241	Forbidden	Forbidden	A	
	Fibers or Fabrics impregnated with weakly nitrated nitrocellulose, n.o.s.	4.1	UN1353	III	4.1	A1	None	213	240	25 kg	100 kg	D	
	Films, nitrocellulose base, from which gelatine has been removed; film scrap, see Celluloid scrap.												
	Films, nitrocellulose base, gelatine coated (except scrap).	4.1	UN1324	III	4.1		None	183	None	25 kg	100 kg	D	91
	Fire extinguisher charges, corrosive liquid.	8	UN1774	II	8	N41	154	202	None	1 L	30 L	A	
	Fire extinguisher charges, expelling, explosive, see Cartridges, power device.												
	Fire extinguishers containing compressed or liquefied gas.	2.2	UN1044		2.2	18	309	309	None	75 kg	150 kg	A	
	Firelighters, solid with flammable liquid.	4.1	UN2623	III	4.1	A1, A19	None	213	None	25 kg	100 kg	A	
	Fireworks	1.1G	UN0333	II	1.1G	108	None	62	None	Forbidden	Forbidden	B	
	Fireworks	1.2G	UN0334	II	1.2G	108	None	62	None	Forbidden	Forbidden	B	
	Fireworks	1.3G	UN0335	II	1.3G	108	None	62	None	Forbidden	Forbidden	B	
	Fireworks	1.4G	UN0336	II	1.4G	108	None	62	None	Forbidden	75 kg	A	24E
	Fireworks	1.4S	UN0337	II	1.4S	108	None	62	None	25 kg	100 kg	A	
W	Fish meal, stabilized or Fish scrap, stabilized.	9	UN2216	III	None		155	218	218	No limit	No limit	A	88
	Fish meal, unstablized or Fish scrap, unstabilized.	4.2	UN1374	II	4.2	A1, A19	None	212	241	15 kg	50 kg	A	119, 120
	Fissile radioactive materials, see Radioactive material, fissile, n.o.s.												
	Flammable compressed gas, see Compressed or Liquefied gas, flammable, etc.												
	Flammable compressed gas (small receptacles not fitted with a dispersion device, not refillable), see Receptacles, etc.												
	Flammable gas in lighters, see Lighters or lighter refills, cigarettes, containing flammable gas.												
G	Flammable liquid, toxic, corrosive, n.o.s.	3	UN3286	I	3, 6.1, 8.		None	201	243	Forbidden	2.5 L	E	21, 40, 100
				II	3, 6.1, 8.	T14	None	202	243	1 L	5 L	B	21, 40, 100

§172.101 Hazardous Materials Table—Continued

Symbols	Hazardous materials descriptions and proper shipping names	Hazard class or Division	Identification Numbers	PG	Label Codes	Special provisions	Packaging (§173.***) Exceptions (8A)	Packaging (§173.***) Non-bulk (8B)	Packaging (§173.***) Bulk (8C)	Quantity limitations Passenger aircraft/rail (9A)	Quantity limitations Cargo aircraft only (9B)	Vessel stowage Location (10A)	Vessel stowage Other (10B)
(1)	(2)	(3)	(4)	(5)	(6)	(7)	(8A)	(8B)	(8C)	(9A)	(9B)	(10A)	(10B)
G	Flammable liquids, corrosive, n.o.s.	3	UN2924	I	3, 8	T42	None	201	243	0.5 L	2.5 L	E	40
				II	3, 8	T15, T26	None	202	243	1 L	5 L	B	40
				III	3, 8	B1, T15, T26	150	203	242	5 L	60 L	A	40
G	Flammable liquids, n.o.s	3	UN1993	I	3	T42	None	201	243	1 L	30 L	E	
				II	3	T8, T31	150	202	242	5 L	60 L	B	
				III	3	B1, B52, T7, T30	150	203	242	60 L	220 L	A	
G	Flammable liquids, toxic, n.o.s	3	UN1992	I	3, 6.1	T42	None	201	243	Forbidden	30L	E	40
				II	3, 6.1	T18	None	202	243	1 L	60 L	B	40
				III	3, 6.1	B1, T18	150	203	242	60 L	220 L	A	
G	Flammable solid, corrosive, inorganic, n.o.s.	4.1	UN3180	II	4.1, 8	A1, B106	151	212	242	15 kg	50 kg	D	40
				III	4.1, 8	A1, B106	151	213	242	25 kg	100 kg	D	40
G	Flammable solid, inorganic, n.o.s.	4.1	UN3178	II	4.1	A1	151	212	240	15 kg	50 kg	B	
				III	4.1	A1	151	213	240	25 kg	100 kg	B	
G	Flammable solid, organic, molten, n.o.s.	4.1	UN3176	II	4.1	T9	151	212	240	Forbidden	Forbidden	C	
				III	4.1	T9	151	213	240	Forbidden	Forbidden	C	
G	Flammable solid, oxidizing, n.o.s	4.1	UN3097	II	4.1, 5.1	131	None	214	214	Forbidden	Forbidden	E	40
G		4.1		III	4.1, 5.1	131	None	214	214	Forbidden	Forbidden	D	40
G	Flammable solid, toxic, inorganic, n.o.s.	4.1	UN3179	II	4.1, 6.1	A1, B106	151	212	242	15 kg	50 kg	B	40
				III	4.1, 6.1	A1, B106	151	213	242	25 kg	100 kg	B	40
G	Flammable solids, corrosive, organic, n.o.s.	4.1	UN2925	II	4.1, 8	A1, B106	None	212	242	15 kg	50 kg	D	40
				III	4.1, 8	A1, B106	151	213	242	25 kg	100 kg	D	40
G	Flammable solids, organic, n.o.s	4.1	UN1325	II	4.1	A1	151	212	240	15 kg	50 kg	B	
				III	4.1	A1	151	213	240	25 kg	100 kg	B	
G	Flammable solids, toxic, organic, n.o.s.	4.1	UN2926	II	4.1, 6.1	A1, B106	None	212	242	15 kg	50 kg	B	40
				III	4.1, 6.1	A1, B106	151	213	242	25 kg	100 kg	B	40
	Flares, aerial	1.3G	UN0093	II	1.3G		None	62	None	Forbidden	75 kg	B	
	Flares, aerial	1.4G	UN0403	II	1.4G		None	62	None	Forbidden	75 kg	A	
	Flares, aerial	1.4S	UN0404	II	1.4S		None	62	None	25 kg	100 kg	A	24E
	Flares, aerial	1.1G	UN0420	II	1.1G		None	62	None	Forbidden	Forbidden	B	
	Flares, aerial	1.2G	UN0421	II	1.2G		None	62	None	Forbidden	Forbidden	B	
	Flares, airplane, see Flares, aerial.												
	Flares, signal, see Cartridges, signal.												
	Flares, surface	1.3G	UN0092	II	1.3G		None	62	None	Forbidden	75 kg	B	
	Flares, surface	1.1G	UN0418	II	1.1G		None	62	None	Forbidden	Forbidden	B	
	Flares, surface	1.2G	UN0419	II	1.2G		None	62	None	Forbidden	Forbidden	B	
	Flares, water-activated, see Contrivances, water-activated, etc.												
	Flash powder	1.1G	UN0094	II	1.1G		None	62	None	Forbidden	Forbidden	E	1E, 5E
	Flash powder	1.3G	UN0305	II	1.3G		None	62	None	Forbidden	Forbidden	E	1E, 5E
	Flue dusts, poisonous, see Arsenical dust.												
	Fluoric acid, see Hydrofluoric acid, solution, etc.												
	Fluorine, compressed	2.3	UN1045		2.3, 5.1, 8.	1	None	302	None	Forbidden	Forbidden	D	40, 89, 90
	Fluoroacetic acid	6.1	UN2642	I	6.1	B100	None	211	242	1 kg	15kg	E	
	Fluoroanilines	6.1	UN2941	III	6.1	T8	153	203	241	60 L	220 L	A	
	Fluorobenzene	3	UN2387	II	3	B101, T8	150	202	242	5 L	60 L	B	
	Fluoroboric acid	8	UN1775	II	8	A6, A7, B2, B15, N3, N34, T15, T27	154	202	242	1 L	30 L	A	
	Fluorophosphoric acid anhydrous.	8	UN1776	II	8	A6, A7, B2, N3, N34, T9, T27	None	202	242	1 L	30 L	A	
	Fluorosilicates, n.o.s	6.1	UN2856	III	6.1		153	213	240	100 kg	200 kg	A	26
	Fluorosilicic acid	8	UN1778	II	8	A6, A7, B2, B15, N3, N34, T12, T27	None	202	242	1 L	30 L	A	
	Fluorosulfonic acid	8	UN1777	I	8	A3, A6, A7, A10, B6, B10, N3, T9, T27	None	201	243	0.5 L	2.5 L	D	40
	Fluorotoluenes	3	UN2388	II	3	T8	150	202	242	5 L	60 L	B	40
	Forbidden materials. See 173.21		Forbidden										
	Formaldehyde, solutions, flammable.	3	UN1198	III	3, 8	B1, T8	150	203	242	5 L	60 L	A	40
	Formaldehyde, solutions, with not less than 25 percent formaldehyde.	8	UN2209	III	8	T1	154	203	241	5 L	60 L	A	
	Formalin, see Formaldehyde, solutions.												
	Formic acid	8	UN1779	II	8	B2, B28, T8	154	202	242	1 L	30 L	A	40
	Fracturing devices, explosive, without detonators for oil wells.	1.1D	UN0099	II	1.1D		None	62	None	Forbidden	Forbidden	B	
	Fuel, aviation, turbine engine	3	UN1863	I	3	T7	150	201	243	1 L	30 L	E	
				II	3	T1	150	202	242	5 L	60 L	B	
				III	3	B1, T1	150	203	242	60 L	220 L	A	
D	Fuel oil (No. 1, 2, 4, 5, or 6)	3	NA1993	III	3	B1	150	203	242	60 L	220 L	A	
	Fulminate of mercury (dry)		Forbidden										
	Fulminate of mercury, wet, see Mercury fulminate, etc.												
	Fulminating gold		Forbidden										

Appendix A

§ 172.101 Hazardous Materials Table—Continued

Symbols	Hazardous materials descriptions and proper shipping names	Hazard class or Division	Identification Numbers	PG	Label Codes	Special provisions	Packaging (§173.***)			Quantity limitations		Vessel stowage	
							Exceptions	Non-bulk	Bulk	Passenger aircraft/rail	Cargo aircraft only	Location	Other
(1)	(2)	(3)	(4)	(5)	(6)	(7)	(8A)	(8B)	(8C)	(9A)	(9B)	(10A)	(10B)
	Fulminating mercury	Forbidden											
	Fulminating platinum	Forbidden											
	Fulminating silver	Forbidden											
	Fulminic acid	Forbidden											
	Fumaryl chloride	8	UN1780	II	8	B2, T8, T26	154	202	242	1 L	30 L	C	8, 40
	Furaldehydes	6.1	UN1199	II	6.1, 3	T15	None	202	243	5 L	60 L	A	
	Furan	3	UN2389	I	3	T18	None	201	243	1 L	30 L	E	40
	Fumigated lading, see §§ 172.302(g), 173.9 and 176.76(h).												
	Furfuryl alcohol	6.1	UN2874	III	6.1	T2	153	203	241	60 L	220 L	A	26, 74
	Furfurylamine	3	UN2526	III	3, 8	B1, T1	150	203	242	5 L	60 L	A	40
	Fuse, detonating, metal clad, see Cord, detonating, metal clad.												
	Fuse, detonating, mild effect, metal clad, see Cord, detonating, mild effect, metal clad.												
	Fuse, igniter *tubular metal clad*	1.4G	UN0103	II	1.4G		None	62	None	Forbidden	75 kg	A	24E
	Fuse, non-detonating *instantaneous or quickmatch.*	1.3G	UN0101	II	1.3G		None	62	None	Forbidden	Forbidden	B	
	Fuse, safety	1.4S	UN0105	II	1.4S		None	62	None	25 kg	100 kg	A	
D	Fusee (*railway or highway*)	4.1	NA1325	II	4.1		None	184	None	15 kg	50 kg	B	
	Fusel oil	3	UN1201	II	3	T1	150	202	242	5 L	60 L	B	
				III	3	B1, T1	150	203	242	60 L	220 L	A	
	Fuses, tracer, see Tracers for ammunition.												
	Fuzes, combination, percussion and time, see Fuzes, detonating (UN 0257, UN 0367); Fuzes, igniting (UN 0317, UN 0368).												
	Fuzes, detonating	1.1B	UN0106	II	1.1B		None	62	None	Forbidden	Forbidden	B	2E, 6E
	Fuzes, detonating	1.2B	UN0107	II	1.2B		None	62	None	Forbidden	Forbidden	B	2E, 6E
	Fuzes, detonating	1.4B	UN0257	II	1.4B	116	None	62	None	Forbidden	75 kg	A	24E
	Fuzes, detonating	1.4S	UN0367	II	1.4S	116	None	62	None	25 kg	100 kg	A	
	Fuzes, detonating, *with protective features.*	1.1D	UN0408	II	1.1D		None	62	None	Forbidden	Forbidden	B	
	Fuzes, detonating, *with protective features.*	1.2D	UN0409	II	1.2D		None	62	None	Forbidden	Forbidden	B	
	Fuzes, detonating, *with protective features.*	1.4D	UN0410	II	1.4D	116	None	62	None	Forbidden	75 kg	A	24E
	Fuzes, igniting	1.3G	UN0316	II	1.3G		None	62	None	Forbidden	Forbidden	B	
	Fuzes, igniting	1.4G	UN0317	II	1.4G		None	62	None	Forbidden	75 kg	A	24E
	Fuzes, igniting	1.4S	UN0368	II	1.4S		None	62	None	25 kg	100 kg	A	
	Galactsan trinitrate	Forbidden											
	Gallium	8	UN2803	III	8		None	162	240	20 kg	20 kg	B	48
	Gas cartridges, (*flammable*) *without a release device, non-re-fillable.*	2.1	UN2037		2.1		306	304	None	1 kg	15 kg	B	40
	Gas generator assemblies (*aircraft*), *containing a non-flammable non-toxic gas and a propellant cartridge.*	2.2			2.2		None	335	None	75 kg	150 kg	A	
D	Gas identification set	2.3	NA9035		2.3	6	None	194	None	Forbidden	Forbidden	D	
	Gas oil *or Diesel fuel or Heating oil, light.*	3	UN1202	III	3	B1, T7, T30	150	203	242	60 L	220 L	A	
G	Gas, refrigerated liquid, flammable, n.o.s. (*cryogenic liquid*).	2.1	UN3312		2.1		None	316	318	Forbidden	Forbidden	D	40
G	Gas, refrigerated liquid, n.o.s. (*cryogenic liquid*).	2.2	UN3158		2.2		320	316	318	50 kg	500 kg	D	
G	Gas, refrigerated liquid, oxidizing, n.o.s. (*cryogenic liquid*).	2.2	UN3311		2.2, 5.1		320	316	318	Forbidden	Forbidden	D	
	Gas sample, non-pressurized, flammable, n.o.s., *not refrigerated liquid.*	2.1	UN3167		2.1		306	302, 304.	None	1 L	5 L	D	
	Gas sample, non-pressurized, toxic, flammable, n.o.s., *not refrigerated liquid.*	2.3	UN3168		2.3, 2.1		306	302	None	Forbidden	1 L	D	
	Gas sample, non-pressurized, toxic, n.o.s., *not refrigerated liquid.*	2.3	UN3169		2.3		306	302, 304.	None	Forbidden	1 L	D	
D	Gasohol *gasoline mixed with ethyl alcohol, with not more than 20 percent alcohol.*	3	NA1203	II	3		150	202	242	5 L	60 L	E	
	Gasoline	3	UN1203	II	3	B33, B101, T8	150	202	242	5 L	60 L	E	
	Gasoline, casinghead, see Gasoline.												
	Gelatine, blasting, see Explosive, blasting, type A.												
	Gelatine dynamites, see Explosive, blasting, type A.												
	Germane	2.3	UN2192		2.3, 2.1	2	None	192	245	Forbidden	Forbidden	D	40
	Glycerol-1,3-dinitrate	Forbidden											
	Glycerol gluconate trinitrate	Forbidden											
	Glycerol lactate trinitrate	Forbidden											
	Glycerol alpha-monochlorohydrin	6.1	UN2689	III	6.1	T2	153	203	241	60 L	220 L	A	

§172.101 HAZARDOUS MATERIALS TABLE—Continued

Symbols	Hazardous materials descriptions and proper shipping names	Hazard class or Division	Identification Numbers	PG	Label Codes	Special provisions	Packaging (§173.***)			Quantity limitations		Vessel stowage	
							Exceptions	Non-bulk	Bulk	Passenger aircraft/rail	Cargo aircraft only	Location	Other
(1)	(2)	(3)	(4)	(5)	(6)	(7)	(8A)	(8B)	(8C)	(9A)	(9B)	(10A)	(10B)
	Glyceryl trinitrate, see Nitroglycerin, etc.												
	Glycidaldehyde	3	UN2622	II	3, 6.1	T8	150	202	243	1 L	60 L	A	40
D	Grenades, empty primed	1.4S	NA0349	II	None		None	62	None	25 kg	100 kg	A	
	Grenades, *hand or rifle, with bursting charge.*	1.1D	UN0284	II	1.1D			62	None	Forbidden	Forbidden	B	
	Grenades, *hand or rifle, with bursting charge.*	1.2D	UN0285	II	1.2D			62	None	Forbidden	Forbidden	B	
	Grenades, *hand or rifle, with bursting charge.*	1.1F	UN0292	II	1.1F			62	None	Forbidden	Forbidden	E	
	Grenades, *hand or rifle, with bursting charge.*	1.2F	UN0293	II	1.2F			62	None	Forbidden	Forbidden	E	
	Grenades, illuminating, see Ammunition, illuminating, etc.												
	Grenades, practice, hand or rifle	1.4S	UN0110	II	1.4S			62	None	25 kg	100 kg	A	
	Grenades, practice, hand or rifle	1.3G	UN0318	II	1.3G			62	None	Forbidden	Forbidden	B	
	Grenades, practice, hand or rifle	1.2G	UN0372	II	1.2G			62	None	Forbidden	Forbidden	B	
	Grenades practice *Hand or rifle*	1.4G	UN0452	II	1.4G			62	None	Forbidden	75 kg	A	24E
	Grenades, smoke, see Ammunition, smoke, etc.												
	Guanidine nitrate	5.1	UN1467	III	5.1	A1	152	213	240	25 kg	100 kg	A	73
	Guanyl nitrosaminoguanylidene hydrazine (dry).	Forbidden											
	Guanyl nitrosaminoguanylidene hydrazine, *wetted with not less than 30 percent water, by mass.*	1.1A	UN0113	II	1.1A	111, 117	None	62	None	Forbidden	Forbidden	E	2E, 6E
	Guanyl nitrosaminoguanyltetrazene (dry).	Forbidden											
	Guanyl nitrosaminoguanyltetrazene, *wetted or Tetrazene, wetted with not less than 30 percent water or mixture of alcohol and water, by mass.*	1.1A	UN0114	II	1.1A	111, 117	None	62	None	Forbidden	Forbidden	E	2E, 6E
	Gunpowder, compressed or Gunpowder in pellets, see Black powder (UN 0028).												
	Gunpowder, granular or as a meal, see Black powder (UN 0027).												
	Hafnium powder, dry	4.2	UN2545	I	4.2	B100	None	211	242	Forbidden	Forbidden	D	
				II	4.2	A19, A20, B101, B106, N34	None	212	241	15 kg	50 kg	D	
				III	4.2	B100	None	213	241	25 kg	100 kg	D	
	Hafnium powder, wetted *with not less than 25 percent water (a visible excess of water must be present) (a) mechanically produced, particle size less than 53 microns; (b) chemically produced, particle size less than 840 microns.*	4.1	UN1326	II	4.1	A6, A19, A20, N34	None	212	241	15 kg	50 kg	E	
	Hand signal device, see Signal devices, hand.												
	Hazardous substances, liquid or solid, n.o.s., see Environmentally hazardous substances, etc.												
GD	Hazardous waste, liquid, n.o.s	9	NA3082	III	9		155	203	241	No limit	No limit	A	
GD	Hazardous waste, solid, n.o.s	9	NA3077	III	9	B54	155	213	240	No limit	No limit	A	
	Helium, compressed	2.2	UN1046		2.2		306	302	302, 314.	75 kg	150 kg	A	85
	Helium-oxygen mixture, see Rare gases and oxygen mixtures.												
	Helium, refrigerated liquid *(cryogenic liquid).*	2.2	UN1963		2.2		320	316	318	50 kg	500 kg	B	
	Heptafluoropropane *or Refrigerant gas R 227.*	2.2	UN3296		2.2		306	304	314, 315.	75 kg	150kg	A	
	n-Heptaldehyde	3	UN3056	III	3	B1, T1	150	203	242	60 L	220 L	A	
	Heptanes	3	UN1206	II	3	T2	150	202	242	5 L	60 L	B	
	n-Heptene	3	UN2278	II	3	B101, T8	150	202	242	5 L	60 L	B	
	Hexachloroacetone	6.1	UN2661	III	6.1	T8	153	203	241	60 L	220 L	B	12, 40
	Hexachlorobenzene	6.1	UN2729	III	6.1		153	203	241	60 L	220 L	A	
	Hexachlorobutadiene	6.1	UN2279	III	6.1	T7	153	203	241	60 L	220 L	A	
	Hexachlorocyclopentadiene	6.1	UN2646	I	6.1	2, B9, B14, B32, B74, B77, T38, T43, T45	None	227	244	Forbidden	Forbidden	D	40
	Hexachlorophene	6.1	UN2875	III	6.1		153	213	240	100 kg	200 kg	A	
	Hexadecyltrichlorosilane	8	UN1781	II	8	A7, B2, B6, N34, T8	None	202	242	Forbidden	30 L	C	40
	Hexadienes	3	UN2458	II	3	B101, T7	None	202	242	5 L	60 L	B	
	Hexaethyl tetraphosphate and compressed gas mixtures.	2.3	UN1612		2.3	3	None	334	None	Forbidden	Forbidden	D	40
	Hexaethyl tetraphosphate *liquid*	6.1	UN1611	II	6.1	N76	None	202	243	5 L	60 L	E	40
	Hexaethyl tetraphosphate, *solid*	6.1	UN1611	II	6.1	N76	None	212	242	25 kg	100 kg	E	40

Appendix A

§172.101 HAZARDOUS MATERIALS TABLE—Continued

Symbols	Hazardous materials descriptions and proper shipping names	Hazard class or Division	Identification Numbers	PG	Label Codes	Special provisions	Packaging (§173.***)			Quantity limitations		Vessel stowage	
							Exceptions	Non-bulk	Bulk	Passenger aircraft/rail	Cargo aircraft only	Location	Other
(1)	(2)	(3)	(4)	(5)	(6)	(7)	(8A)	(8B)	(8C)	(9A)	(9B)	(10A)	(10B)
	Hexafluoroacetone	2.3	UN2420		2.3, 8	2, B9, B14	None	304	314, 315.	Forbidden	Forbidden	D	40
	Hexafluoroacetone hydrate	6.1	UN2552	II	6.1	T14	None	202	243	5 L	60 L	B	40
	Hexafluoroethane, compressed or Refrigerant gas R 116.	2.2	UN2193		2.2		306	304	314, 315.	75 kg	150 kg	A	
	Hexafluorophosphoric acid	8	UN1782	II	8	A6, A7, B2, N3, N34, T9, T27	None	202	242	1 L	30 L	A	
	Hexafluoropropylene compressed or Refrigerant gas R 1216.	2.2	UN1858		2.2		306	304	314, 315.	75 kg	150 kg	A	
	Hexaldehyde	3	UN1207	III	3	B1, T1	150	203	242	60 L	220 L	A	
	Hexamethylene diisocyanate	6.1	UN2281	II	6.1	B101, T14	None	202	243	5 L	60 L	C	13, 40
	Hexamethylene triperoxide diamine (dry).	Forbidden											
	Hexamethylenediamine, solid	8	UN2280	III	8		154	213	240	25 kg	100 kg	A	12
	Hexamethylenediamine solution	8	UN1783	II	8	T8	None	202	242	1 L	30 L	A	
				III	8	T7	154	203	241	5 L	60 L	A	
	Hexamethyleneimine	3	UN2493	II	3, 8	B101, T8	None	202	243	1 L	5 L	B	40
	Hexamethylenetetramine	4.1	UN1328	III	4.1	A1	151	213	240	25 kg	100 kg	A	
	Hexamethylol benzene hexanitrate.	Forbidden											
	Hexanes	3	UN1208	II	3	B101, T8	150	202	242	5 L	60 L	E	
	2,2',4,4',6,6'- Hexanitro-3,3'-dihydroxyazobenzene (dry).	Forbidden											
	Hexanitroazoxy benzene	Forbidden											
	N,N'-(hexanitrodiphenyl) ethylene dinitramine (dry).	Forbidden											
	Hexanitrodiphenyl urea	Forbidden											
	2,2',3,4,4',6-Hexanitrodiphenylamine.	Forbidden											
	Hexanitrodiphenylamine or Dipicrylamine or Hexyl.	1.1D	UN0079	II	1.1D		None	62	None	Forbidden	Forbidden	B	1E, 5E
	2,3',4,4',6,6'-Hexanitrodiphenylether.	Forbidden											
	Hexanitroethane	Forbidden											
	Hexanitrooxanilide	Forbidden											
	Hexanitrostilbene	1.1D	UN0392	II	1.1D		None	62	None	Forbidden	Forbidden	B	1E, 5E
	Hexanoic acid, see Corrosive liquids, n.o.s.												
	Hexanols	3	UN2282	III	3	B1, T1	150	203	242	60 L	220 L	A	
	1-Hexene	3	UN2370	II	3	B101, T8	150	202	242	5 L	60 L	E	
	Hexogen and cyclotetramethylenetetranitramine mixtures, wetted or desensitized see RDX and HMX mixtures, wetted or desensitized etc.												
	Hexogen and HMX mixtures, wetted or desensitized see RDX and HMX mixtures, wetted or desensitized etc.												
	Hexogen and octogen mixtures, wetted or desensitized see RDX and HMX mixtures, wetted or desensitized etc.												
	Hexogen, see Cyclotrimethylenetrinitramine, etc.												
	Hexolite, or Hexotol dry or wetted with less than 15 percent water, by mass.	1.1D	UN0118	II	1.1D		None	62	None	Forbidden	Forbidden	B	1E, 5E
	Hexotonal	1.1D	UN0393	II	1.1D		None	62	None	Forbidden	Forbidden	B	1E, 5E
	Hexyl, see Hexanitrodiphenylamine.												
	Hexyltrichlorosilane	8	UN1784	II	8	A7, B2, B6, N34, T8, T26	None	202	242	Forbidden	30 L	C	40
	High explosives, see individual explosives' entries.												
	HMX, see Cyclotetramethylenetetranitramine, etc.												
	Hydrazine, anhydrous or Hydrazine aqueous solutions with more than 64 percent hydrazine, by mass.	8	UN2029	I	8, 3, 6.1.	A3, A6, A7, A10, B7, B16, B53, T25	None	201	243	Forbidden	2.5 L	D	21, 40, 42, 100
	Hydrazine, aqueous solution with not more than 37 percent hydrazine, by mass.	6.1	UN3293	III	6.1	T7	153	203	241	60 L	220 L	A	
	Hydrazine azide	Forbidden											
	Hydrazine chlorate	Forbidden											
	Hydrazine dicarbonic acid diazide.	Forbidden											
	Hydrazine hydrate or Hydrazine aqueous solutions, with not less than 37 percent but not more than 64 percent hydrazine, by mass.	8	UN2030	II	8, 6.1	B16, B53, B110, T15	None	202	243	Forbidden	30 L	D	40, 42, 82
	Hydrazine perchlorate	Forbidden											
	Hydrazine selenate	Forbidden											

§172.101 HAZARDOUS MATERIALS TABLE—Continued

Symbols	Hazardous materials descriptions and proper shipping names	Hazard class or Division	Identification Numbers	PG	Label Codes	Special provisions	Packaging (§173.***)			Quantity limitations		Vessel stowage	
							Exceptions	Non-bulk	Bulk	Passenger aircraft/rail	Cargo aircraft only	Location	Other
(1)	(2)	(3)	(4)	(5)	(6)	(7)	(8A)	(8B)	(8C)	(9A)	(9B)	(10A)	(10B)
	Hydriodic acid, anhydrous, see Hydrogen iodide, anhydrous.												
	Hydriodic acid	8	UN1787	II	8	A3, A6, B2, N41, T9, T27	154	202	242	1 L	30 L	C	
				III	8	T8, T26	154	203	241	5 L	60 L	C	8
	Hydrobromic acid, anhydrous, see Hydrogen bromide, anhydrous.												
	Hydrobromic acid, *with more than 49 percent hydrobromic acid.*	8	UN1788	II	8	B2, B15, N41, T9, T27	154	202	242	Forbidden	Forbidden	C	
				III	8	T8, T26	154	203	241	Forbidden	Forbidden	C	8
	Hydrobromic acid, *with not more than 49 percent hydrobromic acid.*	8	UN1788	II	8	A3, A6, B2, B15, N41, T9, T27	154	202	242	1 L	30 L	C	
				III	8	T8, T26	154	203	241	5 L	30 L	C	8
	Hydrocarbon gas mixture, compressed, n.o.s.	2.1	UN1964		2.1		306	302	314, 315.	Forbidden	150 kg	E	40
	Hydrocarbon gas mixture, liquefied, n.o.s.	2.1	UN1965		2.1		306	304	314, 315.	Forbidden	150 kg	E	40
	Hydrocarbons, liquid, n.o.s	3	UN3295	I	3	T8, T31	150	201	243	1 L	30 L	E	
				II	3	T8, T31	150	202	242	5 L	60 L	B	
				III	3	B1, T7, T30	150	203	242	60 L	220 L	A	
	Hydrochloric acid, anhydrous, see Hydrogen chloride, anhydrous.												
	Hydrochloric acid	8	UN1789	II	8	A3, A6, B3, B15, N41, T9, T27	154	202	242	1 L	30 L	C	
				III	8	T8, T26	154	203	241	5 L	60 L	C	8
	Hydrocyanic acid, anhydrous, see Hydrogen cyanide etc.												
	Hydrocyanic acid, aqueous solutions *or* Hydrogen cyanide, aqueous solutions *with not more than 20 percent hydrogen cyanide.*	6.1	UN1613	I	6.1	2, B61, B65, B77, B82	None	195	244	Forbidden	Forbidden	D	40
D	Hydrocyanic acid, aqueous solutions *with less than 5 percent hydrogen cyanide.*	6.1	NA1613	II	6.1	T18, T26	None	195	243	Forbidden	5 L	D	40
	Hydrocyanic acid, liquefied, see Hydrogen cyanide, etc.												
	Hydrocyanic acid (prussic), unstabilized.	Forbidden											
	Hydrofluoric acid and Sulfuric acid mixtures.	8	UN1786	I	8, 6.1	A6, A7, B15, B23, N5, N34, T18, T27	None	201	243	Forbidden	2.5 L	D	40, 95
	Hydrofluoric acid, anhydrous, see Hydrogen fluoride, anhydrous.												
	Hydrofluoric acid, *with more than 60 percent strength.*	8	UN1790	I	8, 6.1	A6, A7, B4, B15, B23, N5, N34, T18, T27	None	201	243	0.5 L	2.5 L	D	12, 40
	Hydrofluoric acid, *with not more than 60 percent strength.*	8	UN1790	II	8, 6.1	A6, A7, B15, B110, N5, N34, T18, T27	None	202	243	1 L	30 L	D	12, 40
	Hydrofluoroboric acid, see Fluoroboric acid.												
	Hydrofluorosilicic acid, see Fluorosilicic acid.												
	Hydrogen and Methane mixtures, compressed.	2.1	UN2034		2.1		306	302	302, 314, 315.	Forbidden	150 kg	E	40
	Hydrogen bromide, anhydrous	2.3	UN1048		2.3, 8	3, B14	None	304	314, 315.	Forbidden	25 kg	D	40
	Hydrogen chloride, anhydrous	2.3	UN1050		2.3, 8	3	None	304	None	Forbidden	Forbidden	D	40
	Hydrogen chloride, refrigerated liquid.	2.3	UN2186		2.3, 8	3, B6	None	None	314, 315.	Forbidden	Forbidden	B	40
	Hydrogen, compressed	2.1	UN1049		2.1		306	302	302, 314.	Forbidden	150 kg	E	40, 57
	Hydrogen cyanide, solution in alcohol *with not more than 45 percent hydrogen cyanide.*	6.1	UN3294	I	6.1, 3	2, 25, B9, B14, B32, B74, T38, T43, T45	None	227	244	Forbidden	Forbidden	D	40
	Hydrogen cyanide, stabilized *with less than 3 percent water.*	6.1	UN1051	I	6.1, 3	1, B35, B61, B65, B77, B82	None	195	244	Forbidden	Forbidden	D	40
	Hydrogen cyanide, stabilized, *with less than 3 percent water and absorbed in a porous inert material.*	6.1	UN1614	I	6.1	5	None	195	None	Forbidden	Forbidden	D	25, 40
	Hydrogen fluoride, anhydrous	8	UN1052	I	8, 6.1	3, B7, B46, B71, B77, T24, T27	None	163	243	Forbidden	Forbidden	D	40
	Hydrogen iodide, anhydrous	2.3	UN2197		2.3	3, B14	None	304	314, 315.	Forbidden	Forbidden	D	40
	Hydrogen iodide solution, see Hydriodic acid, solution.												
	Hydrogen peroxide and peroxyacetic acid mixtures, stabilized *with acids, water and not more than 5 percent peroxyacetic acid.*	5.1	UN3149	II	5.1, 8	A2, A3, A6, B53, B104, B110, T14	None	202	243	1 L	5 L	D	25, 66, 75, 106

Appendix A

§172.101 Hazardous Materials Table—Continued

Symbols	Hazardous materials descriptions and proper shipping names	Hazard class or Division	Identification Numbers	PG	Label Codes	Special provisions	(8) Packaging (§173.***)			(9) Quantity limitations		(10) Vessel stowage	
							Exceptions	Non-bulk	Bulk	Passenger aircraft/rail	Cargo aircraft only	Location	Other
(1)	(2)	(3)	(4)	(5)	(6)	(7)	(8A)	(8B)	(8C)	(9A)	(9B)	(10A)	(10B)
	Hydrogen peroxide, aqueous solutions *with more than 40 percent but not more than 60 percent hydrogen peroxide (stabilized as necessary)*.	5.1	UN2014	II	5.1, 8	12, A3, A6, B53, B80, B81, B85, B104, B110, T14, T37	None	202	243	Forbidden	Forbidden	D	25, 66, 75, 106
	Hydrogen peroxide, aqueous solutions *with not less than 8 percent but less than 20 percent hydrogen peroxide (stabilized as necessary)*.	5.1	UN2984	III	5.1	A1, B104, T8, T37	152	203	241	2.5 L	30 L	B	25, 75, 106
	Hydrogen peroxide, aqueous solutions *with not less than 20 percent but not more than 40 percent hydrogen peroxide (stabilized as necessary)*.	5.1	UN2014	II	5.1, 8	A2, A3, A6, B53, B104, B110, T14, T37	None	202	243	1 L	5 L	D	25, 66, 75, 106
	Hydrogen peroxide, stabilized or Hydrogen peroxide aqueous solutions, stabilized *with more than 60 percent hydrogen peroxide*.	5.1	UN2015	I	5.1, 8	12, A3, A6, B53, B80, B81, B85, T15, T37	None	201	243	Forbidden	Forbidden	D	25, 66, 75, 106
	Hydrogen, refrigerated liquid (cryogenic liquid).	2.1	UN1966		2.1		None	316	318, 319.	Forbidden	Forbidden	D	40
	Hydrogen selenide, anhydrous	2.3	UN2202		2.3, 2.1	1	None	192	245	Forbidden	Forbidden	D	40
	Hydrogen sulfate, see Sulfuric acid.												
	Hydrogen sulfide	2.3	UN1053		2.3, 2.1	2, B9, B14	None	304	314, 315.	Forbidden	Forbidden	D	40
	Hydrogendifluorides, n.o.s. solid	8	UN1740	II	8	N3, N34	None	212	240	15 kg	50 kg	A	25, 26, 40
				III	8	N3, N34	154	213	240	25 kg	100 kg	A	25, 26, 40
	Hydrogendifluorides, n.o.s. solutions.	8	UN1740	II	8	N3, N34	None	202	242	1 L	30 L	A	25, 26, 40
				III	8	N3, N34	154	203	241	5 L	60 L	A	25, 26, 40
	Hydroquinone	6.1	UN2662	III	6.1		153	213	240	100 kg	200 kg	A	
	Hydrosilicofluoric acid, see Fluorosilicic acid.												
	Hydroxyl amine iodide	Forbidden											
	Hydroxylamine sulfate	8	UN2865	III	8		154	213	240	25 kg	100 kg	A	
	Hypochlorite solutions	8	UN1791	II	8	A7, B2, B15, N34, T7	154	202	242	1 L	30 L	B	26
				III	8	B104, N34, T7	154	203	241	5 L	60 L	B	26
	Hypochlorites, inorganic, n.o.s	5.1	UN3212	II	5.1		152	212	240	5 kg	25 kg	D	48, 56, 58, 69, 106, 116, 118
	Hyponitrous acid	Forbidden											
	Igniter fuse, metal clad, see Fuse, igniter, tubular, metal clad.												
	Igniters	1.1G	UN0121	II	1.1G		None	62	None	Forbidden	Forbidden	B	
	Igniters	1.2G	UN0314	II	1.2G		None	62	None	Forbidden	Forbidden	B	
	Igniters	1.3G	UN0315	II	1.3G		None	62	None	Forbidden	Forbidden	A	
	Igniters	1.4G	UN0325	II	1.4G		None	62	None	Forbidden	75 kg	A	24E
	Igniters	1.4S	UN0454	II	1.4S		None	62	None	25 kg	100 kg	A	
	3,3'-Iminodipropylamine	8	UN2269	III	8	T8	154	203	241	5 L	60 L	A	
G	Infectious substances, affecting animals only.	6.2	UN2900		6.2		134	196	None	50 mL or 50 g	4 L or 4 kg	B	
G	Infectious substances, affecting humans.	6.2	UN2814		6.2		134	196	None	50 mL or 50 g	4 L or 4 kg	B	
	Inflammable, see Flammable												
	Initiating explosives (dry)	Forbidden											
	Inositol hexanitrate (dry)	Forbidden											
GD	Insecticide gases flammable n.o.s.	2.1	NA1954		2.1		306	304	314, 315.	75 kg	150 kg	D	
G	Insecticide gases, n.o.s	2.2	UN1968		2.2		306	304	314, 315.	75 kg	150 kg	A	
G	Insecticide gases, flammable, n.o.s.	2.1	UN3354		2.1		306	304	314, 315.	Forbidden	150 kg	D	40
G	Insecticide gases, toxic, flammable, n.o.s. *Inhalation hazard Zone A*.	2.3	UN3355		2.3, 2.1	1	None	192	245	Forbidden	Forbidden	D	40
G	Insecticide gases, toxic, flammable, n.o.s. *Inhalation hazard Zone B*.	2.3	UN3355		2.3, 2.1	2, B9, B14	None	302, 305.	314, 315.	Forbidden	Forbidden	D	40
G	Insecticide gases, toxic, flammable, n.o.s. *Inhalation hazard Zone C*.	2.3	UN3355		2.3, 2.1	3, B14	None	302, 305.	314, 315.	Forbidden	Forbidden	D	40
G	Insecticide gases toxic, flammable, n.o.s. *Inhalation hazard Zone D*.	2.3	UN3355		2.3, 2.1	4	None	302, 305.	314, 315.	Forbidden	Forbidden	D	
G	Insecticide gases, toxic, n.o.s	2.3	UN1967		2.3	3	None	193, 334.	245	Forbidden	Forbidden	D	40
	Inulin trinitrate (dry)	Forbidden											
	Iodine azide (dry)	Forbidden											

§172.101 Hazardous Materials Table—Continued

Symbols	Hazardous materials descriptions and proper shipping names	Hazard class or Division	Identification Numbers	PG	Label Codes	Special provisions	Packaging (§173.***) Exceptions	Packaging (§173.***) Non-bulk	Packaging (§173.***) Bulk	Quantity limitations Passenger aircraft/rail	Quantity limitations Cargo aircraft only	Vessel stowage Location	Vessel stowage Other
(1)	(2)	(3)	(4)	(5)	(6)	(7)	(8A)	(8B)	(8C)	(9A)	(9B)	(10A)	(10B)
	Iodine monochloride	8	UN1792	II	8	B6, N41, T8, T26	None	212	240	Forbidden	50 kg	D	40, 66, 74, 89, 90
	Iodine pentafluoride	5.1	UN2495	I	5.1, 6.1, 8.		None	205	243	Forbidden	2.5 L	D	25, 40, 66, 90
	2-Iodobutane	3	UN2390	II	3	T8	150	202	242	5 L	60 L	B	
	Iodomethylpropanes	3	UN2391	II	3	T8	150	202	242	5 L	60 L	B	
	Iodopropanes	3	UN2392	III	3	B1, T8	150	203	242	60 L	220 L	A	
	Iodoxy compounds (dry)	Forbidden											
	Iridium nitratopentamine iridium nitrate.	Forbidden											
	Iron chloride, see Ferric chloride												
	Iron oxide, spent, or Iron sponge, spent obtained from coal gas purification.	4.2	UN1376	III	4.2	B18	None	213	240	Forbidden	Forbidden	E	
	Iron pentacarbonyl	6.1	UN1994	I	6.1, 3	1, B9, B14, B30, B72, B77, T38, T43, T44	None	192	244	Forbidden	Forbidden	D	40
	Iron sesquichloride, see Ferric chloride.												
	Irritating material, see Tear gas substances, etc.												
	Isobutane see also Petroleum gases, liquefied.	2.1	UN1969		2.1	19	306	304	314, 315.	Forbidden	150 kg	E	40
	Isobutanol or Isobutyl alcohol	3	UN1212	III	3	B1, T1	150	203	242	60 L	220 L	A	
	Isobutyl acetate	3	UN1213	II	3	T1	150	202	242	5 L	60 L	B	
	Isobutyl acrylate, inhibited	3	UN2527	III	3	B1, T1	150	203	242	60 L	220 L	A	
	Isobutyl alcohol, see Isobutanol												
	Isobutyl aldehyde, see Isobutyraldehyde.												
D	Isobutyl chloroformate	6.1	NA2742	I	6.1, 3, 8.	2, B9, B14, B32, B74, T38, T43, T45	None	227	244	1 L	30 L	A	12, 13, 22, 25, 40, 48, 100
	Isobutyl formate	3	UN2393	II	3	T1	150	202	242	5 L	60 L	B	
	Isobutyl isobutyrate	3	UN2528	III	3	B1, T1	150	203	242	60 L	220 L	A	
+	Isobutyl isocyanate	3	UN2486	I	3, 6.1	1, B9, B14, B30, B72, T38, T43, T44	None	226	244	Forbidden	Forbidden	D	40
	Isobutyl methacrylate, inhibited	3	UN2283	III	3	B1, T1	150	203	242	60 L	220 L	A	
	Isobutyl propionate	3	UN2394	III	3	B1, T1	150	203	242	60 L	220 L	B	
	Isobutylamine	3	UN1214	II	3, 8	B101, T8	None	202	243	1 L	5 L	B	40
	Isobutylene see also Petroleum gases, liquefied.	2.1	UN1055		2.1	19	306	304	314, 315.	Forbidden	150 kg	E	40
	Isobutyraldehyde or Isobutyl aldehyde.	3	UN2045	II	3	T8	150	202	242	5L	60 L	E	40
	Isobutyric acid	3	UN2529	III	3, 8	B1, T1	150	203	242	5 L	60 L	A	
	Isobutyric anhydride	3	UN2530	III	3, 8	B1, T1	150	203	242	5 L	60 L	A	
	Isobutyronitrile	3	UN2284	II	3, 6.1	T17	None	202	243	1 L	60 L	E	40
	Isobutyryl chloride	3	UN2395	II	3, 8	B100, T9, T26	None	202	243	1 L	5 L	C	40
G	Isocyanates, flammable, toxic, n.o.s or Isocyanate solutions, flammable, toxic, n.o.s. flashpoint less than 23 degrees C.	3	UN2478	II	3, 6.1	5, A3, A7, T15	None	202	243	1 L	60 L	D	40
G	Isocyanates, toxic, flammable, n.o.s. or Isocyanate solutions, toxic, flammable, n.o.s., flash point not less than 23 degrees C but not more than 61 degrees C and boiling point less than 300 degrees C.	6.1	UN3080	II	6.1, 3	T15	None	202	243	5 L	60 L	B	25, 40, 48
G	Isocyanates, toxic, n.o.s. or Isocyanate solutions, toxic, n.o.s., flash point more than 61 degrees C and boiling point less than 300 degrees C.	6.1	UN2206	II	6.1	T15	None	202	243	5 L	60 L	E	25, 40, 48
G	Isocyanates, toxic, n.o.s. or Isocyanate solutions, toxic, n.o.s., flash point more than 61 degrees C and boiling point less than 300 degrees C.	6.1	UN2206	III	6.1	T8	153	203	241	60 L	220 L	E	25, 40, 48
	Isocyanatobenzotrifluorides	6.1	UN2285	II	6.1, 3	5, B101, T14	None	202	243	5 L	60 L	D	25, 40, 48
	Isoheptenes	3	UN2287	II	3	T7	150	202	242	5 L	60 L	B	
	Isohexenes	3	UN2288	II	3	T7	150	202	242	5 L	60 L	E	
	Isooctane, see Octanes												
	Isooctenes	3	UN1216	II	3	T8	150	202	242	5 L	60 L	B	
	Isopentane, see Pentane												
	Isopentanoic acid, see Corrosive liquids, n.o.s.												
	Isopentenes	3	UN2371	I	3	T20	150	201	243	1 L	30 L	E	
	Isophorone diisocyanate	6.1	UN2290	III	6.1	T7	153	203	241	60 L	220 L	B	40
	Isophoronediamine	8	UN2289	III	8	T8	154	203	241	5 L	60 L	A	
	Isoprene, inhibited	3	UN1218	I	3	T20	150	201	243	1 L	30 L	E	
	Isopropanol or Isopropyl alcohol	3	UN1219	II	3	T1	150	202	242	5 L	60 L	B	

Appendix A

§ 172.101 Hazardous Materials Table—Continued

Symbols	Hazardous materials descriptions and proper shipping names	Hazard class or Division	Identification Numbers	PG	Label Codes	Special provisions	Packaging (§173.***) Exceptions (8A)	Packaging (§173.***) Non-bulk (8B)	Packaging (§173.***) Bulk (8C)	Quantity limitations Passenger aircraft/rail (9A)	Quantity limitations Cargo aircraft only (9B)	Vessel stowage Location (10A)	Vessel stowage Other (10B)
(1)	(2)	(3)	(4)	(5)	(6)	(7)	(8A)	(8B)	(8C)	(9A)	(9B)	(10A)	(10B)
	Isopropenyl acetate	3	UN2403	II	3	T1	150	202	242	5 L	60 L	B	
	Isopropenylbenzene	3	UN2303	III	3	B1, T1	150	203	242	60 L	220 L	A	
	Isopropyl acetate	3	UN1220	II	3	T1	150	202	242	5 L	60 L	B	
	Isopropyl acid phosphate	8	UN1793	III	8	T7	154	213	240	25 kg	100 kg	A	
	Isopropyl alcohol, see Isopropanol.												
	Isopropyl butyrate	3	UN2405	III	3	B1, T1	150	203	242	60 L	220 L	A	
	Isopropyl chloroacetate	3	UN2947	III	3	B1, T1	150	203	242	60 L	220 L	A	
	Isopropyl chloroformate	6.1	UN2407	I	6.1, 3, 8.	2, B9, B14, B32, B74, B77, T38, T43, T45	None	227	244	Forbidden	Forbidden	B	40
	Isopropyl 2-chloropropionate	3	UN2934	III	3	B1, T1	150	203	242	60 L	220 L	A	
	Isopropyl isobutyrate	3	UN2406	II	3	T1	150	202	242	5 L	60 L	B	
+	Isopropyl isocyanate	3	UN2483	I	3, 6.1	1, B9, B14, B30, B72, T38, T43, T44	None	226	244	Forbidden	Forbidden	D	40
	Isopropyl mercaptan, see Propanethiols.												
	Isopropyl nitrate	3	UN1222	II	3	T25	150	202	None	5 L	60 L	D	
	Isopropyl phosphoric acid, see Isopropyl acid phosphate.												
	Isopropyl propionate	3	UN2409	II	3	T1	150	202	242	5 L	60 L	B	
	Isopropylamine	3	UN1221	I	3, 8	T20	None	201	243	0.5 L	2.5 L	E	
	Isopropylbenzene	3	UN1918	III	3	B1, T1	150	203	242	60 L	220 L	A	
	Isopropylcumyl hydroperoxide, with more than 72 percent in solution.	Forbidden											
	Isosorbide dinitrate mixture with not less than 60 percent lactose, mannose, starch or calcium hydrogen phosphate.	4.1	UN2907	II	4.1		None	212	None	15 kg	50 kg	E	
	Isosorbide-5-mononitrate	4.1	UN3251	III	4.1	66	151	213	240	Forbidden	Forbidden	D	12
	Isothiocyanic acid	Forbidden											
	Jet fuel, see Fuel aviation, turbine engine.												
D	Jet perforating guns, charged oil well, with detonator.	1.1D	NA0124	II	1.1D	55, 56	None	62	None	Forbidden	Forbidden	A	24E
D	Jet perforating guns, charged oil well, with detonator.	1.4D	NA0494	II	1.4D	55, 56	None	62	None	Forbidden	Forbidden	B	
	Jet perforating guns, charged oil well, without detonator.	1.1D	UN0124	II	1.1D	55	None	62	None	Forbidden	Forbidden	B	
	Jet perforating guns, charged, oil well, without detonator.	1.4D	UN0494	II	1.4D	55, 114	None	62	None	Forbidden	300 kg	A	24E
	Jet perforators, see Charges, shaped, commercial etc.												
	Jet tappers, without detonator, see Charges, shaped commercial, etc.												
	Jet thrust igniters, for rocket motors or Jato, see Igniters.												
	Jet thrust unit (Jato), see Rocket motors.												
	Kerosene	3	UN1223	III	3	B1, T1	150	203	242	60 L	220 L	A	
G	Ketones, liquid, n.o.s	3	UN1224	I	3	T8, T31	None	201	243	1 L	30 L	E	
				II	3	T8, T31	150	202	242	5 L	60 L	B	
				III	3	B1, T7, T30	150	203	242	60 L	220 L	A	
	Krypton, compressed	2.2	UN1056		2.2		306	302	None	75 kg	150 kg	A	
	Krypton, refrigerated liquid (cryogenic liquid).	2.2	UN1970		2.2		320	None	None	50 kg	500 kg	B	
	Lacquer base or lacquer chips, nitrocellulose, dry, see Nitrocellulose, etc. (UN 2557).												
	Lacquer base or lacquer chips, plastic, wet with alcohol or solvent, see Nitrocellulose (UN 2059, UN 2060, UN 2555, UN2556) or Paint etc. (UN1263).												
	Lead acetate	6.1	UN1616	III	6.1		153	213	240	100 kg	200 kg	A	
	Lead arsenates	6.1	UN1617	II	6.1		None	212	242	25 kg	100 kg	A	
	Lead arsenites	6.1	UN1618	II	6.1		None	212	242	25 kg	100 kg	A	
	Lead azide (dry)	Forbidden											
	Lead azide, wetted with not less than 20 percent water or mixture of alcohol and water, by mass.	1.1A	UN0129	II	1.1A	111, 117	None	62	None	Forbidden	Forbidden	E	2E, 6E
	Lead compounds, soluble, n.o.s	6.1	UN2291	III	6.1	138	153	213	240	100 kg	200 kg	A	
	Lead cyanide	6.1	UN1620	II	6.1		None	212	242	25 kg	100 kg	A	26
	Lead dioxide	5.1	UN1872	III	5.1	A1	152	213	240	25 kg	100 kg	A	34
	Lead dross, see Lead sulfate, with more than 3 percent free acid.												
D	Lead mononitroresorcinate	1.1A	NA0473	II	1.1A	111, 117	None	62	None	Forbidden	Forbidden	E	2E, 6E
	Lead nitrate	5.1	UN1469	II	5.1, 6.1		None	212	242	5 kg	25 kg	A	
	Lead nitroresorcinate (dry)	Forbidden											
	Lead perchlorate, solid	5.1	UN1470	II	5.1, 6.1	T8	None	212	242	5 kg	25 kg	A	56, 58, 106
	Lead perchlorate, solution	5.1	UN1470	II	5.1, 6.1	T8	None	202	243	1 L	5 L	A	56, 58, 106

§172.101 HAZARDOUS MATERIALS TABLE—Continued

Symbols	Hazardous materials descriptions and proper shipping names	Hazard class or Division	Identification Numbers	PG	Label Codes	Special provisions	Packaging (§173.***)			Quantity limitations		Vessel stowage	
							Exceptions	Non-bulk	Bulk	Passenger aircraft/rail	Cargo aircraft only	Location	Other
(1)	(2)	(3)	(4)	(5)	(6)	(7)	(8A)	(8B)	(8C)	(9A)	(9B)	(10A)	(10B)
	Lead peroxide, see Lead dioxide												
	Lead phosphite, dibasic	4.1	UN2989	II	4.1		None	212	240	5 kg	25 kg	B	34
				III	4.1		151	213	240	15 kg	50 kg	B	34
	Lead picrate (dry)	Forbidden											
	Lead styphnate (dry)	Forbidden											
	Lead styphnate, wetted or Lead trinitroresorcinate, wetted with not less than 20 percent water or mixture of alcohol and water, by mass.	1.1A	UN0130	II	1.1A	111, 117	None	62	None	Forbidden	Forbidden	E	2E, 6E
	Lead sulfate with more than 3 percent free acid.	8	UN1794	II	8		154	212	240	15 kg	50 kg	A	
	Lead trinitroresorcinate, see Lead styphnate, etc.												
	Life-saving appliances, not self inflating containing dangerous goods as equipment.	9	UN3072		None		None	219	None	No limit	No limit	A	
	Life-saving appliances, self inflating.	9	UN2990		None		None	219	None	No limit	No limit	A	
	Lighter replacement cartridges containing liquefied petroleum gases (and similar devices, each not exceeding 65 grams), see Lighters or lighter refills etc. containing flammable gas.												
D	Lighters for cigars, cigarettes, etc., with lighter fluids.	3	NA1226	II	3	N10	None	21	None	Forbidden	Forbidden	B	
	Lighters, fuse	1.4S	UN0131	II	1.4S		None	62	None	25 kg	100 kg	A	
	Lighters or Lighter refills cigarettes, containing flammable gas.	2.1	UN1057		2.1	N10	None	21, 308	None	1 kg	15 kg	B	40
	Lime, unslaked, see Calcium oxide.												
G	Liquefied gas, flammable, n.o.s	2.1	UN3161		2.1		306	304	314, 315.	Forbidden	150 kg	D	40
G	Liquefied gas, n.o.s	2.2	UN3163		2.2		306	304	314, 315.	75 kg	150 kg	A	
G	Liquefied gas, oxidizing, n.o.s	2.2	UN3157		2.2, 5.1		306	304	314, 315.	75 kg	150 kg	D	
GI	Liquefied gas, toxic, corrosive, n.o.s. Inhalation Hazard Zone A.	2.3	UN3308		2.3, 8	1	None	192	245	Forbidden	Forbidden	D	40
GI	Liquefied gas, toxic, corrosive, n.o.s. Inhalation Hazard Zone B.	2.3	UN3308		2.3, 8	2	None	304	314, 315.	Forbidden	Forbidden	D	40
GI	Liquefied gas, toxic, corrosive, n.o.s. Inhalation Hazard Zone C.	2.3	UN3308		2.3, 8	3	None	304	314, 315.	Forbidden	Forbidden	D	40
GI	Liquefied gas, toxic, corrosive, n.o.s. Inhalation Hazard Zone D.	2.3	UN3308		2.3, 8	4	None	304	314, 315.	Forbidden	Forbidden	D	40
GI	Liquefied gas, toxic, flammable, corrosive, n.o.s. Inhalation Hazard Zone A.	2.3	UN3309		2.3, 2.1, 8.	1	None	192	245	Forbidden	Forbidden	D	17, 40
GI	Liquefied gas toxic, flammable, corrosive, n.o.s. Inhalation Hazard Zone B.	2.3	UN3309		2.3, 2.1, 8.	2	None	304	314, 315.	Forbidden	Forbidden	D	17, 40
GI	Liquefied gas, toxic, flammable, corrosive, n.o.s. Inhalation Hazard Zone C.	2.3	UN3309		2.3, 2.1, 8.	3	None	304	314, 315.	Forbidden	Forbidden	D	17, 40
GI	Liquefied gas, toxic, flammable, corrosive, n.o.s. Inhalation Hazard Zone D.	2.3	UN3309		2.3, 2.1, 8.	4	None	304	314, 315.	Forbidden	Forbidden	D	17, 40
G	Liquefied gas, toxic, flammable, n.o.s. Inhalation Hazard Zone A.	2.3	UN3160		2.3, 2.1	1	None	192	245	Forbidden	Forbidden	D	40
G	Liquefied gas, toxic, flammable, n.o.s. Inhalation Hazard Zone B.	2.3	UN3160		2.3, 2.1	2, B9, B14	None	304	314, 315.	Forbidden	Forbidden	D	40
G	Liquefied gas, toxic, flammable, n.o.s. Inhalation Hazard Zone C.	2.3	UN3160		2.3, 2.1	3, B14	None	304	314, 315.	Forbidden	Forbidden	D	40
G	Liquefied gas, toxic, flammable, n.o.s. Inhalation Hazard Zone D.	2.3	UN3160		2.3, 2.1	4	None	304	314, 315.	Forbidden	Forbidden	D	40
G	Liquefied gas, toxic, n.o.s. Inhalation Hazard Zone A.	2.3	UN3162		2.3	1	None	192	245	Forbidden	Forbidden	D	40
G	Liquefied gas, toxic, n.o.s. Inhalation Hazard Zone B.	2.3	UN3162		2.3	2, B9, B14	None	304	314, 315.	Forbidden	Forbidden	D	40
G	Liquefied gas, toxic, n.o.s. Inhalation Hazard Zone C.	2.3	UN3162		2.3	3, B14	None	304	314, 315.	Forbidden	Forbidden	D	40
G	Liquefied gas, toxic, n.o.s. Inhalation Hazard Zone D.	2.3	UN3162		2.3	4	None	304	314, 315.	Forbidden	Forbidden	D	40
GI	Liquefied gas, toxic, oxidizing, corrosive, n.o.s. Inhalation Hazard Zone A.	2.3	UN3310		2.3, 5.1, 8.	1	None	192	245	Forbidden	Forbidden	D	40, 89, 90

Appendix A

§ 172.101 HAZARDOUS MATERIALS TABLE—Continued

Symbols	Hazardous materials descriptions and proper shipping names	Hazard class or Division	Identification Numbers	PG	Label Codes	Special provisions	Packaging (§173.***) Exceptions	Packaging (§173.***) Non-bulk	Packaging (§173.***) Bulk	Quantity limitations Passenger aircraft/rail	Quantity limitations Cargo aircraft only	Vessel stowage Location	Vessel stowage Other
(1)	(2)	(3)	(4)	(5)	(6)	(7)	(8A)	(8B)	(8C)	(9A)	(9B)	(10A)	(10B)
G	Liquefied gas, toxic, oxidizing, corrosive, n.o.s. *Inhalation Hazard Zone B.*	2.3	UN3310		2.3, 2.1, 8.	2	None	304	314, 315.	Forbidden	Forbidden	D	40, 89, 90
G	Liquefied gas, toxic, oxidizing, corrosive, n.o.s. *Inhalation Hazard Zone C.*	2.3	UN3310		2.3, 2.1, 8.	3	None	304	314, 315.	Forbidden	Forbidden	D	40, 89, 90
G	Liquefied gas, toxic, oxidizing, corrosive, n.o.s. *Inhalation Hazard Zone D.*	2.3	UN3310		2.3, 2.1, 8.	4	None	304	314, 315.	Forbidden	Forbidden	D	40, 89, 90
G	Liquefied gas, toxic, oxidizing, n.o.s. *Inhalation Hazard Zone A.*	2.3	UN3307		2.3, 5.1	1	None	192	245	Forbidden	Forbidden	D	40
G	Liquefied gas, toxic, oxidizing, n.o.s *Inhalation Hazard Zone B.*	2.3	UN3307		2.3, 5.1	2	None	304	314, 315.	Forbidden	Forbidden	D	40
G	Liquefied gas, toxic, oxidizing, n.o.s. *Inhalation Hazard Zone C.*	2.3	UN3307		2.3, 5.1	3	None	304	314, 315.	Forbidden	Forbidden	D	40
G	Liquefied gas, toxic, oxidizing, n.o.s. *Inhalation Hazard Zone D.*	2.3	UN3307		2.3, 5.1	4	None	304	314, 315.	Forbidden	Forbidden	D	40
	Liquefied gases, *non-flammable charged with nitrogen, carbon dioxide or air.*	2.2	UN1058		2.2		306	304	None	75 kg	150 kg	A	
	Liquefied hydrocarbon gas, *see* Hydrocarbon gases, liquefied, n.o.s., *etc.*												
	Liquefied natural gas, *see* Methane, *etc.* (UN 1972).												
	Liquefied petroleum gas *see* Petroleum gases, liquefied.												
	Lithium	4.3	UN1415	I	4.3	A7, A19, B100, N45	None	211	244	Forbidden	15 kg	E	
	Lithium acetylide ethylenediamine complex, see Water reactive solid *etc.*												
	Lithium alkyls	4.2	UN2445	I	4.2, 4.3	B11, T28, T40	None	181	244	Forbidden	Forbidden	D	
	Lithium aluminum hydride	4.3	UN1410	I	4.3	A19, B100,	None	211	242	Forbidden	15 kg	E	
	Lithium aluminum hydride, ethereal.	4.3	UN1411	I	4.3, 3	A2, A3, A11, N34	None	201	244	Forbidden	1 L	D	40
	Lithium batteries, contained in equipment.	9	UN3091	II	9	29	185(i)	185	None	5 kg	5 kg	A	
	Lithium batteries packed with equipment.	9	UN3091	II	9	29	185	185	None	5 kg gross	35 kg gross	A	
	Lithium battery	9	UN3090	II	9	29	185	185	None	5 kg	35 kg gross	A	
	Lithium borohydride	4.3	UN1413	I	4.3	A19, B100, N40	None	211	242	Forbidden	15 kg	E	
	Lithium ferrosilicon	4.3	UN2830	II	4.3	A19, B105, B106	151	212	241	15 kg	50 kg	E	40, 85, 103
	Lithium hydride	4.3	UN1414	I	4.3	A19, B100, N40	None	211	242	Forbidden	15 kg	E	
	Lithium hydride, fused solid	4.3	UN2805	II	4.3	A8, A19, B101, B106	151	212	241	15 kg	50 kg	E	
	Lithium hydroxide, monohydrate *or* Lithium hydroxide, solid.	8	UN2680	II	8		154	212	240	15 kg	50 kg	A	
	Lithium hydroxide, solution	8	UN2679	II	8	B2, T8	154	202	242	1 L	30 L	A	
				III	8	T8	154	203	241	5 L	60 L	A	96
	Lithium hypochlorite, dry *or* Lithium hypochlorite mixtures, dry.	5.1	UN1471	II	5.1	A9, N34	152	212	240	5 kg	25 kg	A	48, 56, 58, 69, 106, 116
	Lithium in cartridges, see Lithium.												
	Lithium nitrate	5.1	UN2722	III	5.1	A1	152	213	240	25 kg	100 kg	A	
	Lithium nitride	4.3	UN2806	I	4.3	A19, B101, B106, N40	None	211	242	Forbidden	15 kg	E	
	Lithium peroxide	5.1	UN1472	II	5.1	A9, N34	152	212	None	5 kg	25 kg	A	13, 75, 106
	Lithium silicon	4.3	UN1417	II	4.3	A19, A20, B105, B106	151	212	241	15 kg	50 kg	A	85, 103
	LNG, *see* Methane *etc.* (UN 1972).												
	London purple	6.1	UN1621	II	6.1		None	212	242	25 kg	100 kg	A	
	LPG, *see* Petroleum gases, liquefied.												
	Lye, see Sodium hydroxide, solutions.												
	Magnesium alkyls	4.2	UN3053	I	4.2, 4.3	B11, T28, T29, T40	None	181	244	Forbidden	Forbidden	D	18
	Magnesium aluminum phosphide.	4.3	UN1419	I	4.3, 6.1	A19, B100, N34, N40	None	211	242	Forbidden	15 kg	E	40, 85
▸	Magnesium arsenate	6.1	UN1622	II	6.1		None	212	242	25 kg	100 kg	A	
	Magnesium bisulfite solution, see Bisulfites, aqueous solutions, n.o.s.												
	Magnesium bromate	5.1	UN1473	II	5.1	A1	152	212	242	5 kg	25 kg	A	56, 58, 106
	Magnesium chlorate	5.1	UN2723	II	5.1		152	212	242	5 kg	25 kg	A	56, 58, 106
	Magnesium diamide	4.2	UN2004	II	4.2	A8, A19, A20	None	212	241	15 kg	50 kg	C	

§172.101 Hazardous Materials Table—Continued

Symbols	Hazardous materials descriptions and proper shipping names	Hazard class or Division	Identification Numbers	PG	Label Codes	Special provisions	(8) Packaging (§173.***)			(9) Quantity limitations		(10) Vessel stowage	
							Exceptions	Non-bulk	Bulk	Passenger aircraft/rail	Cargo aircraft only	Location	Other
(1)	(2)	(3)	(4)	(5)	(6)	(7)	(8A)	(8B)	(8C)	(9A)	(9B)	(10A)	(10B)
	Magnesium diphenyl	4.2	UN2005	I	4.2		None	187	244	Forbidden	Forbidden	C	
	Magnesium dross, wet or hot	Forbidden											
	Magnesium fluorosilicate	6.1	UN2853	III	6.1		153	213	240	100 kg	200 kg	A	26
	Magnesium granules, coated particle size not less than 149 microns.	4.3	UN2950	III	4.3	A1, A19, B108	151	213	240	25 kg	100 kg	A	
	Magnesium hydride	4.3	UN2010	I	4.3	A19, B100, N40	None	211	242	Forbidden	15 kg	E	
	Magnesium or Magnesium alloys with more than 50 percent magnesium in pellets, turnings or ribbons.	4.1	UN1869	III	4.1	A1	151	213	240	25 kg	100 kg	A	39
	Magnesium nitrate	5.1	UN1474	III	5.1	A1	152	213	240	25 kg	100 kg	A	
	Magnesium perchlorate	5.1	UN1475	II	5.1		152	212	242	5 kg	25 kg	A	56, 58, 106
	Magnesium peroxide	5.1	UN1476	II	5.1		152	212	242	5 kg	25 kg	A	13, 75, 106
	Magnesium phosphide	4.3	UN2011	I	4.3, 6.1	A19, N40	None	211	None	Forbidden	15 kg	E	40, 85
	Magnesium, powder or Magnesium alloys, powder.	4.3	UN1418	I	4.3, 4.2	A19, B56	None	211	244	Forbidden	15 kg	A	39
				II	4.3, 4.2	A19, B56, B101, B106	None	212	241	15 kg	50 kg	A	39
				III	4.3, 4.2	A19, B56, B106, B108	None	213	241	25 kg	100 kg	A	39
	Magnesium scrap, see Magnesium, etc. (UN 1869).												
	Magnesium silicide	4.3	UN2624	II	4.3	A19, A20, B105, B106	151	212	241	15 kg	50 kg	B	85, 103
	Magnetized material, see section 173.21.												
D	Maleic acid	8	NA2215	III	8		154	213	240	25 kg	100 kg	A	
	Maleic anhydride	8	UN2215	III	8	T7	154	213	240	25 kg	100 kg	A	
	Malononitrile	6.1	UN2647	II	6.1		None	212	242	25 kg	100 kg	A	12
	Mancozeb (manganese ethylenebisdithiocarbamate complex with zinc) see Maneb.												
	Maneb or Maneb preparations with not less than 60 percent maneb.	4.2	UN2210	III	4.2, 4.3	57, A1, A19, B105	None	213	242	25 kg	100 kg	A	34
	Maneb stabilized or Maneb preparations, stabilized against self-heating.	4.3	UN2968	III	4.3	54, A1, A19, B108	151	213	242	25 kg	100 kg	B	34
	Manganese nitrate	5.1	UN2724	III	5.1	A1	152	213	240	25 kg	100 kg	A	
	Manganese resinate	4.1	UN1330	III	4.1	A1	151	213	240	25 kg	100 kg	A	
	Mannitan tetranitrate	Forbidden											
	Mannitol hexanitrate (dry)	Forbidden											
	Mannitol hexanitrate, wetted or Nitromannite, wetted with not less than 40 percent water, or mixture of alcohol and water, by mass.	1.1D	UN0133	II	1.1D	121	None	62	None	Forbidden	Forbidden	B	1E, 5E
	Marine pollutants, liquid or solid, n.o.s., see Environmentally hazardous substances, liquid or solid, n.o.s.												
	Matches, block, see Matches, 'strike anywhere'.												
	Matches, fusee	4.1	UN2254	III	4.1		186	186	None	Forbidden	Forbidden	A	
	Matches, safety (book, card or strike on box).	4.1	UN1944	III	4.1		186	186	None	25 kg	100 kg	A	
	Matches, strike anywhere	4.1	UN1331	III	4.1		186	186	None	Forbidden	Forbidden	B	
	Matches, wax, Vesta	4.1	UN1945	III	4.1		186	186	None	25 kg	100 kg	B	
	Matting acid, see Sulfuric acid												
	Medicine, liquid, flammable, toxic, n.o.s.	3	UN3248	II	3, 6.1	36	None	202	None	1 L	5 L	B	40
				III	3, 6.1	36	150	203	None	5 L	5 L	A	
	Medicine, liquid, toxic, n.o.s	6.1	UN1851	II	6.1		153	202	243	5 L	5 L	C	40
				III	6.1		153	203	241	5 L	5 L	C	40
	Medicine, solid, toxic, n.o.s	6.1	UN3249	II	6.1	36	153	212	None	5 kg	5 kg	C	40
				III	6.1	36	153	213	None	5 kg	5 kg	C	40
D	Medicines, corrosive, liquid, n.o.s.	8	NA1760	II	8	B3	154	202	242	1 L	30 L	B	40
				III	8		154	203	241	5 L	60 L	A	40
D	Medicines, corrosive, solid, n.o.s	8	NA1759	II	8		154	212	240	15 kg	50 kg	A	
				III	8		154	213	240	25 kg	100 kg	A	
D	Medicines, flammable, liquid, n.o.s.	3	NA1993	I	3		150	201	243	1 L	30 L	E	
				II	3		150	202	242	5 L	60 L	B	
				III	3	B1	150	203	242	60 L	220 L	A	
D	Medicines, flammable, solid, n.o.s.	4.1	NA1325	II	4.1		151	212	240	15 kg	50 kg	B	
D	Medicines, oxidizing substance, solid, n.o.s.	5.1	NA1479	II	5.1		152	212	242	5 kg	25 kg	B	56, 58, 69, 106
	Memtetrahydrophthalic anhydride, see Corrosive liquids, n.o.s.												
	Mercaptans, liquid, flammable, n.o.s. or Mercaptan mixture, liquid, flammable, n.o.s.	3	UN3336	I	3	T23	150	201	243	1 L	30 L	E	95

Appendix A

§ 172.101 HAZARDOUS MATERIALS TABLE—Continued

Symbols	Hazardous materials descriptions and proper shipping names	Hazard class or Division	Identification Numbers	PG	Label Codes	Special provisions	Packaging (§173.***) Exceptions (8A)	Packaging Non-bulk (8B)	Packaging Bulk (8C)	Quantity limitations Passenger aircraft/rail (9A)	Quantity limitations Cargo aircraft only (9B)	Vessel stowage Location (10A)	Vessel stowage Other (10B)
(1)	(2)	(3)	(4)	(5)	(6)	(7)	(8A)	(8B)	(8C)	(9A)	(9B)	(10A)	(10B)
				II	3	T8, T31	150	202	242	5 L	60 L	B	95
				III	3	B1, B52, T7, T30	150	203	241	60 L	220 L	B	95
	Mercaptans, liquid, flammable, toxic, n.o.s. *or* Mercaptan mixtures, liquid, flammable, toxic, n.o.s.	3	UN1228	II	3, 6.1	T13	None	202	243	Forbidden	60 L	B	40, 95
				III	3, 6.1	B1, T8	150	203	242	5 L	220 L	A	40, 95
	Mercaptans, liquid, toxic, flammable, n.o.s. *or* Mercaptan mixtures, liquid, toxic, flammable, n.o.s., flash point not less than 23 degrees C.	6.1	UN3071	II	6.1, 3	T14	None	202	243	5 L	60 L	C	40, 121
	5-Mercaptotetrazol-1-acetic acid	1.4C	UN0448	II	1.4C		None	62	None	Forbidden	75 kg	A	1E, 5E, 24E
	Mercuric arsenate	6.1	UN1623	II	6.1		None	212	242	25 kg	100 kg	A	
	Mercuric chloride	6.1	UN1624	II	6.1		None	212	242	25 kg	100 kg	A	
	Mercuric compounds, see Mercury compounds, *etc.*												
	Mercuric nitrate	6.1	UN1625	II	6.1	N73	None	212	242	25 kg	100kg	A	
+	Mercuric potassium cyanide	6.1	UN1626	I	6.1	N74, N75	None	211	242	5 kg	50 kg	A	26
	Mercuric sulfocyanate, see Mercury thiocyanate.												
	Mercurol, see Mercury nucleate												
	Mercurous azide	Forbidden											
	Mercurous compounds, see Mercury compounds, *etc.*												
	Mercurous nitrate	6.1	UN1627	II	6.1		None	212	242	25 kg	100 kg	A	
A, W	Mercury	8	UN2809	III	8		164	164	240	35 kg	35 kg	B	40, 97
	Mercury acetate	6.1	UN1629	II	6.1		None	212	242	25 kg	100 kg	A	
	Mercury acetylide	Forbidden											
	Mercury ammonium chloride	6.1	UN1630	II	6.1		None	212	242	25 kg	100 kg	A	
	Mercury based pesticides, liquid, flammable, toxic, *flash point less than 23 degrees C.*	3	UN2778	I	3, 6.1		None	201	243	Forbidden	30 L	B	40
				II	3, 6.1		None	202	243	1 L	60 L	B	40
	Mercury based pesticides, liquid, toxic.	6.1	UN3012	I	6.1	T42	None	201	243	1L	30 L	B	40
				II	6.1	T14	None	202	243	5 L	60 L	B	40
				III	6.1	T14	153	203	241	60 L	220 L	A	40
	Mercury based pesticides, liquid, toxic, flammable, *flashpoint not less than 23 degrees C.*	6.1	UN3011	I	6.1, 3	T42	None	201	243	1 L	30 L	B	40
				II	6.1, 3	T14	None	202	243	5 L	60 L	B	40
				III	6.1, 3	T14	153	203	242	60 L	220 L	A	40
	Mercury based pesticides, solid, toxic.	6.1	UN2777	I	6.1		None	211	242	5 kg	50 kg	A	40
				II	6.1		None	212	242	25 kg	100 kg	A	40
				III	6.1		153	213	240	100 kg	200 kg	A	40
	Mercury benzoate	6.1	UN1631	II	6.1		None	212	242	25 kg	100 kg	A	
	Mercury bromides	6.1	UN1634	II	6.1		None	212	242	25 kg	100 kg	A	
	Mercury compounds, liquid, n.o.s.	6.1	UN2024	I	6.1		None	201	243	1 L	30 L	B	40
				II	6.1		None	202	243	5 L	60 L	B	40
				III	6.1		153	203	241	60 L	220 L	B	40
	Mercury compounds, solid, n.o.s	6.1	UN2025	I	6.1		None	211	242	5 kg	50kg	A	
				II	6.1		None	212	242	25 kg	100 kg	A	
				III	6.1		153	213	240	100 kg	200 kg	A	
A	Mercury *contained in manufactured articles.*	8	UN2809	III	8		None	164	None	No limit	No limit	B	40, 97
	Mercury cyanide	6.1	UN1636	II	6.1	N74, N75	None	212	242	25 kg	100 kg	A	26
	Mercury fulminate, wetted with not less than 20 percent water, or mixture of alcohol and water, by mass.	1.1A	UN0135	II	1.1A	111, 117	None	62	None	Forbidden	Forbidden	E	2E, 6E
	Mercury gluconate	6.1	UN1637	II	6.1		None	212	242	25 kg	100 kg	A	
	Mercury iodide, *solid*	6.1	UN1638	II	6.1		None	212	242	25 kg	100 kg	A	
	Mercury iodide aquabasic ammonobasic (Iodide of Millon's base).	Forbidden											
	Mercury iodide, *solution*	6.1	UN1638	II	6.1		None	202	243	5 L	60 L	A	
	Mercury nitride	Forbidden											
	Mercury nucleate	6.1	UN1639	II	6.1		None	212	242	25 kg	100 kg	A	
	Mercury oleate	6.1	UN1640	II	6.1		None	212	242	25 kg	100 kg	A	
	Mercury oxide	6.1	UN1641	II	6.1		None	212	242	25 kg	100 kg	A	
	Mercury oxycyanide	Forbidden											
	Mercury oxycyanide, desensitized.	6.1	UN1642	II	6.1		None	212	242	25 kg	100 kg	A	26, 91
	Mercury potassium iodide	6.1	UN1643	II	6.1		None	212	242	25 kg	100 kg	A	
	Mercury salicylate	6.1	UN1644	II	6.1		None	212	242	25 kg	100 kg	A	
+	Mercury sulfates	6.1	UN1645	II	6.1		None	212	242	25 kg	100 kg	A	
	Mercury thiocyanate	6.1	UN1646	II	6.1		None	212	242	25 kg	100 kg	A	
	Mesityl oxide	3	UN1229	III	3	B1, T1	None	203	242	60 L	220 L	A	
	Metal alkyl halides, water-reactive n.o.s. *or* Metal aryl halides, water-reactive, n.o.s..	4.2	UN3049	I	4.2, 4.3	B9, B11, T28, T29, T40	None	181	244	Forbidden	Forbidden	D	
	Metal alkyl hydrides, water-reactive, n.o.s. *or* Metal aryl hydrides, water-reactive, n.o.s..	4.2	UN3050	I	4.2, 4.3	B9, B11, T28, T29, T40	None	181	244	Forbidden	Forbidden	D	

§ 172.101 HAZARDOUS MATERIALS TABLE—Continued

Symbols	Hazardous materials descriptions and proper shipping names	Hazard class or Division	Identification Numbers	PG	Label Codes	Special provisions	Packaging (§173.***)			Quantity limitations		Vessel stowage	
							Exceptions	Non-bulk	Bulk	Passenger aircraft/rail	Cargo aircraft only	Location	Other
(1)	(2)	(3)	(4)	(5)	(6)	(7)	(8A)	(8B)	(8C)	(9A)	(9B)	(10A)	(10B)
	Metal alkyls, water-reactive, n.o.s. or Metal aryls, water-reactive n.o.s..	4.2	UN2003	I	4.2, 4.3	B11, T42	None	181	244	Forbidden	Forbidden	D	
D	Metal alkyl, solution, n.o.s	3	NA9195	II	3		150	202	242	1 L	4 L	B	
	Metal carbonyls, n.o.s	6.1	UN3281	I	6.1	5	None	201	243	1 L	30 L	B	40
				II	6.1	T14	None	202	243	5 L	60 L	B	40
				III	6.1	T7	153	203	241	60 L	220 L	A	40
	Metal catalyst, dry	4.2	UN2881	I	4.2	N34	None	187	None	Forbidden	Forbidden	C	
				II	4.2	N34	None	187	242	Forbidden	50 kg	C	
				III	4.2	N34	241	187	None	25 kg	100 kg	C	
	Metal catalyst, wetted with a visible excess of liquid.	4.2	UN1378	II	4.2	A2, A8, N34	None	212	None	Forbidden	50 kg	C	
	Metal hydrides, flammable, n.o.s	4.1	UN3182	II	4.1	A1	151	212	240	15 kg	50 kg	E	
				III	4.1	A1	151	213	240	25 kg	100 kg	E	
	Metal hydrides, water reactive, n.o.s.	4.3	UN1409	I	4.3	A19, B100, N34, N40	None	211	242	Forbidden	15 kg	D	
				II	4.3	A19, B101, B106, N34, N40	151	212	242	15 kg	50 kg	D	
	Metal powder, self-heating, n.o.s	4.2	UN3189	II	4.2		None	212	241	15 kg	50 kg	C	
				III	4.2		None	213	241	25 kg	100 kg	C	
	Metal powders, flammable, n.o.s	4.1	UN3089	II	4.1		151	212	240	15 kg	50 kg	B	
				III	4.1		151	213	240	25 kg	100 kg	B	
	Metal salts of methyl nitramine (dry).	Forbidden											
G	Metal salts of organic compounds, flammable, n.o.s.	4.1	UN3181	II	4.1	A1	151	212	240	15 kg	50 kg	B	40
				III	4.1	A1	151	213	240	25 kg	100 kg	B	40
	Metaldehyde	4.1	UN1332	III	4.1	A1	151	213	240	25 kg	100 kg	A	40
G	Metallic substance, water-reactive, n.o.s.	4.3	UN3208	I	4.3	B101, B106	None	211	242	Forbidden	15 kg	E	40
				II	4.3	B101, B106	151	212	242	15 kg	50 kg	E	40
				III	4.3	B105, B108	151	213	241	25 kg	100 kg	E	40
G	Metallic substance, water-reactive, self-heating, n.o.s.	4.3	UN3209	I	4.3, 4.2	B100	None	211	242	Forbidden	15 kg	E	40
				II	4.3, 4.2	B101, B106	None	212	242	15 kg	50 kg	E	40
				III	4.3, 4.2	B101, B106	None	213	242	25 kg	100 kg	E	40
	Methacrylaldehyde, inhibited	3	UN2396	II	3, 6.1	45, T8	None	202	243	1 L	60 L	E	40
	Methacrylic acid, inhibited	8	UN2531	II	8	T8, T47	154	203	241	5 L	60 L	A	
+	Methacrylonitrile, inhibited	3	UN3079	I	3, 6.1	2, B9, B14, B32, B74, T38, T43, T45	None	227	244	Forbidden	30 L	D	12, 40, 48
	Methallyl alcohol	3	UN2614	III	3	B1, T1	150	203	242	60 L	220 L	A	
	Methane and hydrogen, mixtures, see Hydrogen and methane, mixtures, etc.												
	Methane, compressed or Natural gas, compressed (with high methane content).	2.1	UN1971		2.1		306	302	302	Forbidden	150 kg	E	40
	Methane, refrigerated liquid (cryogenic liquid) or Natural gas, refrigerated liquid (cryogenic liquid), with high methane content).	2.1	UN1972		2.1		None	None	318	Forbidden	Forbidden	D	40
	Methanesulfonyl chloride	6.1	UN3246	I	6.1, 8	2, 25, B9, B14, B32, B74, T38, T43, T45	None	227	244	Forbidden	Forbidden	D	40
I	Methanol	3	UN1230	II	3, 6.1	T8	150	202	242	1 L	60 L	B	40
D	Methanol	3	UN1230	II	3	T8	150	202	242	1 L	60 L	B	40
	Methazoic acid	Forbidden											
	4-Methoxy-4-methylpentan-2-one.	3	UN2293	III	3	B1, T1	150	203	242	60 L	220 L	A	
	1-Methoxy-2-propanol	3	UN3092	III	3	B1, T1	150	203	242	60 L	220 L	A	
+	Methoxymethyl isocyanate	3	UN2605	I	3, 6.1	1, B9, B14, B30, B72, T38, T43, T44	None	226	244	Forbidden	Forbidden	D	40
	Methyl acetate	3	UN1231	II	3	B101, T8	150	202	242	5 L	60 L	B	
	Methyl acetylene and propadiene mixtures, stabilized.	2.1	UN1060		2.1		306	304	314, 315.	Forbidden	150 kg	B	40
	Methyl acrylate, inhibited	3	UN1919	II	3	T8	150	202	242	5 L	60 L	B	
	Methyl alcohol, see Methanol												
	Methyl allyl chloride	3	UN2554	II	3	B101, T8	150	202	242	5 L	60 L	E	
	Methyl amyl ketone, see Amyl methyl ketone.												
	Methyl bromide	2.3	UN1062		2.3	3, B14	None	193	314, 315.	Forbidden	25 kg	D	40
	Methyl bromide and chloropicrin mixtures with more than 2 percent chloropicrin, see Chloropicrin and methyl bromide mixtures.												
	Methyl bromide and chloropicrin mixtures with not more than 2 percent chloropicrin, see Methyl bromide.												
	Methyl bromide and ethylene dibromide mixtures, liquid.	6.1	UN1647	I	6.1	2, B9, B14, B32, B74, N65, T38, T43, T45	None	227	244	Forbidden	30 L	C	40
	Methyl bromoacetate	6.1	UN2643	II	6.1	B100, T8	None	202	243	5 L	60 L	D	40

Appendix A

§ 172.101 HAZARDOUS MATERIALS TABLE—Continued

Symbols	Hazardous materials descriptions and proper shipping names	Hazard class or Division	Identification Numbers	PG	Label Codes	Special provisions	Packaging (§173.***) Exceptions (8A)	Packaging (§173.***) Non-bulk (8B)	Packaging (§173.***) Bulk (8C)	Quantity limitations Passenger aircraft/rail (9A)	Quantity limitations Cargo aircraft only (9B)	Vessel stowage Location (10A)	Vessel stowage Other (10B)
(1)	(2)	(3)	(4)	(5)	(6)	(7)	(8A)	(8B)	(8C)	(9A)	(9B)	(10A)	(10B)
	2-Methyl-1-butene	3	UN2459	I	3	T14	None	201	243	1 L	30 L	E	
	2-Methyl-2-butene	3	UN2460	II	3	T14	None	202	242	5 L	60 L	E	
	3-Methyl-1-butene	3	UN2561	I	3	T20	None	201	243	1 L	30 L	E	
	Methyl tert-butyl ether	3	UN2398	II	3	B101, T14	150	202	242	5 L	60 L	E	
	Methyl butyrate	3	UN1237	II	3	T1	150	202	242	5 L	60 L	B	
	Methyl chloride, or Refrigerant gas R 40.	2.1	UN1063		2.1		306	304	314, 315.	5 kg	100 kg	D	40
	Methyl chloride and chloropicrin mixtures, see Chloropicrin and methyl chloride mixtures.												
	Methyl chloride and methylene chloride mixtures.	2.1	UN1912		2.1		306	304	314, 315.	Forbidden	150 kg	D	40
	Methyl chloroacetate	6.1	UN2295	I	6.1, 3	T42	None	201	243	1 L	30 L	D	
	Methyl chlorocarbonate, see Methyl chloroformate.												
	Methyl chloroform, see 1,1,1-Trichloroethane.												
	Methyl chloroformate	6.1	UN1238	I	6.1, 3, 8.	1, B9, B14, B30, B72, N34, T38, T43, T44	None	226	244	Forbidden	Forbidden	D	21, 40, 100
	Methyl chloromethyl ether	6.1	UN1239	I	6.1, 3	1, B9, B14, B30, B72, T38, T43, T44	None	226	244	Forbidden	Forbidden	D	40
	Methyl 2-chloropropionate	3	UN2933	III	3	B1, T7	150	203	242	60 L	220 L	A	
	Methyl dichloroacetate	6.1	UN2299	III	6.1	T1	153	203	241	60 L	220 L	A	
	Methyl ethyl ether, see Ethyl methyl ether.												
	Methyl ethyl ketone, see Ethyl methyl ketone.												
	Methyl ethyl ketone peroxide, in solution with more than 9 percent by mass active oxygen.	Forbidden											
	2-Methyl-5-ethylpyridine	6.1	UN2300	III	6.1	T7	153	203	241	60 L	220 L	A	
	Methyl fluoride, or Refrigerant gas R 41.	2.1	UN2454		2.1		306	304	314, 315.	Forbidden	150 kg	E	40
	Methyl formate	3	UN1243	I	3	T20	150	201	243	1 L	30 L	E	
	2-Methyl-2-heptanethiol	6.1	UN3023	I	6.1, 3	2, B9, B14, B32, B74, T38, T43, T45	None	227	244	Forbidden	Forbidden	D	40, 102
	Methyl iodide	6.1	UN2644	I	6.1	2, B9, B14, B32, B74, T38, T43, T45	None	227	244	Forbidden	Forbidden	A	12, 40
	Methyl isobutyl carbinol	3	UN2053	III	3	B1, T1	150	203	242	60 L	220 L	A	
	Methyl isobutyl ketone	3	UN1245	II	3	T1	150	202	242	5 L	60 L	B	
	Methyl isobutyl ketone peroxide, in solution with more than 9 percent by mass active oxygen.	Forbidden											
	Methyl isocyanate	6.1	UN2480	I	6.1, 3	1, B9, B14, B30, B72, T38, T43, T44	None	226	244	Forbidden	Forbidden	D	26, 40
	Methyl isopropenyl ketone, inhibited.	3	UN1246	II	3	T7	150	202	242	5 L	60 L	B	
	Methyl isothiocyanate	6.1	UN2477	I	6.1, 3	2, B9, B14, B32, B74, T38, T43, T45	None	227	244	Forbidden	Forbidden	A	
	Methyl isovalerate	3	UN2400	II	3	T1	150	202	242	5 L	60 L	B	
	Methyl magnesium bromide, in ethyl ether.	4.3	UN1928	I	4.3, 3		None	201	243	Forbidden	1 L	D	
	Methyl mercaptan	2.3	UN1064		2.3, 2.1	3, B7, B9, B14	None	304	314, 315.	Forbidden	25 kg	D	40
	Methyl mercaptopropionaldehyde, see Thia-4-pentanal.												
	Methyl methacrylate monomer, inhibited.	3	UN1247	II	3	T8	150	202	242	5 L	60 L	B	40
	Methyl nitramine (dry)	Forbidden											
	Methyl nitrate	Forbidden											
	Methyl nitrite	Forbidden											
	Methyl norbornene dicarboxylic anhydride, see Corrosive liquids, n.o.s.												
	Methyl orthosilicate	6.1	UN2606	I	6.1, 3	2, B9, B14, B32, B74, T38, T43, T45	None	227	244	Forbidden	30 L	E	40
D	Methyl parathion liquid	6.1	NA3018	II	6.1	N76, T14	None	202	243	Forbidden	1 L	A	40
D	Methyl parathion solid	6.1	NA2783	II	6.1	N77	None	212	242	25 kg	100 kg	A	40
D	Methyl phosphonic dichloride	6.1	NA9206	I	6.1, 8	2, A3, B9, B14, B32, B74, N34, N43, T38, T43, T45	None	227	244	Forbidden	Forbidden	C	
	Methyl phosphonothioic dichloride, anhydrous, see Corrosive liquid, n.o.s.												
D	Methyl phosphonous dichloride, pyrophoric liquid.	6.1	NA2845	I	6.1, 4.2	2, B9, B14, B16, B32, B74, T38, T43, T45	None	227	244	Forbidden	Forbidden	D	18
	Methyl picric acid (heavy metal salts of).	Forbidden											

§172.101 Hazardous Materials Table—Continued

Symbols	Hazardous materials descriptions and proper shipping names	Hazard class or Division	Identification Numbers	PG	Label Codes	Special provisions	Packaging (§173.***)			Quantity limitations		Vessel stowage	
							Exceptions	Non-bulk	Bulk	Passenger aircraft/rail	Cargo aircraft only	Location	Other
(1)	(2)	(3)	(4)	(5)	(6)	(7)	(8A)	(8B)	(8C)	(9A)	(9B)	(10A)	(10B)
	Methyl propionate	3	UN1248	II	3	B101, T2	150	202	242	5 L	60 L	B	
	Methyl propyl ether	3	UN2612	II	3	T14	150	202	242	5 L	60 L	E	40
	Methyl propyl ketone	3	UN1249	II	3	T1	150	202	242	5 L	60 L	B	
	Methyl sulfate, see Dimethyl sulfate.												
	Methyl sulfide, see Dimethyl sulfide.												
	Methyl trichloroacetate	6.1	UN2533	III	6.1	T1	153	203	241	60 L	220 L	A	
	Methyl trimethylol methane trinitrate.	Forbidden											
	Methyl vinyl ketone, stabilized	6.1	UN1251	I	6.1, 3, 8.	1, 25, B9, B14, B30, B72, T38, T43, T44	None	226	244	Forbidden	Forbidden	B	40
	Methylal	3	UN1234	II	3	T14	None	202	242	5 L	60 L	E	
	Methylamine, anhydrous	2.1	UN1061		2.1		306	304	314, 315.	Forbidden	150 kg	B	40
	Methylamine, aqueous solution	3	UN1235	II	3, 8	B1, T8	150	202	243	1 L	5 L	E	41
	Methylamine dinitramine and dry salts thereof.	Forbidden											
	Methylamine nitroform	Forbidden											
	Methylamine perchlorate (dry)	Forbidden											
	Methylamyl acetate	3	UN1233	III	3	B1, T1	150	203	242	60 L	220 L	A	
	N-Methylaniline	6.1	UN2294	III	6.1	T7	153	203	241	60 L	220L	A	
	alpha-Methylbenzyl alcohol	6.1	UN2937	III	6.1	T1	153	203	241	60 L	220 L	A	
	3-Methylbutan-2-one	3	UN2397	II	3	T1	150	202	242	5 L	60 L	B	
	N-Methylbutylamine	3	UN2945	II	3, 8	T8	None	202	243	1 L	5 L	B	40
	Methylchlorosilane	2.3	UN2534		2.3, 2.1, 8.	2, A2, A3, A7, B9, B14, N34	None	226	314, 315.	Forbidden	Forbidden	D	17, 40
	Methylcyclohexane	3	UN2296	II	3	B1, T1	150	202	242	5 L	60 L	B	
	Methylcyclohexanols, flammable	3	UN2617	III	3	B1, T2	150	203	242	60 L	220 L	A	
	Methylcyclohexanone	3	UN2297	III	3	B1, T1	150	203	242	60 L	220 L	A	
	Methylcyclopentane	3	UN2298	II	3	T8	150	202	242	5 L	60 L	B	
D	Methyldichloroarsine	6.1	NA1556	I	6.1	2	None	192	None	Forbidden	Forbidden	D	40, 95
	Methyldichlorosilane	4.3	UN1242	I	4.3, 8, 3.	A2, A3, A7, B6, B77, N34, T16, T26	None	201	243	Forbidden	1 L	D	21, 28, 40, 49, 100
	Methylene chloride, see Dichloromethane.												
	Methylene glycol dinitrate	Forbidden											
	2-Methylfuran	3	UN2301	II	3	T7	150	202	242	5 L	60 L	E	
	a-Methylglucoside tetranitrate	Forbidden											
	a-Methylglycerol trinitrate	Forbidden											
	5-Methylhexan-2-one	3	UN2302	III	3	B1, T1	150	203	242	60 L	220 L	A	
	Methylhydrazine	6.1	UN1244	I	6.1, 3, 8.	1, B7, B9, B14, B30, B72, B77, N34, T38, T43, T44	None	226	244	Forbidden	Forbidden	D	21, 40, 49, 100
	4-Methylmorpholine or n-methylmorpholine.	3	UN2535	II	3, 8	B6, T8	None	202	243	1 L	5 L	B	40
	Methylpentadienes	3	UN2461	II	3	T7	150	202	242	5 L	60 L	E	
	2-Methylpentan-2-ol	3	UN2560	III	3	B1, T1	150	203	242	60 L	220 L	A	
	Methylpentanes, see Hexanes												
	Methylphenyldichlorosilane	8	UN2437	II	8	T8, T26	154	202	242	1 L	30 L	C	40
	1-Methylpiperidine	3	UN2399	II	3, 8	T8	None	202	243	1 L	5 L	B	
	Methyltetrahydrofuran	3	UN2536	II	3	B101, T7	150	202	242	5 L	60 L	B	
	Methyltrichlorosilane	3	UN1250	I	3, 8	A7, B6, B77, N34, T14, T26	None	201	243	Forbidden	2.5 L	B	40
	alpha-Methylvaleraldehyde	3	UN2367	II	3	B1, T1	150	202	242	5 L	60 L	B	
	Mine rescue equipment containing carbon dioxide, see Carbon dioxide.												
	Mines with bursting charge	1.1F	UN0136	II	1.1F			62	None	Forbidden	Forbidden	E	
	Mines with bursting charge	1.1D	UN0137	II	1.1D			62	None	Forbidden	Forbidden	B	3E, 7E
	Mines with bursting charge	1.2D	UN0138	II	1.2D			62	None	Forbidden	Forbidden	B	3E, 7E
	Mines with bursting charge	1.2F	UN0294	II	1.2F			62	None	Forbidden	Forbidden	E	
	Mixed acid, see Nitrating acid, mixtures etc.												
	Mobility aids, see Wheel chair, electric.												
D	Model rocket motor	1.4C	NA0276	II	1.4C	51	None	62	None	Forbidden	75 kg	A	24E
D	Model rocket motor	1.4S	NA0323	II	1.4S	51	None	62	None	25 kg	100 kg	A	9E
	Molybdenum pentachloride	8	UN2508	III	8	T8, T26	154	213	240	25 kg	100 kg	C	40
	Monochloroacetone (unstabilized).	Forbidden											
	Monochloroethylene, see Vinyl chloride, inhibited.												
	Monoethanolamine, see Ethanolamine, solutions.												
	Monoethylamine, see Ethylamine.												
	Morpholine	3	UN2054	III	3	B1, T1	150	203	242	60 L	220 L	A	
	Morpholine, aqueous, mixture, see Corrosive liquids, n.o.s.												
	Motor fuel anti-knock compounds see Motor fuel anti-knock mixtures.												
+	Motor fuel anti-knock mixtures	6.1	UN1649	I	6.1, 3	14, B9, B90, T26, T39	None	201	244	Forbidden	30 L	D	25, 40

§ 172.101 HAZARDOUS MATERIALS TABLE—Continued

Symbols	Hazardous materials descriptions and proper shipping names	Hazard class or Division	Identification Numbers	PG	Label Codes	Special provisions	(8) Packaging (§173.***)			(9) Quantity limitations		(10) Vessel stowage	
							Exceptions	Non-bulk	Bulk	Passenger aircraft/rail	Cargo aircraft only	Location	Other
(1)	(2)	(3)	(4)	(5)	(6)	(7)	(8A)	(8B)	(8C)	(9A)	(9B)	(10A)	(10B)
	Motor spirit, see Gasoline												
	Muriatic acid, see Hydrochloric acid solution.												
	Musk xylene, see 5-tert-Butyl2,4,6-trinitro-m-xylene.												
	Naphtha see Petroleum distillates n.o.s.												
	Naphthalene, crude or Naphthalene, refined.	4.1	UN1334	III	4.1	A1	151	213	240	25 kg	100 kg	A	
	Naphthalene diozonide	Forbidden											
	Naphthalene, molten	4.1	UN2304	III	4.1	A1, T8	151	213	241	Forbidden	Forbidden	C	
	beta-Naphthylamine	6.1	UN1650	II	6.1	T12, T26	None	212	242	25 kg	100 kg	A	
	alpha-Naphthylamine	6.1	UN2077	III	6.1	T7	153	213	240	100 kg	200 kg	A	
	Naphthylamineperchlorate	Forbidden											
	Naphthylthiourea	6.1	UN1651	II	6.1		None	212	242	25 kg	100 kg	A	
	Naphthylurea	6.1	UN1652	II	6.1		None	212	242	25 kg	100 kg	A	
	Natural gases (with high methane content), see Methane, etc. (UN 1971, UN 1972).												
	Neohexane, see Hexanes												
	Neon, compressed	2.2	UN1065		2.2		306	302	302	75 kg	150 kg	A	
	Neon, refrigerated liquid *(cryogenic liquid).*	2.2	UN1913		2.2		320	316	None	50 kg	500 kg	B	
	New explosive or explosive device, see sections 173.51 and 173.56.												
	Nickel carbonyl	6.1	UN1259	I	6.1, 3	1	None	198	None	Forbidden	Forbidden	D	18, 40
	Nickel cyanide	6.1	UN1653	II	6.1	N74, N75	None	212	242	25 kg	100 kg	A	26
	Nickel nitrate	5.1	UN2725	III	5.1	A1	152	213	240	25 kg	100 kg	A	
	Nickel nitrite	5.1	UN2726	III	5.1	A1	152	213	240	25 kg	100 kg	A	56, 58
	Nickel picrate	Forbidden											
	Nicotine	6.1	UN1654	II	6.1		None	202	243	5 L	60 L	A	
	Nicotine compounds, liquid, n.o.s. or Nicotine preparations, liquid, n.o.s.	6.1	UN3144	I	6.1	A4, T42	None	201	243	1 L	30 L	B	40
				II	6.1	T14	None	202	243	5 L	60 L	B	40
				III	6.1	T7	153	203	241	60 L	220 L	B	40
	Nicotine compounds, solid, n.o.s. or Nicotine preparations, solid, n.o.s.	6.1	UN1655	I	6.1		None	211	242	5 kg	50 kg	B	
				II	6.1		None	212	242	25 kg	100 kg	A	
				III	6.1		153	213	240	100 kg	200 kg	A	
	Nicotine hydrochloride or Nicotine hydrochloride solution.	6.1	UN1656	II	6.1		None	202	243	5 L	60 L	A	
	Nicotine salicylate	6.1	UN1657	II	6.1		None	212	242	25 kg	100 kg	A	
	Nicotine sulfate, *solid*	6.1	UN1658	II	6.1		None	212	242	25 kg	100 kg	A	
	Nicotine sulfate, *solution*	6.1	UN1658	II	6.1	T14	None	202	243	5 L	60 L	A	
	Nicotine tartrate	6.1	UN1659	II	6.1		None	212	242	25 kg	100 kg	A	
	Nitrated paper (unstable)	Forbidden											
	Nitrates, inorganic, aqueous solution, n.o.s.	5.1	UN3218	II	5.1	58, T8	152	202	242	1 L	5 L	B	46
				III	5.1	58, T8	152	203	241	2.5 L	30 L	B	46
	Nitrates, inorganic, n.o.s	5.1	UN1477	II	5.1		152	212	240	5 kg	25 kg	A	46
				III	5.1		152	213	240	25 kg	100 kg	A	46
	Nitrates of diazonium compounds.	Forbidden											
	Nitrating acid mixtures, spent with more than 50 percent nitric acid.	8	UN1826	I	8, 5.1	T12, T27	None	158	243	Forbidden	2.5 L	D	40, 66
	Nitrating acid mixtures spent with not more than 50 percent nitric acid.	8	UN1826	II	8	B2, B100, T12, T27	None	158	242	Forbidden	30 L	D	40
	Nitrating acid mixtures *with more than 50 percent nitric acid.*	8	UN1796	I	8, 5.1	T12, T27	None	158	243	Forbidden	2.5 L	D	40, 66
	Nitrating acid mixtures *with not more than 50 percent nitric acid.*	8	UN1796	II	8	B2, T12, T27	None	158	242	Forbidden	30 L	D	40
	Nitric acid *other than red fuming, with more than 70 percent nitric acid.*	8	UN2031	I	8, 5.1	B47, B53, T9, T27	None	158	243	Forbidden	2.5 L	D	44, 66, 89, 90, 110, 111
	Nitric acid *other than red fuming, with not more than 70 percent nitric acid.*	8	UN2031	II	8	B2, B47, B53, T9, T27	None	158	242	Forbidden	30 L	D	44, 66, 89, 90, 110, 111
+	Nitric acid, red fuming	8	UN2032	I	8, 5.1, 6.1.	2, B9, B32, B74, T38, T43, T45	None	227	244	Forbidden	Forbidden	D	40, 66, 74, 89, 90
	Nitric oxide, compressed	2.3	UN1660		2.3, 5.1, 8.	1, B37, B46, B50, B60, B77	None	337	None	Forbidden	Forbidden	D	40, 89, 90
	Nitric oxide and dinitrogen tetroxide mixtures *or* Nitric oxide and nitrogen dioxide mixtures.	2.3	UN1975		2.3, 5.1, 8.	1, B7, B9, B14, B45, B46, B61, B66, B67, B77	None	337	None	Forbidden	Forbidden	D	40, 89, 90
G	Nitriles, flammable, toxic, n.o.s	3	UN3273	I	3, 6.1		None	201	243	Forbidden	30 L	E	40, 52
				II	3, 6.1	T14	None	202	243	1 L	60 L	B	40, 52
G	Nitriles, toxic, flammable, n.o.s	6.1	UN3275	I	6.1, 3	5	None	201	243	1 L	30 L	B	40

§172.101 HAZARDOUS MATERIALS TABLE—Continued

Symbols (1)	Hazardous materials descriptions and proper shipping names (2)	Hazard class or Division (3)	Identification Numbers (4)	PG (5)	Label Codes (6)	Special provisions (7)	Packaging (§173.***) (8) Exceptions (8A)	Packaging (§173.***) (8) Non-bulk (8B)	Packaging (§173.***) (8) Bulk (8C)	Quantity limitations (9) Passenger aircraft/rail (9A)	Quantity limitations (9) Cargo aircraft only (9B)	Vessel stowage (10) Location (10A)	Vessel stowage (10) Other (10B)
G	Nitriles, toxic, n.o.s.	6.1	UN3276	II	6.1, 3	T14	None	202	243	5 L	60 L	B	40
				I	6.1	5	None	201	243	1 L	30 L	B	
				II	6.1	T14	None	202	243	5 L	60 L	B	
				III	6.1	T7	153	203	241	60 L	220 L	A	
	Nitrites, inorganic, aqueous solution, n.o.s.	5.1	UN3219	II	5.1	T8	152	202	242	1 L	5 L	B	46, 56, 58
				III	5.1	T8	152	203	241	2.5 L	30 L	B	46, 56, 58
	Nitrites, inorganic, n.o.s	5.1	UN2627	II	5.1	33	152	212	None	5 kg	25 kg	A	46, 56, 58
	3-Nitro-4-chlorobenzotrifluoride	6.1	UN2307	II	6.1	T8	None	202	243	5 L	60 L	A	40
	6-Nitro-4-diazotoluene-3-sulfonic acid (dry).	Forbidden											
	Nitro isobutane triol trinitrate	Forbidden											
	N-Nitro-N-methylglycolamide nitrate.	Forbidden											
	2-Nitro-2-methylpropanol nitrate	Forbidden											
	Nitro urea	1.1D	UN0147	II	1.1D		None	62	None	Forbidden	Forbidden	B	1E, 5E
	N-Nitroaniline	Forbidden											
+	Nitroanilines (o-; m-; p-;)	6.1	UN1661	II	6.1	T14	None	212	242	25 kg	100 kg	A	
+	Nitroanisole	6.1	UN2730	III	6.1	T8	153	213	240	100 kg	200 kg	A	
+	Nitrobenzene	6.1	UN1662	II	6.1	T14	None	202	243	5 L	60 L	A	40
	m-Nitrobenzene diazonium perchlorate.	Forbidden											
	Nitrobenzenesulfonic acid	8	UN2305	II	8		154	202	242	1 L	30 L	A	
	Nitrobenzol, see Nitrobenzene												
	5-Nitrobenzotriazol	1.1D	UN0385	II	1.1D		None	62	None	Forbidden	Forbidden	B	1E, 5E, 19E
	Nitrobenzotrifluorides	6.1	UN2306	II	6.1	T8	None	202	243	5 L	60 L	A	40
	Nitrobromobenzenes liquid	6.1	UN2732	III	6.1	T8, T38	153	203	241	60 L	220 L	A	
	Nitrobromobenzenes solid	6.1	UN2732	III	6.1		153	213	240	100 kg	200 kg	A	
	Nitrocellulose, dry or wetted with less than 25 percent water (or alcohol), by mass.	1.1D	UN0340	II	1.1D		None	62	None	Forbidden	Forbidden	B	4E, 27E
	Nitrocellulose membrane filters	4.1	UN3270	II	4.1	43, A1	151	212	240	1 kg	15 kg	D	
	Nitrocellulose, plasticized with not less than 18 percent plasticizing substance, by mass.	1.3C	UN0343	II	1.3C		None	62	None	Forbidden	Forbidden	B	1E, 5E
	Nitrocellulose, solution, flammable with not more than 12.6 percent nitrogen, by mass, and not more than 55 percent nitrocellulose.	3	UN2059	II	3	T8, T31	150	202	242	5 L	60 L	B	
				III	3	B1, T7, T30	150	203	242	60 L	220 L	A	
	Nitrocellulose, unmodified or plasticized with less than 18 percent plasticizing substance, by mass.	1.1D	UN0341	II	1.1D		None	62	None	Forbidden	Forbidden	B	4E, 27E
	Nitrocellulose, wetted with not less than 25 percent alcohol, by mass.	1.3C	UN0342	II	1.3C		None	62	None	Forbidden	Forbidden	B	1E, 5E
	Nitrocellulose with alcohol with not less than 25 percent alcohol by mass, and with not more than 12.6 percent nitrogen, by dry mass.	4.1	UN2556	II	4.1		151	212	None	1 kg	15 kg	D	28
	Nitrocellulose, with not more than 12.6 percent nitrogen, by dry mass, or Nitrocellulose mixture with pigment or Nitrocellulose mixture with plasticizer or Nitrocellulose mixture with pigment and plasticizer.	4.1	UN2557	II	4.1	44	151	212	None	1 kg	15 kg	D	28
	Nitrocellulose with water with not less than 25 percent water, by mass.	4.1	UN2555	II	4.1		151	212	None	15 kg	50 kg	E	28
	Nitrochlorobenzene, see Chloronitrobenzenes etc.												
	Nitrocresols	6.1	UN2446	III	6.1		153	213	240	100 kg	200 kg	A	
	Nitroethane	3	UN2842	III	3	B1, T8	150	203	242	60 L	220 L	A	
	Nitroethyl nitrate	Forbidden											
	Nitroethylene polymer	Forbidden											
	Nitrogen, compressed	2.2	UN1066		2.2		306	302	314, 315.	75 kg	150 kg	A	
	Nitrogen dioxide, see Dinitrogen tetroxide.												
	Nitrogen fertilizer solution, see Fertilizer ammoniating solution etc.												
	Nitrogen, mixtures with rare gases, see Rare gases and nitrogen mixtures.												
	Nitrogen peroxide, see Dinitrogen tetroxide, liquefied.												
	Nitrogen, refrigerated liquid cryogenic liquid.	2.2	UN1977		2.2		320	316	318	50 kg	500 kg	D	

Appendix A

§ 172.101 Hazardous Materials Table—Continued

Symbols	Hazardous materials descriptions and proper shipping names	Hazard class or Division	Identification Numbers	PG	Label Codes	Special provisions	(8) Packaging (§173.***)			(9) Quantity limitations		(10) Vessel stowage	
							Exceptions	Non-bulk	Bulk	Passenger aircraft/rail	Cargo aircraft only	Location	Other
(1)	(2)	(3)	(4)	(5)	(6)	(7)	(8A)	(8B)	(8C)	(9A)	(9B)	(10A)	(10B)
	Nitrogen tetroxide and nitric oxide mixtures, see Nitric oxide and nitrogen tetroxide mixtures.												
	Nitrogen tetroxide, see Dinitrogen tetroxide, liquefied.												
	Nitrogen trichloride	Forbidden											
	Nitrogen trifluoride, compressed	2.2	UN2451		2.2, 5.1		None	302	None	Forbidden	25 kg	D	40
	Nitrogen triiodide	Forbidden											
	Nitrogen triiodide monoamine	Forbidden											
	Nitrogen trioxide	2.3	UN2421		2.3, 5.1, 8	1	None	336	245	Forbidden	Forbidden	D	40, 89, 90
	Nitroglycerin, desensitized with not less than 40 percent nonvolatile water insoluble phlegmatizer, by mass.	1.1D	UN0143	II	1.1D, 6.1	125	None	62	None	Forbidden	Forbidden	B	1E, 4E, 21E
	Nitroglycerin, liquid, not desensitized.	Forbidden											
	Nitroglycerin mixture, desensitized, liquid, flammable, n.o.s. with not more than 30 percent nitroglycerin, by mass.	3	UN3343		3	129	None	214	None	Forbidden	Forbidden	D	
	Nitroglycerin mixture, desensitized, solid, n.o.s. with more than 2 percent but not more than 10 percent nitroglycerin, by mass.	4.1	UN3319	II	4.1	118	None	None	None	Forbidden	0.5 kg	E	
	Nitroglycerin, solution in alcohol, with more than 1 percent but not more than 5 percent nitroglycerin.	3	UN3064	II	3	N8	None	202	None	Forbidden	5 L	E	
	Nitroglycerin, solution in alcohol, with more than 1 percent but not more than 10 percent nitroglycerin.	1.1D	UN0144	II	1.1D		None	62	None	Forbidden	Forbidden	B	1E, 5E, 21E
	Nitroglycerin solution in alcohol with not more than 1 percent nitroglycerin.	3	UN1204	II	3	N34, T25	None	202	None	5 L	60 L	B	
	Nitroguanidine nitrate	Forbidden											
	Nitroguanidine or Picrite, dry or wetted with less than 20 percent water, by mass.	1.1D	UN0282	II	1.1D		None	62	None	Forbidden	Forbidden	B	1E, 5E
	Nitroguanidine, wetted or Picrite, wetted with not less than 20 percent water, by mass.	4.1	UN1336	I	4.1	23, A8, A19, A20, N41	None	211	None	1 kg	15 kg	E	28
	1-Nitrohydantoin	Forbidden											
	Nitrohydrochloric acid	8	UN1798	I	8	A3, B10, N41, T18, T27	None	201	243	Forbidden	2.5 L	D	40, 66, 74, 89, 90
	Nitromannite (dry)	Forbidden											
	Nitromannite, wetted, see Mannitol hexanitrate, etc.												
	Nitromethane	3	UN1261	II	3	T25	150	202	None	Forbidden	60 L	A	
	Nitromuriatic acid, see Nitrohydrochloric acid.												
	Nitronaphthalene	4.1	UN2538	III	4.1	A1	151	213	240	25 kg	100 kg	A	
+	Nitrophenols (o-; m-; p-;)	6.1	UN1663	III	6.1	T8, T38	153	213	240	100 kg	200 kg	A	
	m-Nitrophenyldinitro methane	Forbidden											
	Nitropropanes	3	UN2608	III	3	B1, T1	150	203	242	60 L	220 L	A	
	p-Nitrosodimethylaniline	4.2	UN1369	II	4.2	A19, A20, B101, N34	None	212	241	15 kg	50 kg	D	34
	Nitrostarch, dry or wetted with less than 20 percent water, by mass.	1.1D	UN0146	II	1.1D		None	62	None	Forbidden	Forbidden	B	1E, 5E
	Nitrostarch, wetted with not less than 20 percent water, by mass.	4.1	UN1337	I	4.1	23, A8, A19, A20, N41	None	211	None	1 kg	15 kg	D	28
	Nitrosugars (dry)	Forbidden											
	Nitrosyl chloride	2.3	UN1069		2.3, 8	3, B14	None	304	314, 315.	Forbidden	Forbidden	D	40
	Nitrosylsulfuric acid	8	UN2308	II	8	A3, A6, A7, B2, N34, T9, T27	154	202	242	1 L	30 L	D	40, 66, 74, 89, 90
	Nitrotoluenes, liquid o-; m-; p-;	6.1	UN1664	II	6.1	T14	None	202	243	5 L	60 L	A	
	Nitrotoluenes, solid m-, or p-	6.1	UN1664	II	6.1	T14	None	212	242	25 kg	100 kg	A	
	Nitrotoluidines (mono)	6.1	UN2660	III	6.1		153	213	240	100 kg	200 kg	A	
	Nitrotriazolone or NTO	1.1D	UN0490	II	1.1D		None	62	None	Forbidden	Forbidden	B	1E, 5E
	Nitrous oxide and carbon dioxide mixtures, see Carbon dioxide and nitrous oxide mixtures.												
	Nitrous oxide	2.2	UN1070		2.2, 5.1		306	304	314, 315.	75 kg	150 kg	A	40
	Nitrous oxide, refrigerated liquid	2.2	UN2201		2.2, 5.1	B6	None	304	314, 315.	75 kg	150 kg	B	40
	Nitroxylenes, (o-; m-; p-)	6.1	UN1665	II	6.1	T14	None	202	243	5 L	60 L	A	
	Nitroxylol, see Nitroxylenes												
	Nonanes	3	UN1920	III	3	B1, T1	150	203	242	60 L	220 L	A	

§ 172.101 HAZARDOUS MATERIALS TABLE—Continued

Symbols	Hazardous materials descriptions and proper shipping names	Hazard class or Division	Identification Numbers	PG	Label Codes	Special provisions	Packaging (§173.***)			Quantity limitations		Vessel stowage	
							Exceptions	Non-bulk	Bulk	Passenger aircraft/rail	Cargo aircraft only	Location	Other
(1)	(2)	(3)	(4)	(5)	(6)	(7)	(8A)	(8B)	(8C)	(9A)	(9B)	(10A)	(10B)
	Nonflammable gas, n.o.s., see Compressed or Liquefied gases, etc. (UN 1955, UN 1956).												
	Nonliquefied gases, see Compressed gases, etc.												
	Nonliquefied hydrocarbon gas, see Hydrocarbon gases, compressed, n.o.s.												
	Nonyltrichlorosilane	8	UN1799	II	8	A7, B2, B6, N34, T8, T26	None	202	242	Forbidden	30 L	C	40
	Nordhausen acid, see Sulfuric acid, fuming etc.												
	Octadecyltrichlorosilane	8	UN1800	II	8	A7, B2, B6 N34, T8	None	202	242	Forbidden	30 L	C	40
	Octadiene	3	UN2309	II	3	B1, T1	150	202	242	5 L	60 L	B	
	1,7-Octadine-3,5-diyne-1,8-dimethoxy-9-octadecynoic acid.	Forbidden											
	Octafluorobut-2-ene *or Refrigerant gas R 1318.*	2.2	UN2422		2.2		None	304	314, 315.	75 kg	150 kg	A	
	Octafluorocyclobutane, *or Refrigerant gas R C318.*	2.2	UN1976		2.2		None	304	314, 315.	75 kg	150 kg	A	
	Octafluoropropane *or Refrigerant gas R 218.*	2.2	UN2424		2.2		None	304	314, 315.	75 kg	150 kg	A	
	Octanes	3	UN1262	II	3	T1	150	202	242	5 L	60 L	B	
	Octogen, see Cyclotetramethylene tetranitramine, etc.												
	Octolite *or Octol, dry or wetted with less than 15 percent water, by mass.*	1.1D	UN0266	II	1.1D		None	62	None	Forbidden	Forbidden	B	1E, 5E
	Octonal	1.1D	UN0496		1.1D		None	62	None	Forbidden	Forbidden	B	1E, 5E
	Octyl aldehydes	3	UN1191	III	3	B1, T1	150	203	242	60 L	220 L	A	
	Octyltrichlorosilane	8	UN1801	II	8	A7, B2, B6, N34, T8, T26	None	202	242	Forbidden	30 L	C	40
	Oil gas, compressed	2.3	UN1071		2.3, 2.1	6	None	304	314, 315.	Forbidden	25 kg	D	40
	Oleum, see Sulfuric acid, fuming												
	Organic peroxide type A, liquid or solid.	Forbidden											
G	Organic peroxide type B, liquid	5.2	UN3101	II	5.2, 1	53	152	225	None	Forbidden	Forbidden	D	12, 40
G	Organic peroxide type B, liquid, temperature controlled.	5.2	UN3111	II	5.2, 1	53	None	225	None	Forbidden	Forbidden	D	2, 40
G	Organic peroxide type B, solid	5.2	UN3102	II	5.2, 1	53	152	225	None	Forbidden	Forbidden	D	12, 40
G	Organic peroxide type B, solid, temperature controlled.	5.2	UN3112	II	5.2, 1	53	None	225	None	Forbidden	Forbidden	D	2, 40
	Organic peroxide type C, liquid	5.2	UN3103	II	5.2		152	225	None	5 L	10 L	D	12, 40
G	Organic peroxide type C, liquid, temperature controlled.	5.2	UN3113	II	5.2		None	225	None	Forbidden	Forbidden	D	2, 40
	Organic peroxide type C, solid	5.2	UN3104	II	5.2		152	225	None	5 kg	10 kg	D	12, 40
G	Organic peroxide type C, solid, temperature controlled.	5.2	UN3114	II	5.2		None	225	None	Forbidden	Forbidden	D	2, 40
G	Organic peroxide type D, liquid	5.2	UN3105	II	5.2		152	225	None	5 L	10 L	D	12, 40
G	Organic peroxide type D, liquid, temperature controlled.	5.2	UN3115	II	5.2		None	225	None	Forbidden	Forbidden	D	2, 40
G	Organic peroxide type D, solid	5.2	UN3106	II	5.2		152	225	None	5 kg	10 kg	D	12, 40
G	Organic peroxide type D, solid, temperature controlled.	5.2	UN3116	II	5.2		None	225	None	Forbidden	Forbidden	D	2, 40
G	Organic peroxide type E, liquid	5.2	UN3107	II	5.2		152	225	None	10 L	25 L	D	12, 40
G	Organic peroxide type E, liquid, temperature controlled.	5.2	UN3117	II	5.2		None	225	None	Forbidden	Forbidden	D	2, 40
G	Organic peroxide type E, solid	5.2	UN3108	II	5.2		152	225	None	10 kg	25 kg	D	12, 40
G	Organic peroxide type E, solid, temperature controlled.	5.2	UN3118	II	5.2		None	225	None	Forbidden	Forbidden	D	2, 40
G	Organic peroxide type F, liquid	5.2	UN3109	II	5.2		152	225	225	10 L	25 L	D	12, 40
G	Organic peroxide type F, liquid, temperature controlled.	5.2	UN3119	II	5.2		None	225	225	Forbidden	Forbidden	D	2, 40
G	Organic peroxide type F, solid	5.2	UN3110	II	5.2	T42	152	225	225	10 kg	25 kg	D	12, 40
G	Organic peroxide type F, solid, temperature controlled.	5.2	UN3120	II	5.2		None	225	225	Forbidden	Forbidden	D	2, 40
D	Organic phosphate, mixed with compressed gas *or* Organic phosphate compound, mixed with compressed gas *or* Organic phosphorus compound, mixed with compressed gas.	2.3	NA1955	III	6.1 2.3	B1, T14 3	153 None	203 334	242 None	60 L Forbidden	220 L Forbidden	A D	40 40
	Organic pigments, self-heating	4.2	UN3313	III III	4.2 4.2	B101	None None	213 213	241 241	25 kg 25 kg	100 kg 100 kg	C C	
	Organoarsenic compound, n.o.s	6.1	UN3280	I II III	6.1 6.1 6.1	5 T14 T7	None None 153	211 212 213	242 242 240	5 kg 25 kg 100 kg	50 kg 100 kg 200 kg	B B A	
	Organochlorine pesticides liquid, flammable, toxic, *flash point less than 23 degrees C.*	3	UN2762	I II	3, 6.1 3, 6.1		None None	201 202	243 243	Forbidden 1 L	30 L 60 L	B B	40 40
	Organochlorine pesticides, liquid, toxic.	6.1	UN2996	I	6.1	T42	None	201	243	1 L	30 L	B	40

Appendix A

§172.101 HAZARDOUS MATERIALS TABLE—Continued

Symbols	Hazardous materials descriptions and proper shipping names	Hazard class or Division	Identification Numbers	PG	Label Codes	Special provisions	(8) Packaging (§173.***)			(9) Quantity limitations		(10) Vessel stowage	
							Exceptions	Non-bulk	Bulk	Passenger aircraft/rail	Cargo aircraft only	Location	Other
(1)	(2)	(3)	(4)	(5)	(6)	(7)	(8A)	(8B)	(8C)	(9A)	(9B)	(10A)	(10B)
				II	6.1	T14	None	202	243	5 L	60 L	B	40
				III	6.1	T14	153	203	241	60 L	220 L	A	40
	Organochlorine pesticides, liquid, toxic, flammable, *flashpoint not less than 23 degrees C.*	6.1	UN2995	I	6.1, 3	T42	None	201	243	1 L	30 L	B	40
				II	6.1, 3	T14	None	202	243	5 L	60 L	B	40
				III	6.1	B1, T14	153	203	242	60 L	220 L	A	40
	Organochlorine pesticides, solid toxic.	6.1	UN2761	I	6.1		None	211	242	5 kg	50 kg	A	40
				II	6.1		None	212	242	25 kg	100 kg	A	40
				III	6.1		153	213	240	100 kg	200 kg	A	40
G	Organometallic compound *or* Compound solution *or* Compound dispersion, water-reactive, flammable, n.o.s.	4.3	UN3207	I	4.3, 3	T28	None	201	244	Forbidden	1 L	E	40
G	Organometallic compound *or* Compound solution *or* Compound dispersion, water-reactive, flammable, n.o.s.	4.3	UN3207	II	4.3, 3	T28	None	202	243	1 L	5 L	E	40
G	Organometallic compound *or* Compound solution *or* Compound dispersion, water-reactive, flammable, n.o.s.	4.3	UN3207	III	4.3, 3	T28	None	203	242	5 L	60L	E	40
G	Organometallic compound, toxic n.o.s.	6.1	UN3282	I	6.1	B106	None	211	242	5 kg	50 kg	B	
				II	6.1	T14	None	212	242	25 kg	100 kg	B	
				III	6.1	T7	153	213	240	100 kg	200 kg	A	
	Organophosphorus compound, toxic, flammable, n.o.s.	6.1	UN3279	I	6.1, 3	5	None	201	243	1 L	30 L	B	40
				II	6.1, 3	T14	None	202	243	5 L	60 L	B	40
	Organophosphorus compound, toxic n.o.s.	6.1	UN3278	I	6.1	5	None	201	243	1 L	30 L	B	
				II	6.1	T14	None	202	243	5 L	60 L	B	
				III	6.1	T7	153	203	241	60 L	220 L	A	
	Organophosphorus pesticides, liquid, flammable, toxic, *flash point less than 23 degrees C.*	3	UN2784	I	3, 6.1	T42	None	201	243	Forbidden	30 L	B	40
				II	3, 6.1	T18	None	202	243	1 L	60 L	B	40
	Organophosphorus pesticides, liquid, toxic.	6.1	UN3018	I	6.1	N76, T42	None	201	243	1 L	30 L	B	40
				II	6.1	N76, T14	None	202	243	5 L	60 L	B	40
				III	6.1	N76, T14	153	203	241	60 L	220 L	A	40
	Organophosphorus pesticides, liquid, toxic, flammable, *flashpoint not less than 23 degrees C.*	6.1	UN3017	I	6.1, 3	N76, T42	None	201	243	1 L	30 L	B	40
				II	6.1, 3	N76, T14	None	202	243	5 L	60 L	B	40
				III	6.1, 3	B1, N76, T14	153	203	242	60 L	220 L	A	40
	Organophosphorus pesticides, solid, toxic.	6.1	UN2783	I	6.1	N77	None	211	242	5 kg	50 kg	A	40
				II	6.1	N77	None	212	242	25 kg	100 kg	A	40
				III	6.1	N77	153	213	240	100 kg	200 kg	A	40
	Organotin compounds, liquid, n.o.s.	6.1	UN2788	I	6.1	A3, N33, N34, T42	None	201	243	1 L	30 L	B	40
				II	6.1	A3, N33, N34, T14	None	202	243	5 L	60 L	A	40
				III	6.1	T14	153	203	241	60 L	220 L	A	40
	Organotin compounds, solid, n.o.s.	6.1	UN3146	I	6.1	A5	None	211	242	5 kg	50 kg	B	40
				II	6.1		None	212	242	25 kg	100 kg	A	40
				III	6.1		153	213	240	100 kg	200 kg	A	40
	Organotin pesticides, liquid, flammable, toxic, *flash point less than 23 degrees C.*	3	UN2787	I	3, 6.1		None	201	243	Forbidden	30 L	B	40
				II	3, 6.1		None	202	243	1 L	60 L	B	40
	Organotin pesticides, liquid, toxic.	6.1	UN3020	I	6.1	T42	None	201	243	1 L	30 L	B	40
				II	6.1	T14	None	202	243	5 L	60 L	B	40
				III	6.1	T14	153	203	241	60 L	220 L	A	40
	Organotin pesticides, liquid, toxic, flammable, *flashpoint not less than 23 degrees C.*	6.1	UN3019	I	6.1, 3	T42	None	201	243	1 L	30 L	B	40
				II	6.1, 3	T14	None	202	243	5 L	60 L	B	40
				III	6.1, 3	B1, T14	153	203	242	60 L	220 L	A	40
	Organotin pesticides, solid, toxic	6.1	UN2786	I	6.1		None	211	242	5 kg	50 kg	A	40
				II	6.1		None	212	242	25 kg	100 kg	A	40
				III	6.1		153	213	240	100 kg	200 kg	A	40
	Orthonitroaniline, *see* Nitroanilines *etc.*												
	Osmium tetroxide	6.1	UN2471	I	6.1	A8, B100, N33, N34	None	211	242	5 kg	50 kg	B	40
D	Other regulated substances, liquid, n.o.s.	9	NA3082	III	9		155	203	241	No limit	No limit	A	
D	Other regulated substances, solid, n.o.s.	9	NA3077	III	9	B54	155	213	240	No limit	No limit	A	
G	Oxidizing liquid, corrosive, n.o.s	5.1	UN3098	I	5.1, 8		None	201	244	Forbidden	2.5 L	D	34, 56, 58, 69, 106

§ 172.101 Hazardous Materials Table—Continued

Symbols	Hazardous materials descriptions and proper shipping names	Hazard class or Division	Identification Numbers	PG	Label Codes	Special provisions	Packaging (§173.***) Exceptions	Packaging Non-bulk	Packaging Bulk	Quantity limitations Passenger aircraft/rail	Quantity limitations Cargo aircraft only	Vessel stowage Location	Vessel stowage Other
(1)	(2)	(3)	(4)	(5)	(6)	(7)	(8A)	(8B)	(8C)	(9A)	(9B)	(10A)	(10B)
				II	5.1, 8		None	202	243	1 L	5 L	B	34, 56, 58, 69, 106
				III	5.1, 8		152	203	242	2.5 L	30 L	B	34, 56, 58, 69, 106
G	Oxidizing liquid, n.o.s	5.1	UN3139	I	5.1	127, A2	None	201	243	Forbidden	2.5 L	D	56, 58, 69, 106
				II	5.1	127, A2	152	202	242	1 L	5 L	B	56, 58, 69, 106
				III	5.1	127, A2	152	203	241	2.5 L	30 L	B	56, 58, 69, 106
G	Oxidizing liquid, toxic, n.o.s	5.1	UN3099	I	5.1, 6.1		None	201	244	Forbidden	2.5 L	D	56, 58, 95, 106
				II	5.1, 6.1		None	202	243	1 L	5 L	B	56, 58, 95, 106
				III	5.1, 6.1		152	203	242	2.5 L	30 L	B	56, 58, 95, 106
G	Oxidizing solid, corrosive, n.o.s	5.1	UN3085	I	5.1, 8		None	211	242	1 kg	15 kg	D	13, 34, 56, 58, 69, 106
				II	5.1, 8		None	212	242	5 kg	25 kg	B	13, 34, 56, 58, 69, 106
				III	5.1, 8		152	213	240	25 kg	100 kg	B	13, 34, 56, 58, 69, 106
G	Oxidizing solid, flammable, n.o.s	5.1	UN3137	I	5.1, 4.1		None	214	214	Forbidden	Forbidden		
G	Oxidizing solid, n.o.s	5.1	UN1479	I	5.1		None	211	242	1 kg	15 kg	D	56, 58, 69, 106
				II	5.1		152	212	240	5 kg	25 kg	B	56, 58, 69, 106
				III	5.1		152	213	240	25 kg	100 kg	B	56, 58, 69, 106
G	Oxidizing solid, self-heating, n.o.s.	5.1	UN3100	II	5.1, 4.2		None	214	214	Forbidden	Forbidden		
G	Oxidizing solid, toxic, n.o.s	5.1	UN3087	I	5.1, 6.1		None	211	242	1 kg	15 kg	D	56, 58, 69, 95, 106
				II	5.1, 6.1		None	212	242	5 kg	25 kg	B	56, 58, 69, 95, 106
				III	5.1, 6.1		152	213	240	25 kg	100 kg	B	56, 58, 69, 95, 106
G	Oxidizing solid, water-reactive, n.o.s.	5.1	UN3121		5.1, 4.3		None	214	214	Forbidden	Forbidden		
	Oxygen and carbon dioxide mixtures, see Carbon dioxide and oxygen mixtures.												
	Oxygen, compressed	2.2	UN1072		2.2, 5.1	A52	306	302	314, 315.	75 kg	150 kg	A	
	Oxygen difluoride, compressed	2.3	UN2190		2.3, 5.1, 8.	1	None	304	None	Forbidden	Forbidden	D	13, 40, 89, 90
	Oxygen generator, chemical	5.1	UN3356	II	5.1	60, A51	None	212	None	Forbidden	25 kg gross	D	56, 58, 69, 106
+	Oxygen generator, chemical, spent.	9	NA3356	III	9	61	None	213	None	Forbidden	Forbidden	A	
	Oxygen, mixtures with rare gases, see Rare gases and oxygen mixtures.												
	Oxygen, refrigerated liquid (cryogenic liquid).	2.2	UN1073		2.2, 5.1		320	316	318	Forbidden	Forbidden	D	
	Paint including paint, lacquer, enamel, stain, shellac solutions, varnish, polish, liquid filler, and liquid lacquer base.	3	UN1263	I	3	T8, T31	150	201	243	1 L	30 L	E	
				II	3	B52, T7, T30	150	173	242	5 L	60 L	B	
				III	3	B1, B52, T7, T30	150	173	242	60 L	220 L	A	
	Paint or Paint related material	8	UN3066	II	8	B2, T14	154	173	242	1 L	30 L	A	
				III	8	B52, T7	154	173	241	5 L	60 L	A	
	Paint related material including paint thinning, drying, removing, or reducing compound.	3	UN1263	I	3	T8, T31	150	201	243	1 L	30 L	E	
				II	3	B52, T7, T30	150	173	242	5 L	60 L	B	
				III	3	B1, B52, T7, T30	150	173	242	60 L	220 L	A	
	Paper, unsaturated oil treated incompletely dried (including carbon paper).	4.2	UN1379	III	4.2	B101, B106	None	213	241	Forbidden	Forbidden	A	
	Paraformaldehyde	4.1	UN2213	III	4.1	A1	151	213	240	25 kg	100 kg	A	
	Paraldehyde	3	UN1264	III	3	B1, T1	150	203	242	60 L	220 L	A	
	Paranitroaniline, solid, see Nitroanilines etc.												
D	Parathion	6.1	NA2783	I	6.1	T42	None	201	243	Forbidden	1 L	A	40
				II	6.1	T14	None	202	243	Forbidden	5 L	A	40
D	Parathion and compressed gas mixture.	2.3	NA1967		2.3	3	None	334	245	Forbidden	Forbidden	E	40

Appendix A

§172.101 Hazardous Materials Table—Continued

Symbols	Hazardous materials descriptions and proper shipping names	Hazard class or Division	Identification Numbers	PG	Label Codes	Special provisions	Packaging (§173.***) Exceptions	Packaging (§173.***) Non-bulk	Packaging (§173.***) Bulk	Quantity limitations Passenger aircraft/rail	Quantity limitations Cargo aircraft only	Vessel stowage Location	Vessel stowage Other
(1)	(2)	(3)	(4)	(5)	(6)	(7)	(8A)	(8B)	(8C)	(9A)	(9B)	(10A)	(10B)
	Paris green, solid, see Copper acetoarsenite.												
A,W	*PCB, see* Polychlorinated biphenyls.												
+	Pentaborane	4.2	UN1380	I	4.2, 6.1	1	None	205	245	Forbidden	Forbidden	D	
	Pentachloroethane	6.1	UN1669	II	6.1	T14	None	202	243	5 L	60 L	A	40
	Pentachlorophenol	6.1	UN3155	II	6.1		None	212	242	25 kg	100 kg	A	
	Pentaerythrite tetranitrate (dry)	Forbidden											
	Pentaerythrite tetranitrate mixture, desensitized, solid, n.o.s. *with more than 10 percent but not more than 20 percent PETN, by mass.*	4.1	UN3344	II	4.1	118	None	214	None	Forbidden	Forbidden	E	40
	Pentaerythrite tetranitrate *or* Pentaerythritol tetranitrate *or* PETN, *with not less than 7 percent wax by mass.*	1.1D	UN0411	II	1.1D		None	62	None	Forbidden	Forbidden	B	1E, 5E
	Pentaerythrite tetranitrate, wetted *or* Pentaerythritol tetranitrate, wetted, *or* PETN, wetted *with not less than 25 percent water, by mass, or* Pentaerythrite tetranitrate, *or* Pentaerythritol tetranitrate *or* PETN, desensitized *with not less than 15 percent phlegmatizer by mass.*	1.1D	UN0150	II	1.1D	121	None	62	None	Forbidden	Forbidden	B	1E, 5E
	Pentaerythritol tetranitrate, see Pentaerythrite tetranitrate, etc.												
	Pentafluoroethane *or Refrigerant gas R 125.*	2.2	UN3220		2.2		306	304	314, 315.	75 kg	150 kg	A	
	Pentamethylheptane	3	UN2286	III	3	B1, T1	150	203	242	60 L	220 L	A	
	Pentane-2,4-dione	3	UN2310	III	3, 6.1	B1, T1	150	203	242	60 L	220 L	A	
	Pentanes	3	UN1265	I	3	T20	150	201	243	1 L	30 L	E	
				II	3	T20	150	202	242	5 L	60 L	E	
	Pentanitroaniline (dry)	Forbidden											
	Pentanols	3	UN1105	II	3	T1	150	202	242	5 L	60 L	B	
				III	3	B1, B3, T1	150	203	242	60 L	220 L	A	
	1-Pentene (*n-amylene*)	3	UN1108	I	3	T14	150	201	243	1 L	30 1	E	
	1-Pentol	8	UN2705	II	8	B2, T8	154	202	242	1 L	30 L	B	38
	Pentolite, dry or wetted with less than 15 percent water, by mass.	1.1D	UN0151	II	1.1D		None	62	None	Forbidden	Forbidden	B	1E, 5E
	Pepper spray, see Aerosols, etc. or Self-defense spray, non-pressurized.												
	Perchlorates, inorganic, aqueous solution, n.o.s.	5.1	UN3211	II	5.1	T8	152	202	242	1 L	5 L	B	46, 56, 58
				III	5.1	T8	152	203	241	2.5 L	30 L	B	56, 58, 69, 106
	Perchlorates, inorganic, n.o.s	5.1	UN1481	II	5.1		152	212	242	5 kg	25 kg	A	46, 56
				III	5.1		152	213	240	25 kg	100 kg	A	46, 56
	Perchloric acid, with more than 72 percent acid by mass.	Forbidden											
	Perchloric acid *with more than 50 percent but not more than 72 percent acid, by mass.*	5.1	UN1873	I	5.1, 8	A2, A3, N41, T9, T27	None	201	243	Forbidden	2.5 L	D	66
	Perchloric acid *with not more than 50 percent acid by mass.*	8	UN1802	II	8, 5.1	N41, T9	None	202	243	Forbidden	30 L	C	66
	Perchloroethylene, see Tetrachloroethylene.												
	Perchloromethyl mercaptan	6.1	UN1670	I	6.1	2, A3, A7, B9, B14, B32, B74, N34, T38, T43, T45	None	227	244	Forbidden	Forbidden	D	40
	Perchloryl fluoride	2.3	UN3083		2.3, 5.1	2, B9, B14	None	302	314, 315.	Forbidden	Forbidden	D	40
	Percussion caps, see Primers, cap type.												
	Perfluoro-2-butene, see Octafluorobut-2-ene.												
	Perfluoro(ethyl vinyl ether)	2.1	UN3154		2.1		306	302, 304, 305.	314, 315.	Forbidden	150 kg	E	40
	Perfluoro(methyl vinyl ether)	2.1	UN3153		2.1		306	302, 304, 305.	314, 315.	Forbidden	150 kg	E	40
	Perfumery products *with flammable solvents.*	3	UN1266	II	3	T7, T30	150	202	242	15 L	60 L	B	
				III	3	B1, T7, T30	150	203	242	60 L	220 L	A	
	Permanganates, inorganic, aqueous solution, n.o.s.	5.1	UN3214	II	5.1	26, T8	152	202	242	1 L	5 L	D	56, 58, 69, 106, 107
	Permanganates, inorganic, n.o.s	5.1	UN1482	II	5.1	26, A30	152	212	242	5 kg	25 kg	D	56, 58, 69, 106, 107

§ 172.101 HAZARDOUS MATERIALS TABLE—Continued

Symbols	Hazardous materials descriptions and proper shipping names	Hazard class or Division	Identification Numbers	PG	Label Codes	Special provisions	Packaging (§173.***)			Quantity limitations		Vessel stowage	
							Exceptions	Non-bulk	Bulk	Passenger aircraft/rail	Cargo aircraft only	Location	Other
(1)	(2)	(3)	(4)	(5)	(6)	(7)	(8A)	(8B)	(8C)	(9A)	(9B)	(10A)	(10B)
				III	5.1	26, A30	152	213	240	25 kg	100 kg	D	56, 58, 69, 106, 107
	Peroxides, inorganic, n.o.s	5.1	UN1483	II	5.1	A7, A20, N34	None	212	242	5 kg	25 kg	A	13, 75, 106
				III	5.1	A7, A20, N34	152	213	240	25 kg	100 kg	A	13, 75, 106
	Peroxyacetic acid, with more than 43 percent and with more than 6 percent hydrogen peroxide.	Forbidden											
	Persulfates, inorganic, aqueous solution, n.o.s.	5.1	UN3216	III	5.1	T2	152	203	241	2.5 L	30 L	A	
	Persulfates, inorganic, n.o.s	5.1	UN3215	III	5.1		152	213	240	25 kg	100 kg	A	
G	Pesticides, liquid, flammable, toxic, *flashpoint less than 23 degrees C.*	3	UN3021	I	3, 6.1	B5	None	201	243	Forbidden	30 L	B	
				II	3, 6.1		None	202	243	1 L	60 L	B	
G	Pesticides, liquid, toxic, flammable, n.o.s. *flashpoint not less than 23 degrees C.*	6.1	UN2903	I	6.1, 3	T42	None	201	243	1 L	30 L	B	40
				II	6.1, 3	T14	None	202	243	5 L	60 L	B	40
				III	6.1, 3	B1, T14	153	203	242	60 L	220 L	A	40
G	Pesticides, liquid, toxic, n.o.s	6.1	UN2902	I	6.1	T42	None	201	243	1 L	30 L	B	40
				II	6.1	T14	None	202	243	5 L	60 L	B	40
				III	6.1	T14	153	203	241	60 L	220 L	A	40
G	Pesticides, solid, toxic, n.o.s	6.1	UN2588	I	6.1		None	211	242	5 kg	50 kg	B	40
				II	6.1		None	212	242	25 kg	100 kg	A	40
				III	6.1		153	213	240	100 kg	200 kg	A	40
	PETN, see Pentaerythrite tetranitrate.												
	PETN/TNT, see Pentolite, etc												
	Petrol, see Gasoline												
	Petroleum crude oil	3	UN1267	I	3	T8, T31	None	201	243	1 L	30 L	E	
				II	3	T8, T31	150	202	242	5 L	60 L	B	
				III	3	B1, T7, T30	150	203	242	60 L	220 L	A	
	Petroleum distillates, n.o.s. or Petroleum products, n.o.s.	3	UN1268	I	3	T8, T31	150	201	243	1 L	30 L	E	
				II	3	T8, T31	150	202	242	5 L	60 L	B	
				III	3	B1, T7, T30	150	203	242	60 L	220 L	A	
	Petroleum gases, liquefied or Liquefied petroleum gas.	2.1	UN1075		2.1		306	304	314, 315.	Forbidden	150 kg	E	40
D	Petroleum oil	3	NA1270	I	3	T8, T31	None	201	243	1 L	30 L	E	
				II	3	T8, T31	150	202	243	5 L	60 L	B	
				III	3	B1, T7, T30	150	203	242	60 L	220 L	A	
	Phenacyl bromide	6.1	UN2645	II	6.1	B106	None	212	242	25 kg	100 kg	B	40
+	Phenetidines	6.1	UN2311	III	6.1	T7	153	203	241	60 L	220 L	A	
	Phenol, molten	6.1	UN2312	II	6.1	B14, B100, T8	None	202	243	Forbidden	Forbidden	B	40
+	Phenol, solid	6.1	UN1671	II	6.1	N78, T14	None	212	242	25 kg	100 kg	A	
	Phenol solutions	6.1	UN2821	II	6.1	T14	None	202	243	5 L	60 L	A	
				III	6.1	T7	153	203	241	60 L	220 L	A	
	Phenolsulfonic acid, liquid	8	UN1803	II	8	B2, N41, T8	154	202	242	1 L	30 L	C	14
	Phenoxyacetic acid derivative pesticide, liquid, flammable, toxic *flashpoint less than 23°C.*.	3	UN3346	I	3, 6.1	T23	None	201	243	Forbidden	30 L	B	40
				II	3, 6.1	T14	None	202	243	1 L	60 L	B	40
	Phenoxyacetic acid derivative pesticide, liquid, toxic.	6.1	UN3348	I	6.1	T24, T26	None	201	243	1 L	30 L	B	40
				II	6.1	T14	153	202	243	5 L	60 L	B	40
				III	6.1	T14	153	203	241	60 L	220 L	A	40
	Phenoxyacetic acid derivative pesticide, liquid, toxic, flammable, *flashpoint not less than 23°C.*	6.1	UN3347	I	6.1, 3	T24, T26	None	201	243	1 L	30 L	B	40
				II	6.1, 3	T14	153	202	243	5 L	60 L	B	40
				III	6.1, 3	T14	153	203	241	60 L	220 L	A	40
	Phenoxyacetic acid derivative pesticide, solid, toxic.	6.1	UN3345	I	6.1		None	211	242	5 kg	50 kg	A	40
				II	6.1		153	212	242	25 kg	100 kg	A	40
				III	6.1		153	213	240	100 kg	200 kg	A	40
	Phenyl chloroformate	6.1	UN2746	II	6.1, 8	T12	None	202	243	1 L	30 L	A	12, 13, 21, 25, 40, 100
	Phenyl isocyanate	6.1	UN2487	I	6.1, 3	2, B9, B14, B32, B74, B77, N33, N34, T38, T43, T45	None	227	244	Forbidden	Forbidden	D	20, 40, 95
	Phenyl mercaptan	6.1	UN2337	I	6.1, 3	2, B9, B14, B32, B74, B77, T38, T43, T45	None	227	244	Forbidden	Forbidden	B	26, 40
	Phenyl phosphorus dichloride	8	UN2798	II	8	B2, B15, T8, T26	154	202	242	Forbidden	30 L	B	40
	Phenyl phosphorus thiodichloride.	8	UN2799	II	8	B2, B15, T8, T26	154	202	242	Forbidden	30 L	B	40
	Phenyl urea pesticides, liquid, toxic.	6.1	UN3002	I	6.1	T42	None	201	243	1 L	30 L	B	40
				II	6.1	T14	None	202	243	5 L	60 L	B	40
				III	6.1	T14	153	203	241	60 L	220 L	A	40

Appendix A

§ 172.101 HAZARDOUS MATERIALS TABLE—Continued

Symbols	Hazardous materials descriptions and proper shipping names	Hazard class or Division	Identification Numbers	PG	Label Codes	Special provisions	Packaging (§173.***)			Quantity limitations		Vessel stowage	
							Exceptions	Non-bulk	Bulk	Passenger aircraft/rail	Cargo aircraft only	Location	Other
(1)	(2)	(3)	(4)	(5)	(6)	(7)	(8A)	(8B)	(8C)	(9A)	(9B)	(10A)	(10B)
	Phenylacetonitrile, liquid	6.1	UN2470	III	6.1	T8	153	203	241	60 L	220 L	A	26
	Phenylacetyl chloride	8	UN2577	II	8	B2, T8, T26	154	202	242	1 L	30 L	C	40
	Phenylcarbylamine chloride	6.1	UN1672	I	6.1	2, B9, B14, B32, B74, T38, T43, T45	None	227	244	Forbidden	Forbidden	D	40
	m-Phenylene diaminediperchlorate (dry).	Forbidden											
+	Phenylenediamines (o-; m-; p-;)	6.1	UN1673	III	6.1		153	213	240	100 kg	200 kg	A	
	Phenylhydrazine	6.1	UN2572	II	6.1	T8	None	202	243	5 L	60 L	A	40
	Phenylmercuric acetate	6.1	UN1674	II	6.1		None	212	242	25 kg	100 kg	A	
	Phenylmercuric compounds, n.o.s.	6.1	UN2026	I	6.1		None	211	242	5 kg	50 kg	A	
				II	6.1		None	212	242	25 kg	100 kg	A	
				III	6.1		153	213	240	100 kg	200 kg	A	
	Phenylmercuric hydroxide	6.1	UN1894	II	6.1		None	212	242	25 kg	100 kg	A	
	Phenylmercuric nitrate	6.1	UN1895	II	6.1		None	212	242	25 kg	100 kg	A	
	Phenyltrichlorosilane	8	UN1804	II	8	A7, B6, N34, T8	None	202	242	Forbidden	30 L	C	40
	Phosgene	2.3	UN1076		2.3, 8	1, B7, B46	None	Forbidden	Forbidden	Forbidden	Forbidden	D	40
	9-Phosphabicyclononanes or Cyclooctadiene phosphines.	4.2	UN2940	II	4.2	A19	None	212	241	15 kg	50 kg	A	
	Phosphine	2.3	UN2199		2.3, 2.1	1	None	192	245	Forbidden	Forbidden	D	40
	Phosphoric acid	8	UN1805	III	8	A7, N34, T7	154	203	241	5 L	60 L	A	
	Phosphoric acid triethyleneimine, see Tris-(1-aziridiyl)phosphine oxide, solution.												
	Phosphoric anhydride, see Phosphorus pentoxide.												
	Phosphorous acid	8	UN2834	III	8	T7	154	213	240	25 kg	100 kg	A	48
	Phosphorus, amorphous	4.1	UN1338	III	4.1	A1, A19, B1, B9, B26	None	213	243	25 kg	100 kg	A	74
	Phosphorus bromide, see Phosphorus tribromide.												
	Phosphorus chloride, see Phosphorus trichloride.												
	Phosphorus heptasulfide, free from yellow or white phosphorus.	4.1	UN1339	II	4.1	A20, N34	None	212	240	15 kg	50 kg	B	74
	Phosphorus oxybromide	8	UN1939	II	8	B8, B106, N41, N43	None	212	240	Forbidden	50 kg	C	12, 40
	Phosphorus oxybromide, molten	8	UN2576	II	8	B2, B8, N41, N43, T8, T27	None	202	242	Forbidden	Forbidden	C	40
+	Phosphorus oxychloride	8	UN1810	II	8, 6.1	2, A7, B9, B14, B32, B74, B77, N34, T38, T43, T45	None	227	244	Forbidden	30 L	C	40
	Phosphorus pentabromide	8	UN2691	II	8	A7, B106, N34	154	212	240	Forbidden	50 kg	B	12, 40
	Phosphorus pentachloride	8	UN1806	II	8	A7, B106, N34	None	212	240	Forbidden	50 kg	C	40
	Phosphorus pentafluoride, compressed.	2.3	UN2198		2.3, 8	2, B9, B14	None	302, 304.	314, 315.	Forbidden	Forbidden	D	40
	Phosphorus pentasulfide, free from yellow or white phosphorus.	4.3	UN1340	II	4.3, 4.1	A20, B59, B101, B106	151	212	242	15 kg	50 kg	B	74
	Phosphorus pentoxide	8	UN1807	II	8	A7, N34	154	212	240	15 kg	50 kg	A	
	Phosphorus sesquisulfide, free from yellow or white phosphorus.	4.1	UN1341	II	4.1	A20, N34	None	212	240	15 kg	50 kg	B	74
	Phosphorus tribromide	8	UN1808	II	8	A3, A6, A7, B2, B25, N34, N43, T8	None	202	242	Forbidden	30 L	C	40
	Phosphorus trichloride	6.1	UN1809	I	6.1, 8	2, B9, B14, B15, B32, B74, B77, N34, T38, T43, T45	None	227	244	Forbidden	Forbidden	C	40
	Phosphorus trioxide	8	UN2578	III	8		154	213	240	25 kg	100 kg	A	12
	Phosphorus trisulfide, free from yellow or white phosphorus.	4.1	UN1343	II	4.1	A20, N34	None	212	240	15 kg	50 kg	B	74
	Phosphorus, white dry or Phosphorus, white, under water or Phosphorus white, in solution or Phosphorus, yellow dry or Phosphorus, yellow, under water or Phosphorus, yellow, in solution.	4.2	UN1381	I	4.2, 6.1	B9, B26, N34, T15, T26, T33	None	188	243	Forbidden	Forbidden	E	
	Phosphorus white, molten	4.2	UN2447	I	4.2, 6.1	B9, B26, N34, T15, T26, T29	None	188	243	Forbidden	Forbidden	D	
	Phosphorus (white or red) and a chlorate, mixtures of.	Forbidden											
	Phosphoryl chloride, see Phosphorus oxychloride.												
	Phthalic anhydride with more than .05 percent maleic anhydride.	8	UN2214	III	8	T7	154	213	240	25 kg	100 kg	A	
	Picolines	3	UN2313	III	3	B1, T8	150	203	242	60 L	220 L	A	40
	Picric acid, see Trinitrophenol, etc.												
D	Picric acid, wet, with not less than 10 percent water.	4.1	NA1344	I	4.1	A19, A20, N41	None	211	None	Forbidden	Forbidden	D	

§172.101 HAZARDOUS MATERIALS TABLE—Continued

Symbols	Hazardous materials descriptions and proper shipping names	Hazard class or Division	Identification Numbers	PG	Label Codes	Special provisions	Packaging (§173.***) (8)			Quantity limitations (9)		Vessel stowage (10)	
							Exceptions (8A)	Non-bulk (8B)	Bulk (8C)	Passenger aircraft/rail (9A)	Cargo aircraft only (9B)	Location (10A)	Other (10B)
(1)	(2)	(3)	(4)	(5)	(6)	(7)	(8A)	(8B)	(8C)	(9A)	(9B)	(10A)	(10B)
	Picrite, see Nitroguanidine, etc ..												
	Picryl chloride, see Trinitrochlorobenzene.												
	Pine oil	3	UN1272	III	3	B1, T1	150	203	242	60 L	220 L	A	
	alpha-Pinene	3	UN2368	III	3	B1, T1	150	203	242	60 L	220 L	A	
	Piperazine	8	UN2579	III	8	T7	154	213	240	25 kg	100 kg	A	12
	Piperidine	8	UN2401	I	8, 3	T17	None	201	243	0.5 L	2.5 L	B	
	Pivaloyl chloride, see Trimethylacetyl chloride.												
	Plastic molding compound in dough, sheet or extruded rope form evolving flammable vapor.	9	UN3314	III	9	32	155	221	221	100 kg	200 kg	A	85, 87
	Plastic solvent, n.o.s., see Flammable liquids, n.o.s.												
	Plastics, nitrocellulose-based, self-heating, n.o.s.	4.2	UN2006	III	4.2		None	213	None	Forbidden	Forbidden	C	
	Poisonous gases, n.o.s., see Compressed or liquefied gases, flammable or toxic, n.o.s.												
	Polyalkylamines, n.o.s., see Amines, etc.												
A, W	Polychlorinated biphenyls, liquid	9	UN2315	II	9	9, 81	155	202	241	100 L	220 L	A	34
A, W	Polychlorinated biphenyls, solid	9	UN2315	II	9	9, 81	155	212	240	100 kg	200 kg	A	34
	Polyester resin kit	3	UN3269		3	40	152	225	None	5 kg	5 kg	B	
	Polyhalogenated biphenyls, liquid or Polyhalogenated terphenyls liquid.	9	UN3151	II	9		155	204	241	100 L	220 L	A	34
	Polyhalogenated biphenyls, solid or Polyhalogenated terphenyls, solid.	9	UN3152	II	9		155	204	241	100 kg	200 kg	A	34
	Polymeric beads, expandable, evolving flammable vapor.	9	UN2211	III	None	32	155	221	221	100 kg	200 kg	A	85, 87
	Potassium	4.3	UN2257	I	4.3	A19, A20, B27, B100, N6, N34, T15, T26	None	211	244	Forbidden	15 kg	D	
	Potassium arsenate	6.1	UN1677	II	6.1		None	212	242	25 kg	100 kg	A	
	Potassium arsenite	6.1	UN1678	II	6.1		None	212	242	25 kg	100 kg	A	
	Potassium bisulfite solution, see Bisulfites, inorganic, aqueous solutions, n.o.s.												
	Potassium borohydride	4.3	UN1870	I	4.3	A19, B100, N40	None	211	242	Forbidden	15 kg	E	
	Potassium bromate	5.1	UN1484	II	5.1		152	212	242	5 kg	25 kg	A	56, 58, 106
	Potassium carbonyl	Forbidden											
	Potassium chlorate	5.1	UN1485	II	5.1	A9, N34	152	212	242	5 kg	25 kg	A	56, 58, 106
	Potassium chlorate, aqueous solution.	5.1	UN2427	II	5.1	A2, T8	152	202	241	1 L	5 L	B	56, 58, 106
				III	5.1	A2, T8	152	203	241	2.5 L	30 L	B	56, 58, 69, 106
	Potassium chlorate mixed with mineral oil, see Explosive, blasting, type C.												
	Potassium cuprocyanide	6.1	UN1679	II	6.1		None	212	242	25 kg	100kg	A	26
	Potassium cyanide	6.1	UN1680	I	6.1	B69, B77, N74, N75, T18, T26	None	211	242	5 kg	50 kg	B	52
	Potassium dichloro isocyanurate or Potassium dichloro-s-triazinetrione, see Dichloroisocyanuric acid, dry or Dichloroisocyanuric acid salts etc.												
	Potassium dithionite or Potassium hydrosulfite.	4.2	UN1929	II	4.2	A8, A19, A20	None	212	241	15 kg	50 kg	E	13
	Potassium fluoride	6.1	UN1812	III	6.1	T8	153	213	240	100 kg	200 kg	A	26
	Potassium fluoroacetate	6.1	UN2628	I	6.1		None	211	242	5 kg	50 kg	E	
	Potassium fluorosilicate	6.1	UN2655	III	6.1		153	213	240	100 kg	200 kg	A	26
	Potassium hydrate, see Potassium hydroxide, solid.												
	Potassium hydrogen fluoride, see Potassium hydrogen difluoride.												
	Potassium hydrogen fluoride solution, see Corrosive liquid, n.o.s.												
	Potassium hydrogen sulfate	8	UN2509	II	8	A7, N34	154	212	240	15 kg	50 kg	A	
	Potassium hydrogendifluoride, solid.	8	UN1811	II	8, 6.1	B106, N3, N34, T8	154	212	240	15 kg	50 kg	A	25, 26, 40
	Potassium hydrogendifluoride, solution.	8	UN1811	II	8, 6.1	N3, N34, T8	154	202	243	1 L	30 L	A	26, 40, 95
	Potassium hydrosulfite, see Potassium dithionite.												
	Potassium hydroxide, liquid, see Potassium hydroxide solution.												
	Potassium hydroxide, solid	8	UN1813	II	8		154	212	240	15 kg	50 kg	A	
	Potassium hydroxide, solution	8	UN1814	II	8	B2, T8	154	202	242	1 L	30 L	A	
				III	8	T7	154	203	241	5 L	60 L	A	

Appendix A

§ 172.101 Hazardous Materials Table—Continued

Symbols	Hazardous materials descriptions and proper shipping names	Hazard class or Division	Identification Numbers	PG	Label Codes	Special provisions	(8) Packaging (§173.***)			(9) Quantity limitations		(10) Vessel stowage	
							Exceptions	Non-bulk	Bulk	Passenger aircraft/rail	Cargo aircraft only	Location	Other
(1)	(2)	(3)	(4)	(5)	(6)	(7)	(8A)	(8B)	(8C)	(9A)	(9B)	(10A)	(10B)
	Potassium hypochlorite, solution, see Hypochlorite solutions, etc.												
	Potassium, metal alloys	4.3	UN1420	I	4.3	A19, A20, B27	None	211	244	Forbidden	15 kg	D	
	Potassium metal, liquid alloy, see Alkali metal alloys, liquid.												
	Potassium metavanadate	6.1	UN2864	II	6.1		None	212	242	25 kg	100 kg	A	
	Potassium monoxide	8	UN2033	II	8		154	212	240	15 kg	50 kg	A	
	Potassium nitrate	5.1	UN1486	III	5.1	A1, A29	152	213	240	25 kg	100 kg	A	
	Potassium nitrate and sodium nitrite mixtures.	5.1	UN1487	II	5.1	B78	152	212	240	5 kg	25 kg	A	56, 58
	Potassium nitrite	5.1	UN1488	II	5.1		152	212	242	5 kg	25 kg	A	56, 58
	Potassium perchlorate, solid	5.1	UN1489	II	5.1		152	212	242	5 kg	25 kg	A	56, 58, 106
	Potassium perchlorate, solution	5.1	UN1489	II	5.1		152	202	242	1 L	5 L	A	56, 58, 106
	Potassium permanganate	5.1	UN1490	II	5.1		152	212	240	5 kg	25 kg	D	56, 58, 69, 106, 107
	Potassium peroxide	5.1	UN1491	I	5.1	A20, N34	None	211	None	Forbidden	15 kg	B	13, 75, 106
	Potassium persulfate	5.1	UN1492	III	5.1	A1, A29	152	213	240	25 kg	100 kg	A	
	Potassium phosphide	4.3	UN2012	I	4.3, 6.1	A19, N40	None	211	None	Forbidden	15 kg	E	40, 85
	Potassium selenate, see Selenates or Selenites.												
	Potassium selenite, see Selenates or Selenites.												
	Potassium sodium alloys	4.3	UN1422	I	4.3	A19, B27, N34, N40, T15, T26	None	211	244	Forbidden	15 kg	D	
	Potassium sulfide, anhydrous *or* Potassium sulfide *with less than 30 percent water of crystallization.*	4.2	UN1382	II	4.2	A19, A20, B16, B106, N34	None	212	241	15 kg	50 kg	A	
	Potassium sulfide, hydrated *with not less than 30 percent water of crystallization.*	8	UN1847	II	8		154	212	240	15 kg	50 kg	A	26
	Potassium superoxide	5.1	UN2466	I	5.1	A20	None	211	None	Forbidden	15 kg	B	13, 75, 106
	Powder cake, wetted *or* Powder paste, wetted *with not less than 17 percent alcohol by mass.*	1.1C	UN0433	II	1.1C		None	62	None	Forbidden	Forbidden	B	1E, 5E
	Powder cake, wetted *or* Powder paste, wetted *with not less than 25 percent water, by mass.*	1.3C	UN0159	II	1.3C		None	62	None	Forbidden	Forbidden	B	1E, 5E
	Powder paste, see Powder cake, etc.												
	Powder, smokeless	1.1C	UN0160	II	1.1C		None	62	None	Forbidden	Forbidden	B	10E, 26E
	Powder, smokeless	1.3C	UN0161	II	1.3C		None	62	None	Forbidden	Forbidden	B	10E, 26E
	Power device, explosive, see Cartridges, power device.												
	Primers, cap type	1.4S	UN0044	II	None		None	62	None	25 kg	100 kg	A	
	Primers, cap type	1.1B	UN0377	II	1.1B		None	62	None	Forbidden	Forbidden	B	2E, 6E
	Primers, cap type	1.4B	UN0378	II	1.4B		None	62	None	Forbidden	75 kg	A	24E
	Primers, small arms, see Primers, cap type.												
	Primers, tubular	1.3G	UN0319	II	1.3G		None	62	None	Forbidden	Forbidden	B	
	Primers, tubular	1.4G	UN0320	II	1.4G		None	62	None	Forbidden	75 kg	A	24E
	Primers, tubular	1.4S	UN0376	II	None		None	62	None	25 kg	100 kg	A	
	Printing ink, *flammable*	3	UN1210	I	3	T8, T31	150	173	243	1 L	30 L	E	
				II	3	T7, T30	150	173	242	5 L	60 L	B	
				III	3	B1, T7, T30	150	173	242	60 L	220 L	A	
	Projectiles, illuminating, see Ammunition, illuminating, etc.												
	Projectiles, *inert with tracer*	1.4S	UN0345	II	1.4S			62	None	25 kg	100 kg	A	3E, 7E, 9E
	Projectiles, *inert, with tracer*	1.3G	UN0424	II	1.3G			62	None	Forbidden	Forbidden	B	3E, 7E
	Projectiles, *inert, with tracer*	1.4G	UN0425	II	1.4G			62	None	Forbidden	75 kg	A	3E, 7E, 24E
	Projectiles, *with burster or expelling charge.*	1.2D	UN0346	II	1.2D			62	None	Forbidden	Forbidden	B	3E, 7E
	Projectiles, *with burster or expelling charge.*	1.4D	UN0347	II	1.4D			62	None	Forbidden	75 kg	A	3E, 7E, 24E
	Projectiles, *with burster or expelling charge.*	1.2F	UN0426	II	1.2F			62	None	Forbidden	Forbidden	E	
	Projectiles, *with burster or expelling charge.*	1.4F	UN0427	II	1.4F			62	None	Forbidden	Forbidden	E	
	Projectiles, *with burster or expelling charge.*	1.2G	UN0434	II	1.2G			62	None	Forbidden	Forbidden	B	3E, 7E
	Projectiles, *with burster or expelling charge.*	1.4G	UN0435	II	1.4G			62	None	Forbidden	75 kg	A	3E, 7E, 24E
	Projectiles, *with bursting charge*	1.1F	UN0167	II	1.1F			62	None	Forbidden	Forbidden	E	
	Projectiles, *with bursting charge*	1.1D	UN0168	II	1.1D			62	None	Forbidden	Forbidden	B	3E, 7E
	Projectiles, *with bursting charge*	1.2D	UN0169	II	1.2D			62	None	Forbidden	Forbidden	B	3E, 7E
	Projectiles, *with bursting charge*	1.2F	UN0324	II	1.2F			62	None	Forbidden	Forbidden	E	

§172.101 HAZARDOUS MATERIALS TABLE—Continued

Symbols (1)	Hazardous materials descriptions and proper shipping names (2)	Hazard class or Division (3)	Identification Numbers (4)	PG (5)	Label Codes (6)	Special provisions (7)	(8) Packaging (§173.***)			(9) Quantity limitations		(10) Vessel stowage	
							Exceptions (8A)	Non-bulk (8B)	Bulk (8C)	Passenger aircraft/rail (9A)	Cargo aircraft only (9B)	Location (10A)	Other (10B)
	Projectiles, *with bursting charge*	1.4D	UN0344	II	1.4D			62	None	Forbidden	75 kg	A	3E, 7E, 24E
	Propadiene, inhibited	2.1	UN2200		2.1		None	304	314, 315.	Forbidden	150 kg	B	40
	Propadiene mixed with methyl acetylene, see Methyl acetylene and propadiene mixtures, stabilized.												
	Propane *see also Petroleum gases, liquefied.*	2.1	UN1978		2.1	19	306	304	314, 315.	Forbidden	150 kg	E	40
	Propanethiols	3	UN2402	II	3	T8	150	202	242	5 L	60 L	E	95, 102
	n-Propanol *or Propyl alcohol, normal.*	3	UN1274	II	3	B1, T1	150	202	242	5 L	60 L	B	
				III	3	B1, T1	150	203	242	60 L	220 L	A	
D	Propargyl alcohol	3	NA1986	II	3, 6.1		None	202	243	Forbidden	1 L	B	40
	Propellant, liquid	1.3C	UN0495	II	1.3C	37	None	62	None	Forbidden	Forbidden	B	
	Propellant, liquid	1.1C	UN0497	II	1.1C	37	None	62	None	Forbidden	Forbidden	B	
	Propellant, solid	1.1C	UN0498	II	1.1C		None	62	None	Forbidden	Forbidden	A	
	Propellant, solid	1.3C	UN0499	II	1.3C		None	62	None	Forbidden	Forbidden	A	
	Propionaldehyde	3	UN1275	II	3	T14	150	202	242	5 L	60 L	E	
	Propionic acid	8	UN1848	III	8	T7	154	203	241	5 L	60 L	A	
	Propionic anhydride	8	UN2496	III	8	T2	154	203	241	5 L	60 L	A	
	Propionitrile	3	UN2404	II	3, 6.1	T14	None	202	243	Forbidden	60 L	E	40
	Propionyl chloride	3	UN1815	II	3, 8	B100, T8, T26	None	202	243	1 L	5 L	B	40
	n-Propyl acetate	3	UN1276	II	3	T1	150	202	242	5 L	60 L	B	
	Propyl alcohol, see Propanol												
	n-Propyl benzene	3	UN2364	III	3	B1, T1	150	203	242	60 L	220 L	A	
	Propyl chloride	3	UN1278	II	3	N34, T14	None	202	242	Forbidden	60 L	E	
	n-Propyl chloroformate	6.1	UN2740	I	6.1, 3, 8.	2, A3, A6, A7, B9, B14, B32, B74, B77, N34, T38, T43, T45	None	227	244	Forbidden	2.5 L	B	21, 40, 100
	Propyl formates	3	UN1281	II	3	T8	150	202	242	5 L	60 L	B	
	n-Propyl isocyanate	6.1	UN2482	I	6.1, 3	1, A7, B9, B14, B30, B72, T38, T43, T44	None	226	244	Forbidden	Forbidden	D	40
	Propyl mercaptan, see Propanethiols.												
	n-Propyl nitrate	3	UN1865	II	3	T25	150	202	None	5 L	60 L	D	
	Propylamine	3	UN1277	II	3, 8	N34, T14	None	202	243	1 L	5 L	E	40
	Propylene *see also Petroleum gases, liquefied.*	2.1	UN1077		2.1	19	306	304	314, 315.	Forbidden	150 kg	E	40
	Propylene chlorohydrin	6.1	UN2611	II	6.1, 3	T9	None	202	243	5 L	60 L	A	12, 40, 48
	Propylene oxide	3	UN1280	I	3	A3, N34, T20, T29	None	201	243	1 L	30 L	E	40
	Propylene tetramer	3	UN2850	III	3	B1, T1	150	203	242	60 L	220 L	A	
	1,2-Propylenediamine	8	UN2258	II	8, 3	A3, A6, N34, T8	None	202	243	1 L	30 L	A	40
	Propyleneimine, inhibited	3	UN1921	I	3, 6.1	A3, N34, T24	None	201	243	1 L	30 L	B	40
	Propyltrichlorosilane	8	UN1816	II	8, 3	A7, B2, B6, N34, T8, T26	None	202	243	Forbidden	30 L	C	40
	Prussic acid, see Hydrogen cyanide.												
	Pyrethroid pesticide, liquid, toxic, flammable, *flashpoint less than 23 °C.*	3	UN3350	I	3, 6.1	T24, T26	None	201	243	Forbidden	30 L	B	40
				II	3, 6.1	T14	None	202	243	1 L	60 L	B	40
	Pyrethroid pesticide, liquid toxic	6.1	UN3352	I	6.1		None	211	242	1 L	30 L	A	40
				II	6.1		153	212	242	5 L	60 L	A	40
				III	6.1		153	213	240	60 L	220 L	A	40
	Pyrethroid pesticide, liquid, flammable, toxic, *flashpoint not less than 23 degrees C.*	6.1	UN3351	I	6.1, 3	T24, T26	None	201	243	1 L	30 L	B	40
				II	6.1, 3	T14	153	202	243	5 L	60 L	B	40
				III	6.1, 3	T14	153	203	241	60 L	220 L	B	40
	Pyrethroid pesticide, solid, toxic	6.1	UN3349	I	6.1		None	211	242	5 kg	50 kg	A	40
				II	6.1		153	212	242	25 kg	100 kg	A	40
				III	6.1		153	213	240	100 kg	200 kg	A	40
	Pyridine	3	UN1282	II	3	T8	None	202	242	5 L	60 L	B	21, 100
	Pyridine perchlorate	Forbidden											
G	Pyrophoric liquid, inorganic, n.o.s.	4.2	UN3194	I	4.2		None	181	244	Forbidden	Forbidden	D	18
G	Pyrophoric liquids, organic, n.o.s	4.2	UN2845	I	4.2	B11, T42	None	181	244	Forbidden	Forbidden	D	18
G	Pyrophoric metals, n.o.s., *or* Pyrophoric alloys, n.o.s.	4.2	UN1383	I	4.2	B11	None	187	242	Forbidden	Forbidden	D	
G	Pyrophoric organometallic compound, water-reactive, n.o.s.	4.2	UN3203	I	4.2, 4.3	T28, T40	None	187	242	Forbidden	Forbidden	D	
G	Pyrophoric solid, inorganic, n.o.s	4.2	UN3200	I	4.2		None	187	242	Forbidden	Forbidden	D	
G	Pyrophoric solids, organic, n.o.s	4.2	UN2846	I	4.2		None	187	242	Forbidden	Forbidden	D	
	Pyrosulfuryl chloride	8	UN1817	II	8	B2, T9, T27	154	202	242	1 L	30 L	C	40
	Pyroxylin solution or solvent, see Nitrocellulose.												
	Pyrrolidine	3	UN1922	II	3, 8	T1	None	202	243	1 L	5 L	B	40
•	Quebrachitol pentanitrate	Forbidden											
	Quicklime, see Calcium oxide												
	Quinoline	6.1	UN2656	III	6.1	T8	153	203	241	60 L	220 L	A	12
	R 114, *see* Dichlorotetrafluoroethane.												
	R 115, *see* Chloropentafluoroethane.												

Appendix A

§172.101 Hazardous Materials Table—Continued

Symbols	Hazardous materials descriptions and proper shipping names	Hazard class or Division	Identification Numbers	PG	Label Codes	Special provisions	(8) Packaging (§173.***)			(9) Quantity limitations		(10) Vessel stowage	
							Exceptions	Non-bulk	Bulk	Passenger aircraft/rail	Cargo aircraft only	Location	Other
(1)	(2)	(3)	(4)	(5)	(6)	(7)	(8A)	(8B)	(8C)	(9A)	(9B)	(10A)	(10B)
	R 116, see Hexafluoroethane												
	R 124, see Chlorotetrafluoroethane.												
	R 133a, see Chlorotrifluoroethane.												
	R 152a, see Difluoroethane												
	R 500, see Dichlorodifluoromethane and difluorethane, etc.												
	R 502, see Chlorodifluoromethane and chloropentafluoroethane mixture, etc.												
	R 503, see Chlorotrifluoromethane and trifluoromethane, etc.												
	R 12, see Dichlorodifluoromethane.												
	R 12B1, see Chlorodifluorobromomethane.												
	R 13, see Chlorotrifluoromethane.												
	R 13B1, see Bromotrifluoromethane.												
	R 14, see Tetrafluoromethane												
	R 21, see Dichlorofluoromethane.												
	R 22, see Chlorodifluoromethane.												
	Radioactive material, excepted package-articles manufactured from natural or depleted uranium or natural thorium.	7	UN2910		None		422, 426.	422, 426.	422, 426.			A	
	Radioactive material, excepted package-empty package or empty packaging.	7	UN2910		empty		428	428	428			A	
	Radioactive material, excepted package-instruments or articles.	7	UN2910		None		422, 424.	422, 424.	422, 424.			A	
	Radioactive material, excepted package-limited quantity of material.	7	UN2910		None		421, 422.	421, 422.	421, 422.			A	
	Radioactive material, fissile, n.o.s.	7	UN2918		7		453	417	417			A	40, 95
	Radioactive material, low specific activity, n.o.s. or Radioactive material, LSA, n.o.s.	7	UN2912		7		421, 428.	427	427			A	
	Radioactive material, n.o.s	7	UN2982		7		421, 428.	415, 416.	415, 416.			A	40, 95
	Radioactive material, special form, n.o.s.	7	UN2974		7		421, 424.	415, 416.	415, 416.			A	
	Radioactive material, surface contaminated object or Radioactive material, SCO.	7	UN2913		7		421, 424, 426.	427	427			A	
	Railway torpedo, see Signals, railway track, explosive.												
	Rare gases and nitrogen mixtures, compressed.	2.2	UN1981		2.2		306	302	None	75 kg	150 kg	A	
	Rare gases and oxygen mixtures, compressed.	2.2	UN1980		2.2		306	302	None	75 kg	150 kg	A	
	Rare gases mixtures, compressed.	2.2	UN1979		2.2		306	302	None	75 kg	150 kg	A	
	RC 318, see Octafluorocyclobutane.												
	RDX and cyclotetramethylenetetranitramine, wetted or desensitized see RDX and HMX mixtures, wetted or desensitized.												
	RDX and HMX mixtures, wetted with not less than 15 percent water by mass or RDX and HMX mixtures, desensitized with not less than 10 percent phlegmatizer by mass.	1.1D	UN0391	II	1.1D		None	62	None	Forbidden	Forbidden	B	1E, 5E
	RDX and Octogen mixtures, wetted or desensitized see RDX and HMX mixtures, wetted or desensitized etc.												
	RDX, see Cyclotrimethylene trinitramine, etc.												
	Receptacles, small, containing gas (gas cartridges) flammable, without release device, not refillable and not exceeding 1 L capacity.	2.1	UN2037		2.1		306	304	None	1 kg	15 kg	B	40

§ 172.101 Hazardous Materials Table—Continued

Symbols	Hazardous materials descriptions and proper shipping names	Hazard class or Division	Identification Numbers	PG	Label Codes	Special provisions	Packaging (§173.***)			Quantity limitations		Vessel stowage	
							Exceptions	Non-bulk	Bulk	Passenger aircraft/rail	Cargo aircraft only	Location	Other
(1)	(2)	(3)	(4)	(5)	(6)	(7)	(8A)	(8B)	(8C)	(9A)	(9B)	(10A)	(10B)
	Receptacles, small, containing gas (gas cartridges) *non-flammable, without release device, not refillable and not exceeding 1 L capacity.*	2.2	UN2037		2.2		306	304	None	1 kg	15 kg	B	40
	Red phosphorus, see Phosphorus, amorphous.												
	Refrigerant gas R 404A	2.2	UN3337		2.2		306	304	314, 315.	75 kg	150 kg	A	
	Refrigerant gas R 407A	2.2	UN3338		2.2		306	304	314, 315.	75 kg	150 kg	A	
	Refrigerant gas R 407B	2.2	UN3339		2.2		306	304	314, 315.	75 kg	150 kg	A	
	Refrigerant gas R 407C	2.2	UN3340		2.2		306	304	314, 315.	75 kg	150 kg	A	
G	Refrigerant gases, n.o.s	2.2	UN1078		2.2		306	304	314, 315.	75 kg	150 kg	A	
D	Refrigerant gases, n.o.s. or Dispersant gases, n.o.s.	2.1	NA1954		2.1		306	304	314, 315.	Forbidden	150 kg*	D	40
D	Refrigerating machine	3	NA1993	III	3		174	174	None	10 L	10 L	A	
D	Refrigerating machines, containing flammable, non-poisonous, liquefied gas.	2.1	NA1954		2.1		306	306	306	Forbidden	25 kg	C	40
G	Refrigerating machines, containing non-flammable, nontoxic, liquefied gas or ammonia solution (UN2672).	2.2	UN2857		2.2		306, 307.	306	306, 307.	450 kg	450 kg	A	
D	Regulated medical waste	6.2	UN3291	II	6.2	A13, A14	134	197	None	Forbidden	Forbidden	E	
	Release devices, explosive	1.4S	UN0173	II	1.4S		None	62	None	25 kg	100 kg	A	
	Resin solution, *flammable*	3	UN1866	I	3	B52, T8, T31	150	201	243	1 L	30 L	E	
				II	3	B52, T7, T30	150	173	242	5 L	60 L	B	
				III	3	B1, B52, T7, T30	150	173	242	60 L	220 L	A	
	Resorcinol	6.1	UN2876	III	6.1		153	213	240	100 kg	200 kg	A	
	Rifle grenade, see Grenades, hand or rifle, etc.												
	Rifle powder, see Powder, smokeless (UN 0160).												
	Rivets, explosive	1.4S	UN0174	II	1.4S		None	62	None	25 kg	100 kg	A	
	Road asphalt or tar liquid, see Tars, liquid, etc.												
	Rocket motors	1.3C	UN0186	II	1.3C		109	None	62	None	Forbidden	220 kg	B	
	Rocket motors	1.1C	UN0280	II	1.1C		109	None	62	None	Forbidden	Forbidden	B	
	Rocket motors	1.2C	UN0281	II	1.2C		109	None	62	None	Forbidden	Forbidden	B	
	Rocket motors, liquid fueled	1.2J	UN0395	II	1.2J		109	None	62	None	Forbidden	Forbidden	E	7E, 16E, 23E
	Rocket motors, liquid fueled	1.3J	UN0396	II	1.3J		109	None	62	None	Forbidden	Forbidden	E	7E, 16E, 23E
	Rocket motors with hypergolic liquids *with or without an expelling charge.*	1.3L	UN0250	II	1.3L		109	None	62	None	Forbidden	Forbidden	E	2E, 8E, 11E, 17E
	Rocket motors with hypergolic liquids *with or without an expelling charge.*	1.2L	UN0322	II	1.2L		109	None	62	None	Forbidden	Forbidden	E	2E, 8E, 11E, 17E
	Rockets, line-throwing	1.2G	UN0238	II	1.2G			None	62	None	Forbidden	Forbidden	B	
	Rockets, line-throwing	1.3G	UN0240	II	1.3G			None	62	None	Forbidden	75 kg	B	
	Rockets, line-throwing	1.4G	UN0453	II	1.4G			None	62	None	Forbidden	75 kg	A	24E
	Rockets, liquid fueled *with bursting charge.*	1.1J	UN0397	II	1.1J			None	62	None	Forbidden	Forbidden	E	7E, 16E, 23E
	Rockets, liquid fueled *with bursting charge.*	1.2J	UN0398	II	1.2J			None	62	None	Forbidden	Forbidden	E	7E, 16E, 23E
	Rockets, *with bursting charge*	1.1F	UN0180	II	1.1F			None	62	None	Forbidden	Forbidden	E	
	Rockets, *with bursting charge*	1.1E	UN0181	II	1.1E			None	62	None	Forbidden	Forbidden	B	
	Rockets, *with bursting charge*	1.2E	UN0182	II	1.2E			None	62	None	Forbidden	Forbidden	E	
	Rockets, *with bursting charge*	1.2F	UN0295	II	1.2F			None	62	None	Forbidden	Forbidden	E	
	Rockets, *with expelling charge*	1.2C	UN0436	II	1.2C			None	62	None	Forbidden	Forbidden	B	
	Rockets, *with expelling charge*	1.3C	UN0437	II	1.3C			None	62	None	Forbidden	Forbidden	B	
	Rockets, *with expelling charge*	1.4C	UN0438	II	1.4C			None	62	None	Forbidden	75 kg	A	24E
	Rockets, *with inert head*	1.3C	UN0183	II	1.3C			None	62	None	Forbidden	Forbidden	B	
	Rosin oil	3	UN1286	II	3	T7	150	202	242	5 L	60 L	B	
				III	3	B1, T1	150	203	242	60 L	220 L	A	
	Rubber solution	3	UN1287	II	3	T7, T30	150	202	242	5 L	60 L	B	
				III	3	B1, T7, T30	150	203	242	60 L	220 L	A	
	Rubidium	4.3	UN1423	I	4.3	22, A7, A19, B100, N34, N40, N45	None	211	242	Forbidden	15 kg	D	
	Rubidium hydroxide	8	UN2678	II	8	T8	154	212	240	15 kg	50 kg	A	
	Rubidium hydroxide solution	8	UN2677	II	8	B2, T8	154	202	242	1 L	30 L	A	
				III	8	T7	154	203	241	5 L	60 L	A	
	Safety fuse, see Fuse, safety												
G	Samples, explosive, other than initiating explosives.		UN0190	II		113	None	62	None	Forbidden	Forbidden	E	12E
	Sand acid, see Fluorosilicic acid												

Appendix A

§172.101 Hazardous Materials Table—Continued

Symbols	Hazardous materials descriptions and proper shipping names	Hazard class or Division	Identification Numbers	PG	Label Codes	Special provisions	(8) Packaging (§173.***)			(9) Quantity limitations		(10) Vessel stowage	
							Exceptions	Non-bulk	Bulk	Passenger aircraft/rail	Cargo aircraft only	Location	Other
(1)	(2)	(3)	(4)	(5)	(6)	(7)	(8A)	(8B)	(8C)	(9A)	(9B)	(10A)	(10B)
	Seed cake, containing vegetable oil solvent extractions and expelled seeds, with not more than 10 percent of oil and when the amount of moisture is higher than 11 percent, with not more than 20 percent of oil and moisture combined..	4.2	UN1386	III	None ..	N7	None	213	241	Forbidden	Forbidden	A	13
I	Seed cake with more than 1.5 percent oil and not more than 11 percent moisture.	4.2	UN1386	III	None ..	N7	None	213	241	Forbidden	Forbidden	E	13
I	Seed cake with not more than 1.5 percent oil and not more than 11 percent moisture.	4.2	UN2217	III	None ..	N7	None	213	241	Forbidden	Forbidden	A	13
	Selenates or Selenites	6.1	UN2630	I	6.1		None	211	242	5 kg	50 kg	E	
	Selenic acid	8	UN1905	I	8	N34	None	211	242	Forbidden	25 kg	A	
	Selenium compound, n.o.s	6.1	UN3283	I	6.1		None	211	242	5 kg	50 kg	B	
				II	6.1	T14	None	212	242	25 kg	100 kg	B	
				III	6.1	T7	153	213	240	100 kg	200 kg	A	
	Selenium disulfide	6.1	UN2657	II	6.1		None	212	242	25 kg	100 kg	A	
	Selenium hexafluoride	2.3	UN2194		2.3, 8	1	None	302	None ..	Forbidden	Forbidden	D	40
	Selenium nitride	Forbidden											
D	Selenium oxide	6.1	NA2811	I	6.1		None	211	242	5 kg	50 kg	B	
	Selenium oxychloride	8	UN2879	I	8, 6.1	A3, A6, A7, N34, T12, T27	None	201	243	0.5 L	2.5 L	E	40
	Self-defense spray, aerosol, see Aerosols, etc.												
+AD	Self-defense spray, non-pressurized.	9	NA3334	III	9	A37	155	203	None ..	No limit	No limit	A	
G	Self-heating liquid, corrosive, inorganic, n.o.s.	4.2	UN3188	II	4.2, 8		None	202	243	1 L	5 L	C	
				III	4.2, 8		None	203	241	5 L	60 L	C	
G	Self-heating liquid, corrosive, organic, n.o.s.	4.2	UN3185	II	4.2, 8		None	202	243	1 L	5 L	C	
				III	4.2, 8		None	203	241	5 L	60 L	C	
G	Self-heating liquid, inorganic, n.o.s.	4.2	UN3186	II	4.2		None	202	242	1 L	5 L	C	
				III	4.2		None	203	241	5 L	60 L	C	
G	Self-heating liquid, organic, n.o.s	4.2	UN3183	II	4.2		None	202	242	1 L	5 L	C	
				III	4.2		None	203	241	5 L	60 L	C	
G	Self-heating liquid, toxic, inorganic, n.o.s.	4.2	UN3187	II	4.2, 6.1		None	202	243	1 L	5 L	C	
				III	4.2, 6.1		None	203	241	5 L	60 L	C	
G	Self-heating liquid, toxic, organic, n.o.s.	4.2	UN3184	II	4.2, 6.1		None	202	243	1 L	5 L	C	
				III	4.2, 6.1		None	203	241	5 L	60 L	C	
G	Self-heating solid, corrosive, inorganic, n.o.s.	4.2	UN3192	II	4.2, 8		None	212	242	15 kg	50 kg	C	
				III	4.2, 8		None	213	242	25 kg	100 kg	C	
G	Self-heating, solid, corrosive, organic, n.o.s.	4.2	UN3126	II	4.2, 8		None	212	242	15 kg	50 kg	C	
				III	4.2, 8		None	213	242	25 kg	100 kg	C	
G	Self-heating solid, inorganic, n.o.s.	4.2	UN3190	II	4.2		None	212	241	15 kg	50 kg	C	
				III	4.2		None	213	241	25 kg	100 kg	C	
G	Self-heating, solid, organic, n.o.s	4.2	UN3088	II	4.2	B101	None	212	241	15 kg	50 kg	C	
				III	4.2	B101	None	213	241	25 kg	100 kg	C	
G	Self-heating, solid, oxidizing, n.o.s.	4.2	UN3127		4.2, 5.1		None	214	214	Forbidden	Forbidden		
G	Self-heating solid, toxic, inorganic, n.o.s.	4.2	UN3191	II	4.2, 6.1		None	212	242	15 kg	50 kg	C	
				III	4.2, 6.1		None	213	242	25 kg	100 kg	C	
G	Self-heating, solid, toxic, organic, n.o.s.	4.2	UN3128	II	4.2, 6.1		None	212	242	15 kg	50 kg	C	
				III	4.2, 6.1		None	213	242	25 kg	100 kg	C	
	Self-propelled vehicle, see Engines or Batteries etc.												
G	Self-reactive liquid type B	4.1	UN3221	II	4.1	53	None	224	None ..	Forbidden	Forbidden	D	61
G	Self-reactive liquid type B, temperature controlled.	4.1	UN3231	II	4.1	53	None	224	None ..	Forbidden	Forbidden	D	2, 61
G	Self-reactive liquid type C	4.1	UN3223	II	4.1		None	224	None ..	5 L	10 L	D	61
G	Self-reactive liquid type C, temperature controlled.	4.1	UN3233	II	4.1		None	224	None ..	Forbidden	Forbidden	D	2, 61
G	Self-reactive liquid type D	4.1	UN3225	II	4.1		None	224	None ..	5 L	10 L	D	61
G	Self-reactive liquid type D, temperature controlled.	4.1	UN3235	II	4.1		None	224	None ..	Forbidden	Forbidden	D	2, 61
G	Self-reactive liquid type E	4.1	UN3227	II	4.1		None	224	None ..	10 L	25 L	D	61
G	Self-reactive liquid type E, temperature controlled.	4.1	UN3237	II	4.1		None	224	None ..	Forbidden	Forbidden	D	2, 61
G	Self-reactive liquid type F	4.1	UN3229	II	4.1		None	224	None ..	10 L	25 L	D	61
G	Self-reactive liquid type F, temperature controlled.	4.1	UN3239	II	4.1		None	224	None ..	Forbidden	Forbidden	D	2, 61
G	Self-reactive solid type B	4.1	UN3222	II	4.1	53	None	224	None ..	Forbidden	Forbidden	D	61
G	Self-reactive solid type B, temperature controlled.	4.1	UN3232	II	4.1	53	None	224	None ..	Forbidden	Forbidden	D	2, 61
G	Self-reactive solid type C	4.1	UN3224	II	4.1		None	224	None ..	5 kg	10 kg	D	61
G	Self-reactive solid type C, temperature controlled.	4.1	UN3234	II	4.1		None	224	None ..	Forbidden	Forbidden	D	2, 61
G	Self-reactive solid type D	4.1	UN3226	II	4.1		None	224	None ..	5 kg	10 kg	D	61

§ 172.101 Hazardous Materials Table—Continued

Symbols	Hazardous materials descriptions and proper shipping names	Hazard class or Division	Identification Numbers	PG	Label Codes	Special provisions	Packaging (§173.***) Exceptions	Packaging (§173.***) Non-bulk	Packaging (§173.***) Bulk	Quantity limitations Passenger aircraft/rail	Quantity limitations Cargo aircraft only	Vessel stowage Location	Vessel stowage Other
(1)	(2)	(3)	(4)	(5)	(6)	(7)	(8A)	(8B)	(8C)	(9A)	(9B)	(10A)	(10B)
G	Self-reactive solid type D, temperature controlled.	4.1	UN3236	II	4.1		None	224	None	Forbidden	Forbidden	D	2, 61
G	Self-reactive solid type E	4.1	UN3228	II	4.1		None	224	None	10 kg	25 kg	D	61
G	Self-reactive solid type E, temperature controlled.	4.1	UN3238	II	4.1		None	224	None	Forbidden	Forbidden	D	2, 61
G	Self-reactive solid type F	4.1	UN3230	II	4.1		None	224	None	10 kg	25 kg	D	61
G	Self-reactive solid type F, temperature controlled.	4.1	UN3240	II	4.1		None	224	None	Forbidden	Forbidden	D	2, 61
	Shale oil	3	UN1288	I	3	T7	None	201	243	1 L	30 L	B	
				II	3	T7, T30	150	202	242	5 L	60 L	B	
				III	3	B1, T7, T30	150	203	242	60 L	220 L	A	
	Shaped charges, commercial, see Charges, shaped, commercial etc.												
	Signal devices, hand	1.4G	UN0191	II	1.4G		None	62	None	Forbidden	75 kg	A	24E
	Signal devices, hand	1.4S	UN0373	II	1.4S		None	62	None	25 kg	100 kg	A	
	Signals, distress, ship	1.1G	UN0194	II	1.1G		None	62	None	Forbidden	Forbidden	B	
	Signals, distress, ship	1.3G	UN0195	II	1.3G		None	62	None	Forbidden	75 kg	B	
	Signals, highway, see Signal devices, hand; Fireworks, type D.												
	Signals, railway track, explosive	1.1G	UN0192	II	1.1G		None	62	None	Forbidden	Forbidden	B	
	Signals, railway track, explosive	1.4S	UN0193	II	1.4S		None	62	None	25 kg	100 kg	A	
	Signals, railway track, explosive	1.3G	UN0492		1.3G		None	62	None	Forbidden	Forbidden	E	1E, 8E
	Signals, railway track, explosive	1.4G	UN0493		1.4G		None	62	None	Forbidden	75 kg	A	24E
	Signals, ship distress, water-activated, see Contrivances, water-activated, etc.												
	Signals, smoke	1.1G	UN0196	II	1.1G		None	62	None	Forbidden	Forbidden	B	
	Signals, smoke	1.4G	UN0197	II	1.4G		None	62	None	Forbidden	75 kg	A	24E
	Signals, smoke	1.2G	UN0313	II	1.2G		None	62	None	Forbidden	Forbidden	B	
	Signals, smoke	1.3G	UN0487	II	1.3G		None	62	None	Forbidden	Forbidden	B	
	Silane, compressed	2.1	UN2203		2.1		None	302	None	Forbidden	Forbidden	E	40, 57, 104
	Silicofluoric acid, see Fluorosilicic acid.												
	Silicon chloride, see Silicon tetrachloride.												
	Silicon powder, amorphous	4.1	UN1346	III	4.1	A1	None	213	240	25 kg	100 kg	A	
	Silicon tetrachloride	8	UN1818	II	8	A3, A6, B2, B6, T18, T26, T29	154	202	242	1 L	30 L	C	40
	Silicon tetrafluoride, compressed	2.3	UN1859		2.3, 8	2	None	302	None	Forbidden	Forbidden	D	40
	Silver acetylide (dry)	Forbidden											
	Silver arsenite	6.1	UN1683	II	6.1		None	212	242	25 kg	100 kg	A	
	Silver azide (dry)	Forbidden											
	Silver chlorite (dry)	Forbidden											
	Silver cyanide	6.1	UN1684	II	6.1		None	212	242	25 kg	100 kg	A	26, 40
	Silver fulminate (dry)	Forbidden											
	Silver nitrate	5.1	UN1493	II	5.1		152	212	242	5 kg	25 kg	A	
	Silver oxalate (dry)	Forbidden											
	Silver picrate (dry)	Forbidden											
	Silver picrate, wetted *with not less than 30 percent water, by mass*.	4.1	UN1347	I	4.1		None	211	None	Forbidden	Forbidden	D	28, 36
	Sludge, acid	8	UN1906	II	8	A3, A7, B2, N34, T9, T27	None	202	242	Forbidden	30 L	C	14
D	Smokeless powder for small arms *(100 pounds or less)*.	4.1	NA3178	I	4.1	16	None	171	None	Forbidden	7.3 kg	A	
	Soda lime *with more than 4 percent sodium hydroxide*.	8	UN1907	III	8		154	213	240	25 kg	100 kg	A	
	Sodium	4.3	UN1428	I	4.3	A7, A8, A19, A20, B9, B48, B68, N34, T15, T29, T46	None	211	244	Forbidden	15 kg	D	
	Sodium aluminate, solid	8	UN2812	III	8		154	213	240	25 kg	100 kg	A	
	Sodium aluminate, solution	8	UN1819	II	8	B2, T8	154	202	242	1 L	30 L	A	
				III	8	T7	154	203	241	5 L	60 L	A	
	Sodium aluminum hydride	4.3	UN2835	II	4.3	A8, A19, A20, B100	151	212	242	Forbidden	50 kg	E	
	Sodium ammonium vanadate	6.1	UN2863	II	6.1		None	212	242	25 kg	100 kg	A	
	Sodium arsanilate	6.1	UN2473	III	6.1		153	213	240	100 kg	200 kg	A	
	Sodium arsenate	6.1	UN1685	II	6.1		None	212	242	25 kg	100 kg	A	
	Sodium arsenite, aqueous solutions.	6.1	UN1686	II	6.1	T15	None	202	243	5 L	60 L	A	
				III	6.1	T15	153	203	241	60 L	220 L	A	
	Sodium arsenite, solid	6.1	UN2027	II	6.1		None	212	242	25 kg	100 kg	A	
	Sodium azide	6.1	UN1687	II	6.1		None	212	242	25 kg	100 kg	A	36, 52, 91
	Sodium bifluoride, see Sodium hydrogendifluoride.												
	Sodium bisulfite, solution, see Bisulfites, aqueous solutions, n.o.s.												
	Sodium borohydride	4.3	UN1426	I	4.3	B100, N40	None	211	242	Forbidden	15 kg	E	

Appendix A

§172.101 HAZARDOUS MATERIALS TABLE—Continued

Symbols	Hazardous materials descriptions and proper shipping names	Hazard class or Division	Identification Numbers	PG	Label Codes	Special provisions	Packaging (§173.***)			Quantity limitations		Vessel stowage	
							Exceptions	Non-bulk	Bulk	Passenger aircraft/rail	Cargo aircraft only	Location	Other
(1)	(2)	(3)	(4)	(5)	(6)	(7)	(8A)	(8B)	(8C)	(9A)	(9B)	(10A)	(10B)
	Sodium borohydride and sodium hydroxide solution, *with not more than 12 percent sodium borohydride and not more than 40 percent sodium hydroxide by mass.*	8	UN3320	II	8	B2, N34, T8	154	202	242	1 L	30 L	A	26
				III	8	B2, N34, T7	154	203	241	5 L	60 L	A	
	Sodium bromate	5.1	UN1494	II	5.1		152	212	242	5 kg	25 kg	A	56, 58, 106
	Sodium cacodylate	6.1	UN1688	II	6.1		None	212	242	25 kg	100 kg	A	26
	Sodium chlorate	5.1	UN1495	II	5.1	A9, N34, T8	152	212	240	5 kg	25 kg	A	56, 58, 106
	Sodium chlorate, aqueous solution.	5.1	UN2428	II	5.1	A2, B6, T8	152	202	241	1 L	5 L	B	56, 58, 106
				III	5.1	A2, T8	152	203	241	2.5 L	30 L	B	56, 58, 69, 106
	Sodium chlorate mixed with dinitrotoluene, *see* Explosive blasting, type C.												
	Sodium chlorite	5.1	UN1496	II	5.1	A9, N34, T8	None	212	242	5 kg	25 kg	A	56, 58, 106
	Sodium chloroacetate	6.1	UN2659	III	6.1		153	213	240	100 kg	200 kg	A	
	Sodium cuprocyanide, solid	6.1	UN2316	I	6.1		None	211	242	5 kg	50 kg	A	26
	Sodium cuprocyanide, solution	6.1	UN2317	I	6.1	T8, T26	None	201	243	1 L	30 L	B	26, 40
	Sodium cyanide	6.1	UN1689	I	6.1	B69, B77, N74, N75, T42	None	211	242	5 kg	50 kg	B	52
	Sodium dichloroisocyanurate or Sodium dichloro-s-triazinetrione, *see* Dichloroisocyanuric acid etc.												
	Sodium dinitro-o-cresolate, *dry or wetted with less than 15 percent water, by mass.*	1.3C	UN0234	II	1.3C		None	62	None	Forbidden	Forbidden	B	1E, 5E
	Sodium dinitro-o-cresolate, *wetted with not less than 15 percent water, by mass.*	4.1	UN1348	I	4.1, 6.1	23, A8, A19, A20, N41	None	211	None	1 kg	15 kg	E	28, 36
	Sodium dithionite *or* Sodium hydrosulfite.	4.2	UN1384	II	4.2	A19, A20, B106	None	212	241	15 kg	50 kg	E	13
	Sodium fluoride	6.1	UN1690	III	6.1	T8	153	213	240	100 kg	200 kg	A	26
	Sodium fluoroacetate	6.1	UN2629	I	6.1		None	211	242	5 kg	50 kg	E	
	Sodium fluorosilicate	6.1	UN2674	III	6.1		153	213	240	100 kg	200 kg	A	26
	Sodium hydrate, *see* Sodium hydroxide, solid.												
	Sodium hydride	4.3	UN1427	I	4.3	A19, B100, N40	None	211	242	Forbidden	15 kg	E	
	Sodium hydrogendifluoride, *solid*	8	UN2439	II	8	B106, N3, N34	154	212	240	15 kg	50 kg	A	12, 25, 26, 40
	Sodium hydrogendifluoride *solution.*	8	UN2439	II	8	N3, N34	154	202	242	1 L	30 L	A	12, 25, 26, 40
D	Sodium hydrosulfide, solution	8	NA2922	II	8, 6.1	B2	154	202	243	1 L	30 L	B	40, 95
	Sodium hydrosulfide, *with less than 25 percent water of crystallization.*	4.2	UN2318	II	4.2	A7, A19, A20	None	212	241	15 kg	50 kg	A	
	Sodium hydrosulfide *with not less than 25 percent water of crystallization.*	8	UN2949	II	8	A7	154	212	240	15 kg	50 kg	A	26
	Sodium hydrosulfite, *see* Sodium dithionite.												
	Sodium hydroxide, solid	8	UN1823	II	8		154	212	240	15 kg	50 kg	A	
	Sodium hydroxide solution	8	UN1824	II	8	B2, N34, T8	154	202	242	1 L	30 L	A	
				III	8	N34, T7	154	203	241	5 L	60 L	A	
	Sodium hypochlorite, solution, *see* Hypochlorite solutions etc.												
	Sodium metal, liquid alloy, *see* Alkali metal alloys, liquid, n.o.s.												
	Sodium methylate	4.2	UN1431	II	4.2, 8	A19	None	212	242	15 kg	50 kg	B	
	Sodium methylate solutions *in alcohol.*	3	UN1289	II	3, 8	T8, T31	None	202	243	1 L	5 L	B	
				III	3, 8	B1, T7, T30	150	203	242	5 L	60 L	A	
	Sodium monoxide	8	UN1825	II	8		154	212	240	15 kg	50 kg	A	
	Sodium nitrate	5.1	UN1498	III	5.1	A1, A29	152	213	240	25 kg	100 kg	A	
	Sodium nitrate and potassium nitrate mixtures.	5.1	UN1499	III	5.1	A1, A29	152	213	240	25 kg	100 kg	A	
	Sodium nitrite	5.1	UN1500	III	5.1, 6.1	A1, A29	152	213	240	25 kg	100 kg	A	56, 58
	Sodium pentachlorophenate	6.1	UN2567	II	6.1		None	212	242	25 kg	100 kg	A	
	Sodium perchlorate	5.1	UN1502	II	5.1		152	212	242	5 kg	25 kg	A	56, 58, 106
	Sodium permanganate	5.1	UN1503	II	5.1		152	212	242	5 kg	25 kg	D	56, 58, 69, 106, 107
	Sodium peroxide	5.1	UN1504	I	5.1	A20, N34	None	211	None	Forbidden	15 kg	B	13, 75, 106
	Sodium peroxoborate, *anhydrous.*	5.1	UN3247	II	5.1		152	212	240	5 kg	25 kg	A	13, 25, 106
	Sodium persulfate	5.1	UN1505	III	5.1	A1	152	213	240	25 kg	100 kg	A	
	Sodium phosphide	4.3	UN1432	I	4.3, 6.1	A19, N40	None	211	None	Forbidden	15 kg	E	40, 85

§ 172.101 HAZARDOUS MATERIALS TABLE—Continued

Symbols	Hazardous materials descriptions and proper shipping names	Hazard class or Division	Identification Numbers	PG	Label Codes	Special provisions	Packaging (§173.***)			Quantity limitations		Vessel stowage	
							Exceptions	Non-bulk	Bulk	Passenger aircraft/rail	Cargo aircraft only	Location	Other
(1)	(2)	(3)	(4)	(5)	(6)	(7)	(8A)	(8B)	(8C)	(9A)	(9B)	(10A)	(10B)
	Sodium picramate, dry or wetted with less than 20 percent water, by mass.	1.3C	UN0235	II	1.3C		None	62	None	Forbidden	Forbidden	B	1E, 5E
	Sodium picramate, wetted with not less than 20 percent water, by mass.	4.1	UN1349	I	4.1	23, A8, A19, N41	None	211	None	Forbidden	15 kg	E	28, 36
	Sodium picryl peroxide	Forbidden											
	Sodium potassium alloys, see Potassium sodium alloys.												
	Sodium selenate, see Selenates or Selenites.												
D	Sodium selenite	6.1	NA2630	II	6.1		None	212	242	25 kg	100 kg	E	
	Sodium sulfide, anhydrous or Sodium sulfide with less than 30 percent water of crystallization.	4.2	UN1385	II	4.2	A19, A20, B106, N34	None	212	241	15 kg	50 kg	A	
	Sodium sulfide, hydrated with not less than 30 percent water.	8	UN1849	II	8	T8	154	212	240	15 kg	50 kg	A	26
	Sodium superoxide	5.1	UN2547	I	5.1	A20, N34	None	211	None	Forbidden	15 kg	E	13, 75, 106
	Sodium tetranitride	Forbidden											
G	Solids containing corrosive liquid, n.o.s.	8	UN3244	II	8	49	154	212	240	15 kg	50 kg	B	40
G	Solids containing flammable liquid, n.o.s.	4.1	UN3175	II	4.1	47	151	212	240	15 kg	50 kg	B	
G	Solids containing toxic liquid, n.o.s.	6.1	UN3243	II	6.1	48	None	212	240	25 kg	100 kg	B	40
	Sounding devices, explosive	1.2F	UN0204	II	1.2F		None	62	None	Forbidden	Forbidden	E	
	Sounding devices, explosive	1.1F	UN0296	II	1.1F		None	62	None	Forbidden	Forbidden	E	
	Sounding devices, explosive	1.1D	UN0374	II	1.1D		None	62	None	Forbidden	Forbidden	B	
	Sounding devices, explosive	1.2D	UN0375	II	1.2D		None	62	None	Forbidden	Forbidden	B	
	Spirits of salt, see Hydrochloric acid.												
	Squibs, see Igniters etc												
	Stannic chloride, anhydrous	8	UN1827	II	8	B2, T8, T26	154	202	242	1 L	30 L	C	
	Stannic chloride, pentahydrate	8	UN2440	III	8		154	213	240	25 kg	100 kg	A	
	Stannic phosphide	4.3	UN1433	I	4.3, 6.1	A19, B100, N40	None	211	242	Forbidden	15 kg	E	40, 85
	Steel swarf, see Ferrous metal borings, etc.												
	Stibine	2.3	UN2676		2.3, 2.1	1	None	304	None	Forbidden	Forbidden	D	40
	Storage batteries, wet, see Batteries, wet etc.												
	Strontium arsenite	6.1	UN1691	II	6.1		None	212	242	25 kg	100 kg	A	
	Strontium chlorate	5.1	UN1506	II	5.1	A1, A9, N34	152	212	242	5 kg	25 kg	A	56, 58, 106
	Strontium nitrate	5.1	UN1507	III	5.1	A1, A29	152	213	240	25 kg	100 kg	A	
	Strontium perchlorate	5.1	UN1508	II	5.1		152	212	242	5 kg	25 kg	A	56, 58, 106
	Strontium peroxide	5.1	UN1509	II	5.1		152	212	242	5 kg	25 kg	A	13, 75, 106
	Strontium phosphide	4.3	UN2013	I	4.3, 6.1	A19, N40	None	211	None	Forbidden	15 kg	E	40, 85
	Strychnine or Strychnine salts	6.1	UN1692	I	6.1		None	211	242	5 kg	50 kg	A	40
	Styphnic acid, see Trinitroresorcinol, etc.												
	Styrene monomer, inhibited	3	UN2055	III	3	B1, T1	150	203	242	60 L	220 L	A	
G	Substances, explosive, n.o.s	1.1L	UN0357	II	1.1L	101	None	62	None	Forbidden	Forbidden	E	2E, 8E, 11E, 17E
G	Substances, explosive, n.o.s	1.2L	UN0358	II	1.2L	101	None	62	None	Forbidden	Forbidden	E	2E, 8E, 11E, 17E
G	Substances, explosive, n.o.s	1.3L	UN0359	II	1.3L	101	None	62	None	Forbidden	Forbidden	E	2E, 8E, 11E, 17E
G	Substances, explosive, n.o.s	1.1A	UN0473	II	1.1A	101, 111	None	62	None	Forbidden	Forbidden	E	2E, 6E
G	Substances, explosive, n.o.s	1.1C	UN0474	II	1.1C	101	None	62	None	Forbidden	Forbidden	B	1E, 5E
G	Substances, explosive, n.o.s	1.1D	UN0475	II	1.1D	101	None	62	None	Forbidden	Forbidden	B	1E, 5E
G	Substances, explosive, n.o.s	1.1G	UN0476	II	1.1G	101	None	62	None	Forbidden	Forbidden	E	1E, 8E
G	Substances, explosive, n.o.s	1.3C	UN0477	II	1.3C	101	None	62	None	Forbidden	Forbidden	B	1E, 5E
G	Substances, explosive, n.o.s	1.3G	UN0478	II	1.3G	101	None	62	None	Forbidden	Forbidden	E	1E, 8E
G	Substances, explosive, n.o.s	1.4C	UN0479	II	1.4C	101	None	62	None	Forbidden	75 kg	A	1E, 5E
G	Substances, explosive, n.o.s	1.4D	UN0480	II	1.4D	101	None	62	None	Forbidden	75 kg	A	1E, 5E, 24E
G	Substances, explosive, n.o.s	1.4S	UN0481	II	1.4S	101	None	62	None	25 kg	75 kg	A	
G	Substances, explosive, n.o.s	1.4G	UN0485	II	1.4G	101	None	62	None	Forbidden	75kg	E	1E, 8E
G	Substances, explosive, very insensitive, n.o.s., or Substances, EVI, n.o.s.	1.5D	UN0482	II	1.5D	101	None	62	None	Forbidden	Forbidden	B	1E, 5E
	Substituted nitrophenol pesticides, liquid, flammable, toxic, flash point less than 23 degrees C.	3	UN2780	I	3, 6.1		None	201	243	Forbidden	30 L	B	40
				II	3, 6.1		None	202	243	1 L	60 L	B	40
	Substituted nitrophenol pesticides, liquid, toxic.	6.1	UN3014	I	6.1	T42	None	201	243	1 L	30 L	B	40
				II	6.1	T14	None	202	243	5 L	60 L	B	40
				III	6.1	T14	153	203	241	60 L	220 L	A	40

Appendix A

§172.101 Hazardous Materials Table—Continued

Symbols	Hazardous materials descriptions and proper shipping names	Hazard class or Division	Identification Numbers	PG	Label Codes	Special provisions	Packaging (§173.***)			Quantity limitations		Vessel stowage	
							Exceptions	Non-bulk	Bulk	Passenger aircraft/rail	Cargo aircraft only	Location	Other
(1)	(2)	(3)	(4)	(5)	(6)	(7)	(8A)	(8B)	(8C)	(9A)	(9B)	(10A)	(10B)
	Substituted nitrophenol pesticides, liquid, toxic, flammable flashpoint not less than 23 degrees C.	6.1	UN3013	I	6.1, 3	T42	None	201	243	1 L	30 L	B	40
				II	6.1, 3	T14	None	202	243	5 L	60 L	B	40
				III	6.1, 3	B1, T14	153	203	242	60 L	220 L	A	40
	Substituted nitrophenol pesticides, solid, toxic.	6.1	UN2779	I	6.1		None	211	242	5 kg	50 kg	A	40
				II	6.1		None	212	242	25 kg	100 kg	A	40
				III	6.1		153	213	240	100 kg	200 kg	A	40
	Sucrose octanitrate (dry)	Forbidden											
	Sulfamic acid	8	UN2967	III	8		154	213	240	25 kg	100 kg	A	
D	Sulfur	9	NA1350	III	9	30	None	None	240	No limit	No limit	A	19, 74
I	Sulfur	4.1	UN1350	III	9	30, T1	None	None	240	No limit	No limit	A	19, 74
	Sulfur and chlorate, loose mixtures of.	Forbidden											
	Sulfur chlorides	8	UN1828	I	8	5, A3, B10, B77, N34, T18, T27	None	201	243	Forbidden	2.5 L	C	40
	Sulfur dichloride, see Sulfur chlorides.												
	Sulfur dioxide	2.3	UN1079		2.3, 8	3, B14	None	304	314, 315.	Forbidden	25 kg	D	40
	Sulfur dioxide solution, see Sulfurous acid.												
	Sulfur hexafluoride	2.2	UN1080		2.2		306	304	314, 315.	75 kg	150 kg	A	
D	Sulfur, molten	9	NA2448	III	9	T9, T38	None	213	247	Forbidden	Forbidden	C	61
I	Sulfur, molten	4.1	UN2448	III	4.1	T9, T38	None	213	247	Forbidden	Forbidden	C	61
	Sulfur tetrafluoride	2.3	UN2418		2.3, 8	1	None	302	245	Forbidden	Forbidden	D	40
+	Sulfur trioxide, inhibited or Sulfur trioxide, stabilized.	8	UN1829	I	8, 6.1	2, A7, B9, B14, B32, B49, B74, B77, N34, T38, T43, T45	None	227	244	Forbidden	25 kg	A	40
+, D	Sulfur trioxide, uninhibited	8	NA1829	I	8, 6.1	2, A7, B9, B14, B32, B49, B74, B77, N34, T38, T43, T45	None	227	244	Forbidden	25 kg	C	10, 40
	Sulfuretted hydrogen, see Hydrogen sulfide, liquefied.												
	Sulfuric acid, fuming with less than 30 percent free sulfur trioxide.	8	UN1831	I	8	A3, A7, B84, N34, T18, T27	None	201	243	Forbidden	2.5 L	C	14, 40
+	Sulfuric acid, fuming with 30 percent or more free sulfur trioxide.	8	UN1831	I	8, 6.1	2, A3, A6, A7, B9, B14, B32, B74, B77, B84, N34, T38, T43, T45	None	227	244	Forbidden	Forbidden	C	14, 40
	Sulfuric acid, spent	8	UN1832	II	8	A3, A7, B2, B83, B84, N34, T9, T27	None	202	242	Forbidden	30 L	C	14
	Sulfuric acid with more than 51 percent acid.	8	UN1830	II	8	A3, A7, B3, B83, B84, N34, T9, T27	154	202	242	1 L	30 L	C	14
	Sulfuric acid with not more than 51% acid.	8	UN2796	II	8	A3, A7, B2, B15, N6, N34, T9, T27	154	202	242	1 L	30 L	B	
	Sulfuric and hydrofluoric acid mixtures, see Hydrofluoric and sulfuric acid mixtures.												
	Sulfuric anhydride, see Sulfur trioxide, inhibited.												
	Sulfurous acid	8	UN1833	II	8	B3, T8	154	202	242	1 L	30 L	B	40
+	Sulfuryl chloride	8	UN1834	I	8, 6.1	1, A3, B6, B9, B10, B14, B30, B74, B77, N34, T38, T43, T44	None	226	244	Forbidden	Forbidden	C	40
	Sulfuryl fluoride	2.3	UN2191		2.3	4	None	304	314, 315.	Forbidden	25 kg	D	40
	Tars, liquid including road asphalt and oils, bitumen and cut backs.	3	UN1999	II	3	B13, T7, T30	150	202	242	5 L	60 L	B	
				III	3	B1, B13, T7, T30	150	203	242	60 L	220 L	A	
	Tear gas candles	6.1	UN1700	II	6.1, 4.1		None	340	None	Forbidden	50 kg	D	40
	Tear gas cartridges, see Ammunition, tear-producing, etc.												
D	Tear gas devices with more than 2 percent tear gas substances, by mass.	6.1	NA1693	I	6.1		None	340	None	Forbidden	Forbidden	D	40
				II	6.1		None	340	None	Forbidden	Forbidden	D	40
	Tear gas devices, with not more than 2 percent tear gas substances, by mass, see Aerosols, etc.												
	Tear gas grenades, see Tear gas candles.												
G	Tear gas substances, liquid, n.o.s.	6.1	UN1693	I	6.1		None	201	None	Forbidden	Forbidden	D	40
				II	6.1		None	202	None	Forbidden	5 L	D	40
G	Tear gas substances, solid, n.o.s.	6.1	UN1693	I	6.1		None	211	None	Forbidden	Forbidden	D	40
				II	6.1		None	212	None	Forbidden	25 kg	D	40
	Tellurium compound, n.o.s.	6.1	UN3284	I	6.1		None	211	242	5 kg	50 kg	B	
				II	6.1	T14	None	212	242	25 kg	100 kg	B	

§ 172.101 Hazardous Materials Table—Continued

Symbols	Hazardous materials descriptions and proper shipping names	Hazard class or Division	Identification Numbers	PG	Label Codes	Special provisions	Packaging (§173.***) Exceptions (8A)	Packaging (§173.***) Non-bulk (8B)	Packaging (§173.***) Bulk (8C)	Quantity limitations Passenger aircraft/rail (9A)	Quantity limitations Cargo aircraft only (9B)	Vessel stowage Location (10A)	Vessel stowage Other (10B)
(1)	(2)	(3)	(4)	(5)	(6)	(7)	(8A)	(8B)	(8C)	(9A)	(9B)	(10A)	(10B)
	Tellurium hexafluoride	2.3	UN2195		2.3, 8	1	None	302	None	Forbidden	Forbidden	D	40
	Terpene hydrocarbons, n.o.s	3	UN2319	III	3	B1, T1	150	203	242	60 L	220 L	A	
	Terpinolene	3	UN2541	III	3	B1, T1	150	203	242	60 L	220 L	A	
	Tetraazido benzene quinone	Forbidden											
	Tetrabromoethane	6.1	UN2504	III	6.1	T7	153	203	241	60 L	220 L	A	
	Tetrachloroethane	6.1	UN1702	II	6.1	N36, T14	None	202	243	5 L	60 L	A	40
	Tetrachloroethylene	6.1	UN1897	III	6.1	N36, T1	153	203	241	60 L	220 L	A	40
	Tetraethyl dithiopyrophosphate	6.1	UN1704	II	6.1		None	212	242	25 kg	100 kg	D	40
D	Tetraethyl lead, liquid	6.1	NA1649	I	6.1, 3		None	201	None	Forbidden	Forbidden	E	40
D	Tetraethyl pyrophosphate, *liquid*	6.1	NA3018	I	6.1		None	201	243	Forbidden	1 L	A	40
D	Tetraethyl pyrophosphate *solid*	6.1	NA2783	I	6.1	N77	None	211	242	Forbidden	50 kg	A	40
	Tetraethyl silicate	3	UN1292	III	3	B1, T1	150	203	242	60 L	220 L	A	
	Tetraethylammonium perchlorate (dry).	Forbidden											
	Tetraethylenepentamine	8	UN2320	III	8	T2	154	203	241	5 L	60 L	A	
	1,1,1,2-Tetrafluoroethane *or Refrigerant gas R 134a*.	2.2	UN3159		2.2		306	304	314, 315.	75 kg	150 kg	A	
	Tetrafluoroethylene, inhibited	2.1	UN1081		2.1		306	304	None	Forbidden	150 kg	E	40
	Tetrafluoromethane, compressed *or Refrigerant gas R 14*.	2.2	UN1982		2.2		None	302	None	75 kg	150 kg	A	
	1,2,3,6-Tetrahydrobenzaldehyde	3	UN2498	III	3	B1, T1	150	203	242	60 L	220 L	A	
	Tetrahydrofuran	3	UN2056	II	3	T8	None	202	242	5 L	60 L	B	
	Tetrahydrofurfurylamine	3	UN2943	III	3	B1, T1	150	203	242	60 L	220 L	A	
	Tetrahydrophthalic anhydrides with more than 0.05 percent of maleic anhydride.	8	UN2698	III	8		154	213	240	25 kg	100 kg	A	
	1,2,3,6-Tetrahydropyridine	3	UN2410	II	3	T8	150	202	242	5 L	60 L	B	
	Tetrahydrothiophene	3	UN2412	II	3	T7	150	202	242	5 L	60 L	B	
	Tetramethylammonium hydroxide.	8	UN1835	II	8	B2, T8	154	202	242	1 L	30 L	A	
	Tetramethylene diperoxide dicarbamide.	Forbidden											
	Tetramethylsilane	3	UN2749	I	3	T21, T26	None	201	243	Forbidden	30 L	D	
	Tetranitro diglycerin	Forbidden											
	Tetranitroaniline	1.1D	UN0207	II	1.1D		None	62	None	Forbidden	Forbidden	B	1E, 5E
+	Tetranitromethane	5.1	UN1510	I	5.1, 6.1	2, B9, B14, B32, B74, T38, T43, T45	None	227	None	Forbidden	Forbidden	D	40, 66, 106
	2,3,4,6-Tetranitrophenol	Forbidden											
	2,3,4,6-Tetranitrophenyl methyl nitramine.	Forbidden											
	2,3,4,6-Tetranitrophenylnitramine	Forbidden											
	Tetranitroresorcinol (dry)	Forbidden											
	2,3,5,6-Tetranitroso-1,4-dinitrobenzene.	Forbidden											
	2,3,5,6-Tetranitroso nitrobenzene (dry).	Forbidden											
	Tetrapropylorthotitanate	3	UN2413	III	3	B1, T8	150	203	242	60 L	220 L	A	
	Tetrazene, *see Guanyl nitrosaminoguanyltetrazene*.												
	Tetrazine (dry)	Forbidden											
	Tetrazol-1-acetic acid	1.4C	UN0407	II	1.4C		None	62	None	Forbidden	75 kg	A	1E, 5E, 24E
	Tetrazolyl azide (dry)	Forbidden											
	Tetryl, *see Trinitrophenylmethylnitramine*.												
	Thallium chlorate	5.1	UN2573	II	5.1, 6.1		None	212	242	5 kg	25 kg	A	56, 58, 106
	Thallium compounds, n.o.s	6.1	UN1707	II	6.1		None	212	242	25 kg	100 kg	A	
	Thallium nitrate	6.1	UN2727	II	6.1, 5.1		None	212	242	5 kg	25 kg	A	
D	Thallium sulfate, solid	6.1	NA1707	II	6.1		None	212	242	5 kg	50 kg	A	
	4-Thiapentanal	6.1	UN2785	III	6.1	T8	153	203	241	60 L	220 L	D	25, 49
	Thioacetic acid	3	UN2436	II	3	T8	150	202	242	5 L	60 L	B	
	Thiocarbamate pesticide, liquid, flammable, toxic, *flash point less than 23 degrees C*.	3	UN2772	I	3, 6.1		None	201	243	Forbidden	30 L	B	40
				II	3, 6.1		None	202	243	1 L	60 L	B	40
	Thiocarbamate pesticides, liquid, flammable, toxic, *flash point not less than 23 degrees C*.	6.1	UN3005	I	6.1, 3	T42	None	201	243	1 L	30 L	B	40
				II	6.1, 3	T14	None	202	243	5 L	60 L	B	40
				III	6.1, 3	T13	153	203	242	60 L	220 L	A	40
	Thiocarbamate pesticides, liquid, toxic.	6.1	UN3006	I	6.1	T42	None	201	243	1 L	30 L	B	40
				II	6.1	T14	None	202	243	5 L	60 L	B	40
				III	6.1	T14	153	203	241	60 L	220 L	A	40
	Thiocarbamate pesticides, solid, toxic.	6.1	UN2771	I	6.1		None	211	242	5 kg	50 kg	A	40
				II	6.1		None	212	242	25 kg	100 kg	A	40
				III	6.1		153	213	240	100 kg	200 kg	A	40
	Thiocarbonylchloride, see Thiophosgene.												
	Thioglycol	6.1	UN2966	II	6.1	T8	None	202	243	5 L	60 L	A	
	Thioglycolic acid	8	UN1940	II	8	A7, B2, N34, T8	154	202	242	1 L	30 L	A	
	Thiolactic acid	6.1	UN2936	II	6.1	T8	None	212	242	25 kg	100 kg	A	

Appendix A

§ 172.101 HAZARDOUS MATERIALS TABLE—Continued

Symbols	Hazardous materials descriptions and proper shipping names	Hazard class or Division	Identification Numbers	PG	Label Codes	Special provisions	Packaging (§173.***)			Quantity limitations		Vessel stowage	
							Exceptions	Non-bulk	Bulk	Passenger aircraft/rail	Cargo aircraft only	Location	Other
(1)	(2)	(3)	(4)	(5)	(6)	(7)	(8A)	(8B)	(8C)	(9A)	(9B)	(10A)	(10B)
	Thionyl chloride	8	UN1836	I	8	A7, B6, B10, N34, T18, T27	None	201	243	Forbidden	Forbidden	C	40
	Thiophene	3	UN2414	II	3	B101, T2	150	202	242	5 L	60 L	B	40
+	Thiophosgene	6.1	UN2474	II	6.1	2, A7, B9, B14, B32, B74, N33, N34, T38, T43, T45	None	227	244	Forbidden	60 L	B	26, 40
	Thiophosphoryl chloride	8	UN1837	II	8	A3, A7, B2, B8, B25, B101, N34, T12	None	202	242	Forbidden	30 L	C	40
	Thiourea dioxide	4.2	UN3341	II	4.2		None	212	241	15 kg	50 kg	D	
				III	4.2		None	213	241	25 kg	100 kg	D	
	Thorium metal, pyrophoric	7	UN2975		7, 4.2		None	418	None	Forbidden	Forbidden	D	
	Thorium nitrate, solid	7	UN2976		7, 5.1		None	419	None	Forbidden	15 kg	A	
	Tin chloride, fuming, see Stannic chloride, anhydrous.												
	Tin perchloride or Tin tetrachloride, see Stannic chloride, anhydrous.												
	Tinctures, medicinal	3	UN1293	II	3	T8, T31	150	202	242	5 L	60 L	B	
				III	3	B1, T7, T30	150	203	242	60 L	220 L	A	
	Tinning flux, see Zinc chloride												
	Titanium disulphide	4.2	UN3174	III	4.2		None	213	241	25 kg	100 kg	A	
	Titanium hydride	4.1	UN1871	II	4.1	A19, A20, N34	None	212	241	15 kg	50 kg	E	
	Titanium powder, dry	4.2	UN2546	I	4.2		None	211	242	Forbidden	Forbidden	D	
				II	4.2	A19, A20, N5, N34	None	212	241	15 kg	50 kg	D	
				III	4.2		None	213	241	25 kg	100 kg	D	
	Titanium powder, wetted with not less than 25 percent water (a visible excess of water must be present) (a) mechanically produced, particle size less than 53 microns; (b) chemically produced, particle size less than 840 microns.	4.1	UN1352	II	4.1	A19, A20, N34	None	212	240	15 kg	50 kg	E	
	Titanium sponge granules or Titanium sponge powders.	4.1	UN2878	III	4.1	A1	None	213	240	25 kg	100 kg	D	
D	Titanium sulfate solution	8	NA1760	II	8	B2, B15	None	202	242	1 L	30 L	B	40
+	Titanium tetrachloride	8	UN1838	II	8, 6.1	2, A3, A6, B7, B9, B14, B32, B74, B77, T38, T43, T45	None	227	244	Forbidden	30 L	C	40
	Titanium trichloride mixtures	8	UN2869	II	8	A7, B106, N34	154	212	240	15 kg	50 kg	A	40
				III	8	A7, N34	154	213	240	25 kg	100 kg	A	40
	Titanium trichloride, pyrophoric or Titanium trichloride mixtures, pyrophoric.	4.2	UN2441	I	4.2, 8	A7, A8, A19, A20, N34	None	181	244	Forbidden	Forbidden	D	40
	TNT mixed with aluminum, see Tritonal.												
	TNT, see Trinitrotoluene, etc												
	Toluene	3	UN1294	II	3	T1	150	202	242	5 L	60 L	B	
+	Toluene diisocyanate	6.1	UN2078	II	6.1	B110, T14	None	202	243	5 L	60 L	D	25, 40
	Toluene sulfonic acid, see Alkyl, or Aryl sulfonic acid etc.												
+	Toluidines liquid	6.1	UN1708	II	6.1	T14	None	202	243	5 L	60 L	A	
+	Toluidines solid	6.1	UN1708	II	6.1		None	212	242	25 kg	100 kg	A	
	2,4-Toluylenediamine or 2,4-Toluenediamine.	6.1	UN1709	III	6.1	T7	153	213	240	100 kg	200 kg	A	
	Torpedoes, liquid fueled, with inert head.	1.3J	UN0450	II	1.3J			62	None	Forbidden	Forbidden	E	7E, 16E, 23E
	Torpedoes, liquid fueled, with or without bursting charge.	1.1J	UN0449	II	1.1J			62	None	Forbidden	Forbidden	E	7E, 16E, 23E
	Torpedoes with bursting charge	1.1E	UN0329	II	1.1E			62	None	Forbidden	Forbidden	B	
	Torpedoes with bursting charge	1.1F	UN0330	II	1.1F			62	None	Forbidden	Forbidden	B	
	Torpedoes with bursting charge	1.1D	UN0451	II	1.1D			62	None	Forbidden	Forbidden	B	
G	Toxic liquid, corrosive, inorganic, n.o.s.	6.1	UN3289	I and II	6.1, 8	T42	None	201	243	0.5 L	2.5 L	A	
G				II	6.1, 8	T14	None	202	243	1 L	30 L	A	
G	Toxic liquid, corrosive, inorganic, n.o.s. Inhalation Hazard, Packing Group I, Zone A.	6.1	UN3289	I	6.1, 8	1, B9, B14, B30, B72, T38, T43, T44	None	226	244	Forbidden	Forbidden	B	40
G	Toxic liquid, corrosive, inorganic, n.o.s. Inhalation Hazard, Packing Group I, Zone B.	6.1	UN3289	I	6.1, 8	2, B9, B14, B32, B74, T38, T43, T45	None	227	244	Forbidden	Forbidden	B	40
G	Toxic liquid, inorganic, n.o.s	6.1	UN3287	I	6.1	T42	None	201	243	1 L	30 L	A	
G				II	6.1	B110, T14	None	202	243	5 L	60 L	A	
G				III	6.1	T7	153	203	241	60 L	220 L	A	
G	Toxic liquid, inorganic, n.o.s. Inhalation Hazard, Packing Group I, Zone A.	6.1	UN3287	I	6.1	1, B9, B14, B30, B72, T38, T43, T44	None	226	244	Forbidden	Forbidden	B	40
G	Toxic liquid, inorganic, n.o.s. Inhalation Hazard, Packing Group I, Zone B.	6.1	UN3287	I	6.1	2, B9, B14, B32, B74, T38, T43, T45	None	227	244	Forbidden	Forbidden	B	40
G	Toxic liquids, corrosive, organic, n.o.s.	6.1	UN2927	I	6.1, 8	T42	None	201	243	0.5 L	2.5 L	B	40
G				II	6.1, 8	T42	None	202	243	1 L	30 L	B	40

§172.101 Hazardous Materials Table—Continued

Symbols	Hazardous materials descriptions and proper shipping names	Hazard class or Division	Identification Numbers	PG	Label Codes	Special provisions	Packaging (§173.***)			Quantity limitations		Vessel stowage	
							Exceptions (8A)	Non-bulk (8B)	Bulk (8C)	Passenger aircraft/rail (9A)	Cargo aircraft only (9B)	Location (10A)	Other (10B)
(1)	(2)	(3)	(4)	(5)	(6)	(7)	(8A)	(8B)	(8C)	(9A)	(9B)	(10A)	(10B)
G	Toxic liquids, corrosive, organic, n.o.s., inhalation hazard, Packing Group I, Zone A.	6.1	UN2927	I	6.1, 8	1, B9, B14, B30, B72, T38, T43, T44	None	226	244	Forbidden	Forbidden	D	20, 40, 95
G	Toxic liquids, corrosive, organic, n.o.s., inhalation hazard, Packing Group I, Zone B.	6.1	UN2927	I	6.1, 8	2, B9, B14, B32, B74, T38, T43, T45	None	227	244	Forbidden	Forbidden	D	20, 40, 95
G	Toxic liquids, flammable, organic, n.o.s.	6.1	UN2929	I	6.1, 3	T42	None	201	243	1 L	30 L	B	40
G				II	6.1, 3	T15	None	202	243	5 L	60 L	B	40
G	Toxic liquids, flammable, organic, n.o.s., inhalation hazard, Packing Group I, Zone A.	6.1	UN2929	I	6.1, 3	1, B9, B14, B30, B72, T38, T43, T44	None	226	244	Forbidden	Forbidden	D	20, 40, 95
G	Toxic liquids, flammable, organic, n.o.s., inhalation hazard, Packing Group I, Zone B.	6.1	UN2929	I	6.1, 3	2, B9, B14, B32, B74, T38, T43, T45	None	227	244	Forbidden	Forbidden	D	20, 40, 95
G	Toxic, liquids, organic, n.o.s	6.1	UN2810	I	6.1	T42	None	201	243	1 L	30 L	B	40
G				II	6.1	B110, T14	None	202	243	5 L	60 L	B	40
G				III	6.1	T7	153	203	241	60 L	220 L	A	40
G	Toxic, liquids, organic, n.o.s. Inhalation hazard, Packing Group I, Zone A.	6.1	UN2810	I	6.1	1, B9, B14, B30, B72, T38, T43, T44	None	226	244	Forbidden	Forbidden	D	20, 40, 95
G	Toxic, liquids, organic, n.o.s. Inhalation hazard, Packing Group I, Zone B.	6.1	UN2810	I	6.1	2, B9, B14, B32, B74, T38, T43, T45	None	227	244	Forbidden	Forbidden	D	20, 40, 95
G	Toxic liquids, oxidizing, n.o.s	6.1	UN3122	I	6.1, 5.1	A4	None	201	243	Forbidden	2.5 L	C	
G				II	6.1, 5.1		None	202	243	1 L	5 L	C	
G	Toxic liquids, oxidizing, n.o.s. Inhalation hazard, Packing Group I, Zone A.	6.1	UN3122	I	6.1, 5.1	1, B9, B14, B30, B72, T38, T43, T44	None	226	244	Forbidden	Forbidden	C	
G	Toxic liquids, oxidizing, n.o.s. Inhalation Hazard, Packing Group I, Zone B.	6.1	UN3122	I	6.1, 5.1	2, B9, B14, B32, T38, T43, T45	None	227	244	Forbidden	Forbidden	C	
G	Toxic liquids, water-reactive, n.o.s.	6.1	UN3123	I	6.1, 4.3	A4	None	201	243	Forbidden	1 L	E	40
G				II	6.1, 4.3		None	202	243	1 L	5 L	E	40
G	Toxic liquids, water-reactive, n.o.s. Inhalation hazard, packing group I, Zone A.	6.1	UN3123	I	6.1, 4.3	1, B9, B14, B30, B72, T38, T43, T44	None	226	244	Forbidden	Forbidden	E	40
G	Toxic liquids, water-reactive, n.o.s. Inhalation hazard, packing group I, Zone B.	6.1	UN3123	I	6.1, 4.3	2, B9, B14, B32, B74, T38, T43, T45	None	227	244	Forbidden	Forbidden	E	40
G	Toxic solid, corrosive, inorganic, n.o.s.	6.1	UN3290	I	6.1, 8		None	211	242	1 kg	25 kg	A	
				II	6.1, 8		None	212	242	15 kg	50 kg	A	
G	Toxic solid, inorganic, n.o.s	6.1	UN3288	I	6.1		None	211	242	5 kg	50 kg	A	
				II	6.1		None	212	242	25 kg	100 kg	A	
				III	6.1		153	213	240	100 kg	200 kg	A	
G	Toxic solids, corrosive, organic, n.o.s.	6.1	UN2928	I	6.1, 8		None	211	242	1 kg	25 kg	B	40
				II	6.1, 8		None	212	242	15 kg	50 kg	B	40
G	Toxic solids, flammable, organic, n.o.s.	6.1	UN2930	I	6.1, 4.1	B106	None	211	242	1 kg	15 kg	B	
				II	6.1, 4.1	B106	None	212	242	15 kg	50 kg	B	
G	Toxic solids, organic, n.o.s	6.1	UN2811	I	6.1		None	211	242	5 kg	50 kg	B	
				II	6.1		None	212	242	25 kg	100 kg	B	
				III	6.1		153	213	240	100 kg	200 kg	A	
G	Toxic solids, oxidizing, n.o.s	6.1	UN3086	I	6.1, 5.1		None	211	242	1 kg	15 kg	C	
				II	6.1, 5.1		None	212	242	15 kg	50 kg	C	
G	Toxic solids, self-heating, n.o.s.	6.1	UN3124	I	6.1, 4.2	A5, B100	None	211	242	5 kg	15 kg	D	40
				II	6.1, 4.2		None	212	242	15 kg	50 kg	D	40
G	Toxic solids, water-reactive, n.o.s.	6.1	UN3125	I	6.1, 4.3	A5, B100	None	211	242	5 kg	15 kg	D	40
				II	6.1, 4.3	B101	None	212	242	15 kg	50 kg	D	40
D	Toy Caps	1.4S	NA0337	II	1.4S		None	62	None	25 kg	100 kg	A	
	Tracers for ammunition	1.3G	UN0212	II	1.3G		None	62	None	Forbidden	Forbidden	B	9E
	Tracers for ammunition	1.4G	UN0306	II	1.4G		None	62	None	Forbidden	75 kg	A	24E
	Tractors, see Vehicles, self propelled.												
	Tri-(b-nitroxyethyl) ammonium nitrate.	Forbidden											
	Triallyl borate	6.1	UN2609	III	6.1		153	203	241	60 L	220 L	A	
	Triallylamine	3	UN2610	III	3, 8	B1, T1	None	203	242	5 L	60 L	A	13
	Triazine pesticides, flammable, toxic, flash point less than 23 degrees C.	3	UN2764	I	3, 6.1		None	201	243	Forbidden	30 L	B	40
				II	3, 6.1		None	202	243	1 L	60 L	B	40
	Triazine pesticides, liquid, toxic	6.1	UN2998	I	6.1	T42	None	201	243	1 L	30 L	B	40
				II	6.1	T14	None	202	243	5 L	60 L	B	40
				III	6.1	T14	153	203	241	60 L	220 L	A	40
	Triazine pesticides, liquid, toxic, flammable, flashpoint not less than 23 degrees C.	6.1	UN2997	I	6.1, 3	T42	None	201	243	1 L	30 L	B	40
				II	6.1, 3	T14	None	202	243	5 L	60 L	B	40
				III	6.1, 3	T14	153	203	242	60 L	220 L	A	40
	Triazine pesticides, solid, toxic	6.1	UN2763	I	6.1		None	211	242	5 kg	50 kg	A	40
				II	6.1		None	212	242	25 kg	100 kg	A	40
				III	6.1		153	213	240	100 kg	200 kg	A	40
	Tributylamine	6.1	UN2542	II	6.1	B110, T14	None	202	243	5 L	60 L	A	

Appendix A

§ 172.101 HAZARDOUS MATERIALS TABLE—Continued

Symbols	Hazardous materials descriptions and proper shipping names	Hazard class or Division	Identification Numbers	PG	Label Codes	Special provisions	Packaging (§173.***) Exceptions	Packaging (§173.***) Non-bulk	Packaging (§173.***) Bulk	Quantity limitations Passenger aircraft/rail	Quantity limitations Cargo aircraft only	Vessel stowage Location	Vessel stowage Other
(1)	(2)	(3)	(4)	(5)	(6)	(7)	(8A)	(8B)	(8C)	(9A)	(9B)	(10A)	(10B)
	Tributylphosphane	4.2	UN3254	I	4.2		None	211	242	Forbidden	Forbidden	D	
D	mono-(Trichloro) tetra-(monopotassium dichloro)-penta-s-triazinetrione, dry (with more than 39 percent available chlorine).	5.1	NA2468	II	5.1		152	212	240	5 kg	25 kg	A	13
	Trichloro-s-triazinetrione dry, with more than 39 percent available chlorine, see Trichloroisocyanuric acid, dry.												
	Trichloroacetic acid	8	UN1839	II	8	A7, N34	154	212	240	15 kg	50 kg	A	
	Trichloroacetic acid, solution	8	UN2564	II	8	A3, A6, A7, B2, N34, T8	154	202	242	1 L	30 L	B	
				III	8	A3, A6, A7, N34, T7	154	203	241	5 L	60 L	B	8
+	Trichloroacetyl chloride	8	UN2442	II	8, 6.1	2, A3, A7, B9, B14, B32, B74, N34, T38, T43, T45	None	227	244	Forbidden	Forbidden	D	40
	Trichlorobenzenes, liquid	6.1	UN2321	III	6.1	T7	153	203	241	60 L	220 L	A	
	Trichlorobutene	6.1	UN2322	II	6.1	T8	None	202	243	5 L	60 L	A	25, 40
	1,1,1-Trichloroethane	6.1	UN2831	III	6.1	N36, T7	153	203	241	60 L	220 L	A	40
	Trichloroethylene	6.1	UN1710	III	6.1	N36, T1	153	203	241	60 L	220 L	A	40
	Trichloroisocyanuric acid, dry	5.1	UN2468	II	5.1		152	212	240	5 kg	25 kg	A	13
	Trichloromethyl perchlorate	Forbidden											
	Trichlorosilane	4.3	UN1295	I	4.3, 3, 8.	A7, N34, T24, T26	None	201	244	Forbidden	Forbidden	D	21, 28, 40, 49, 100
	Tricresyl phosphate with more than 3 percent ortho isomer.	6.1	UN2574	II	6.1	A3, N33, N34, T8	None	202	243	5 L	60 L	A	
	Triethyl phosphite	3	UN2323	III	3	B1, T1	150	203	242	60 L	220 L	A	
	Triethylamine	3	UN1296	II	3, 8	B101, T8	None	202	243	1 L	5 L	B	40
	Triethylenetetramine	8	UN2259	II	8	B2, T8	154	202	242	1 L	30 L	B	40
	Trifluoroacetic acid	8	UN2699	I	8	A3, A6, A7, B4, N3, N34, T18, T27	None	201	243	0.5 L	2.5 L	B	12, 40
	Trifluoroacetyl chloride	2.3	UN3057		2.3, 8	2, B7, B9, B14	None	304	314, 315.	Forbidden	Forbidden	D	40
	Trifluorochloroethylene, inhibited	2.3	UN1082		2.3, 2.1	3, B14	None	304	314, 315.	Forbidden	Forbidden	D	40
	1,1,1-Trifluoroethane, compressed or Refrigerant gas R 143a.	2.1	UN2035		2.1		306	304	314, 315.	Forbidden	150 kg	B	40
	Trifluoromethane or Refrigerant gas R 23.	2.2	UN1984		2.2		306	304	314, 315.	75 kg	150 kg	A	
	Trifluoromethane, refrigerated liquid.	2.2	UN3136		2.2		306	None	314, 315.	50 kg	500 kg	D	
	2-Trifluoromethylaniline	6.1	UN2942	III	6.1		153	203	241	60 L	220 L	A	
	3-Trifluoromethylaniline	6.1	UN2948	II	6.1	T14	None	202	243	5 L	60 L	A	40
	Triformoxime trinitrate	Forbidden											
	Triisobutylene	3	UN2324	III	3	B1, T7, T30	150	203	242	60 L	220 L	A	
	Triisopropyl borate	3	UN2616	II	3	T8, T31	150	202	242	5 L	60 L	A	
				III	3	B1, T8, T31	150	203	242	60 L	220 L	A	
D	Trimethoxysilane	6.1	NA9269	I	6.1, 3	2, B9, B14, B32, B74, T38, T43, T45	None	227	244	Forbidden	Forbidden	E	40
	Trimethyl borate	3	UN2416	II	3	T14	150	202	242	5 L	60 L	B	
	Trimethyl phosphite	3	UN2329	III	3	B1, T1	150	203	242	60 L	220 L	A	
	1,3,5-Trimethyl-2,4,6-trinitrobenzene.	Forbidden											
	Trimethylacetyl chloride	6.1	UN2438	I	6.1, 8, 3.	2, A3, A6, A7, B3, B9, B14, B32, B74, N34, T38, T43, T45	None	227	244	Forbidden	Forbidden	D	25, 40
	Trimethylamine, anhydrous	2.1	UN1083		2.1		306	304	314, 315.	Forbidden	150 kg	B	40
	Trimethylamine, aqueous solutions with not more than 50 percent trimethylamine by mass.	3	UN1297	I	3, 8	T42	None	201	243	0.5 L	2.5 L	D	40, 41
				II	3, 8	B1, T14	None	202	243	1 L	5 L	B	40, 41
				III	3, 8	B1	150	203	242	5 L	60 L	A	40, 41
	1,3,5-Trimethylbenzene	3	UN2325	III	3	B1, T1	150	203	242	60 L	220 L	A	
	Trimethylchlorosilane	3	UN1298	II	3, 8	A3, A7, B77, N34, T14, T26	None	202	243	1 L	5 L	E	40
	Trimethylcyclohexylamine	8	UN2326	III	8	T2	154	203	241	5 L	60 L	A	
	Trimethylene glycol diperchlorate.	Forbidden											
	Trimethylhexamethylene diisocyanate.	6.1	UN2328	III	6.1	T8	153	203	241	60 L	220 L	B	
	Trimethylhexamethylenediamines.	8	UN2327	III	8	T7	154	203	241	5 L	60 L	A	
	Trimethylol nitromethane trinitrate.	Forbidden											
	Trinitro-meta-cresol	1.1D	UN0216	II	1.1D		None	62	None	Forbidden	Forbidden	B	1E, 5E
	2,4,6-Trinitro-1,3-diazobenzene	Forbidden											
	2,4,6-Trinitro-1,3,5-triazido benzene (dry).	Forbidden											
	Trinitroacetic acid	Forbidden											
	Trinitroacetonitrile	Forbidden											

§172.101 HAZARDOUS MATERIALS TABLE—Continued

Symbols	Hazardous materials descriptions and proper shipping names	Hazard class or Division	Identification Numbers	PG	Label Codes	Special provisions	(8) Packaging (§173.***)			(9) Quantity limitations		(10) Vessel stowage	
							Exceptions	Non-bulk	Bulk	Passenger aircraft/rail	Cargo aircraft only	Location	Other
(1)	(2)	(3)	(4)	(5)	(6)	(7)	(8A)	(8B)	(8C)	(9A)	(9B)	(10A)	(10B)
	Trinitroamine cobalt	Forbidden											
	Trinitroaniline or Picramide	1.1D	UN0153	II	1.1D		None	62	None	Forbidden	Forbidden	B	1E, 5E
	Trinitroanisole	1.1D	UN0213	II	1.1D		None	62	None	Forbidden	Forbidden	B	1E, 5E
	Trinitrobenzene, *dry or wetted with less than 30 percent water, by mass.*	1.1D	UN0214	II	1.1D		None	62	None	Forbidden	Forbidden	B	1E, 5E
	Trinitrobenzene, wetted *with not less than 30 percent water, by mass.*	4.1	UN1354	I	4.1	23, A2, A8, A19, N41	None	211	None	0.5 kg	0.5 kg	E	28
	Trinitrobenzenesulfonic acid	1.1D	UN0386	II	1.1D		None	62	None	Forbidden	Forbidden	E	1E, 5E
	Trinitrobenzoic acid, *dry or wetted with less than 30 percent water, by mass.*	1.1D	UN0215	II	1.1D		None	62	None	Forbidden	Forbidden	B	1E, 5E
	Trinitrobenzoic acid, wetted *with not less than 30 percent water, by mass.*	4.1	UN1355	I	4.1	23, A2, A8, A19, N41	None	211	None	0.5 kg	0.5 kg	E	28
	Trinitrochlorobenzene or Picryl chloride.	1.1D	UN0155	II	1.1D		None	62	None	Forbidden	Forbidden	B	1E, 5E
	Trinitroethanol	Forbidden											
	Trinitroethylnitrate	Forbidden											
	Trinitrofluorenone	1.1D	UN0387	II	1.1D		None	62	None	Forbidden	Forbidden	B	1E, 5E
	Trinitromethane	Forbidden											
	1,3,5-Trinitronaphthalene	Forbidden											
	Trinitronaphthalene	1.1D	UN0217	II	1.1D		None	62	None	Forbidden	Forbidden	B	1E, 5E
	Trinitrophenetole	1.1D	UN0218	II	1.1D		None	62	None	Forbidden	Forbidden	B	1E, 5E
	Trinitrophenol or Picric acid, *dry or wetted with less than 30 percent water, by mass.*	1.1D	UN0154	II	1.1D		None	62	None	Forbidden	Forbidden	B	1E, 5E
	Trinitrophenol, wetted *with not less than 30 percent water, by mass.*	4.1	UN1344	I	4.1	23, A8, A19, N41	None	211	None	1 kg	15 kg	E	28, 36
	2,4,6-Trinitrophenyl guanidine (dry).	Forbidden											
	2,4,6-Trinitrophenyl nitramine	Forbidden											
	2,4,6-Trinitrophenyl trimethylol methyl nitramine trinitrate (dry).	Forbidden											
	Trinitrophenylmethylnitramine or Tetryl.	1.1D	UN0208	II	1.1D		None	62	None	Forbidden	Forbidden	B	1E, 5E
	Trinitroresorcinol or Styphnic acid, *dry or wetted with less than 20 percent water, or mixture of alcohol and water, by mass.*	1.1D	UN0219	II	1.1D		None	62	None	Forbidden	Forbidden	B	1E, 5E
	Trinitroresorcinol, wetted or Styphnic acid, wetted *with not less than 20 percent water, or mixture of alcohol and water by mass.*	1.1D	UN0394	II	1.1D		None	62	None	Forbidden	Forbidden	B	1E, 5E
	2,4,6-Trinitroso-3-methyl nitraminoanisole.	Forbidden											
	Trinitrotetramine cobalt nitrate	Forbidden											
	Trinitrotoluene *and* Trinitrobenzene mixtures *or* TNT and trinitrobenzene mixtures *or* TNT and hexanitrostilbene mixtures *or* Trinitrotoluene and hexanitrostilnene mixtures.	1.1D	UN0388	II	1.1D		None	62	None	Forbidden	Forbidden	B	1E, 5E
	Trinitrotoluene mixtures containing Trinitrobenzene and Hexanitrostilbene *or* TNT mixtures containing trinitrobenzene and hexanitrostilbene.	1.1D	UN0389	II	1.1D		None	62	None	Forbidden	Forbidden	B	1E, 5E
	Trinitrotoluene or TNT, *dry or wetted with less than 30 percent water, by mass.*	1.1D	UN0209	II	1.1D		None	62	None	Forbidden	Forbidden	B	1E, 5E
	Trinitrotoluene, wetted *with not less than 30 percent water, by mass.*	4.1	UN1356	I	4.1	23, A2, A8, A19, N41	None	211	None	0.5 kg	0.5 kg	E	28
	Tripropylamine	3	UN2260	III	3, 8	B1, T8	150	203	242	5 L	60 L	A	40
	Tripropylene	3	UN2057	II	3	T1	150	202	242	5 L	60 L	B	
				III	3	B1, T1	150	203	242	60 L	220 L	A	
	Tris-(1-aziridinyl)phosphine oxide, solution.	6.1	UN2501	II	6.1	T8	None	202	243	5 L	60L	A	
				III	6.1	T7	153	203	241	60 L	220 L	A	
	Tris, bis-bifluoroamino diethoxy propane (TVOPA).	Forbidden											
	Tritonal	1.1D	UN0390	II	1.1D		None	62	None	Forbidden	Forbidden	B	1E, 5E
	Tungsten hexafluoride	2.3	UN2196		2.3, 8	2	None	338	None	Forbidden	Forbidden	D	40
	Turpentine	3	UN1299	III	3	B1, T1	150	203	242	60 L	220 L	A	
	Turpentine substitute	3	UN1300	I	3	T1	None	201	243	1 L	30 L	B	
				II	3	T1	150	202	242	5 L	60 L	B	
				III	3	B1, T1	150	203	242	60 L	220 L	A	
	Undecane	3	UN2330	III	3	B1, T1	150	203	242	60 L	220 L	A	
	Uranium hexafluoride, *fissile excepted or non-fissile.*	7	UN2978		7, 8		423	420, 427.	420, 427.				

Appendix A

§172.101 Hazardous Materials Table—Continued

Symbols	Hazardous materials descriptions and proper shipping names	Hazard class or Division	Identification Numbers	PG	Label Codes	Special provisions	Packaging (§173.***) Exceptions	Packaging (§173.***) Non-bulk	Packaging (§173.***) Bulk	Quantity limitations Passenger aircraft/rail	Quantity limitations Cargo aircraft only	Vessel stowage Location	Vessel stowage Other
(1)	(2)	(3)	(4)	(5)	(6)	(7)	(8A)	(8B)	(8C)	(9A)	(9B)	(10A)	(10B)
	Uranium hexafluoride, fissile (*with more than 1 percent U-235*).	7	UN2977		7, 8		453	417, 420.	417, 420.			A	
	Uranium metal, pyrophoric	7	UN2979		7, 4.2		None	418	None			D	
	Uranyl nitrate hexahydrate solution.	7	UN2980		7, 8		421, 427.	415, 416, 417.	415, 416, 417.			D	
	Uranyl nitrate, solid	7	UN2981		7, 5.1		None	419	None	Forbidden	15 kg	A	
	Urea hydrogen peroxide	5.1	UN1511	III	5.1, 8	A1, A7, A29	152	213	240	25 kg	100kg	A	13
	Urea nitrate, *dry or wetted with less than 20 percent water, by mass*.	1.1D	UN0220	II	1.1D	119	None	62	None	Forbidden	Forbidden	B	1E, 5E
	Urea nitrate, wetted *with not less than 20 percent water, by mass*.	4.1	UN1357	I	4.1	39, A8, A19, N41	None	211	None	1 kg	15 kg	A	28
	Urea peroxide, see Urea hydrogen peroxide.												
	Valeraldehyde	3	UN2058	II	3	T1	150	202	242	5 L	60 L	B	
	Valeric acid, see Corrosive liquids, n.o.s.												
	Valeryl chloride	8	UN2502	II	8, 3	A3, A6, A7, B2, N34, T8	154	202	243	1 L	30 L	C	40
	Vanadium compound, n.o.s	6.1	UN3285	I	6.1		None	211	242	5 kg	50 kg	B	
				II	6.1	T14	None	212	242	25 kg	100 kg	B	
				III	6.1	T7	153	213	240	100 kg	200 kg	A	
	Vanadium oxytrichloride	8	UN2443	II	8	A3, A6, A7, B2, B16, N34, T8, T26	154	202	242	Forbidden	30 L	C	40
	Vanadium pentoxide, *non-fused form*.	6.1	UN2862	III	6.1		153	213	240	100 kg	200 kg	A	40
	Vanadium tetrachloride	8	UN2444	I	8	A3, A6, A7, B4, N34, T8, T26	None	201	243	Forbidden	2.5 L	C	40
	Vanadium trichloride	8	UN2475	III	8		154	213	240	25 kg	100 kg	A	40
	Vanadyl sulfate	6.1	UN2931	II	6.1		None	212	242	25 kg	100 kg	A	
	Vehicle, flammable gas powered	9	UN3166		9	135	220	220	220	Forbidden	No limit	A	
	Vehicle, flammable liquid powered.	9	UN3166		9	135	220	220	220	No limit	No limit	A	
	Very signal cartridge, see Cartridges, signal.												
	Vinyl acetate, inhibited	3	UN1301	II	3	T8	150	202	242	5 L	60 L	B	
	Vinyl bromide, inhibited	2.1	UN1085		2.1		306	304	314, 315.	Forbidden	150 kg	B	40
	Vinyl butyrate, inhibited	3	UN2838	II	3	T7	150	202	242	5 L	60 L	B	
	Vinyl chloride, inhibited *or* Vinyl chloride, stabilized.	2.1	UN1086		2.1	21, B44	306	304	314, 315.	Forbidden	150 kg	B	40
	Vinyl chloroacetate	6.1	UN2589	II	6.1, 3	T14	None	202	243	5 L	60 L	A	
	Vinyl ethyl ether, inhibited	3	UN1302	I	3	A3, B100, T14	None	201	243	1 L	30 L	D	
	Vinyl fluoride, inhibited	2.1	UN1860		2.1		306	304	314, 315.	Forbidden	150 kg	E	40
	Vinyl isobutyl ether, inhibited	3	UN1304	II	3	T8	150	202	242	5 L	60 L	B	
	Vinyl methyl ether, inhibited	2.1	UN1087		2.1	B44	306	304	314, 315.	Forbidden	150 kg	B	
	Vinyl nitrate polymer	Forbidden											
	Vinyltoluenes, inhibited	3	UN2618	III	3	B1, T1	150	203	242	60 L	220 L	A	
	Vinylidene chloride, inhibited	3	UN1303	I	3	T23, T29	150	201	243	1 L	30 L	E	40
	Vinylpyridines, inhibited	6.1	UN3073	II	6.1, 3, 8.	B100, T8	None	202	242	1 L	30 L	B	40
	Vinyltrichlorosilane, inhibited	3	UN1305	I	3, 8	A3, A7, B6, N34, T14, T26	None	201	243	Forbidden	2.5 L	B	40
	Warheads, rocket *with burster or expelling charge*.	1.4D	UN0370	II	1.4D		None	62	None	Forbidden	75 kg	A	3E, 7E, 24E
	Warheads, rocket *with burster or expelling charge*.	1.4F	UN0371	II	1.4F		None	62	None	Forbidden	Forbidden	E	
	Warheads, rocket *with bursting charge*.	1.1D	UN0286	II	1.1D		None	62	None	Forbidden	Forbidden	B	3E, 7E
	Warheads, rocket *with bursting charge*.	1.2D	UN0287	II	1.2D		None	62	None	Forbidden	Forbidden	B	3E, 7E
	Warheads, rocket *with bursting charge*.	1.1F	UN0369	II	1.1F		None	62	None	Forbidden	Forbidden	E	
	Warheads, torpedo *with bursting charge*.	1.1D	UN0221	II	1.1D		None	62	None	Forbidden	Forbidden	B	3E, 7E
G	Water-reactive liquid, corrosive, n.o.s.	4.3	UN3129	I	4.3, 8		None	201	243	Forbidden	1 L	D	
				II	4.3, 8	B106	None	202	243	1 L	5 L	E	85
				III	4.3, 8	B106	None	203	242	5 L	60 L	E	
G	Water-reactive liquid, n.o.s	4.3	UN3148	I	4.3		None	201	244	Forbidden	1 L	E	40
				II	4.3	B106	None	202	243	1 L	5 L	E	40
				III	4.3	B106	None	203	242	5 L	60 L	E	40
G	Water-reactive liquid, toxic, n.o.s	4.3	UN3130	I	4.3, 6.1	A4	None	201	243	Forbidden	1 L	D	
				II	4.3, 6.1	B106	None	202	243	1 L	5 L	E	85
				III	4.3, 6.1	B106	None	203	242	5 L	60 L	E	85
G	Water-reactive solid, corrosive, n.o.s.	4.3	UN3131	I	4.3, 8	B101, B106, N40	None	211	242	Forbidden	15 kg	D	
				II	4.3, 8	B101, B106	151	212	242	15 kg	50 kg	E	85
				III	4.3, 8	B105, B106	151	213	241	25 kg	100 kg	E	85
G	Water-reactive solid, flammable, n.o.s.	4.3	UN3132	I	4.3, 4.1	B101, B106, N40	None	211	242	Forbidden	15 kg	D	
				II	4.3, 4.1	B101, B106	151	212	242	15 kg	50 kg	E	
				III	4.3, 4.1	B105, B106	151	213	241	25 kg	100 kg	E	
G	Water-reactive solid, n.o.s	4.3	UN2813	I	4.3	B101, B106, N40	None	211	242	Forbidden	15 kg	E	40

§172.101 HAZARDOUS MATERIALS TABLE—Continued

Symbols	Hazardous materials descriptions and proper shipping names	Hazard class or Division	Identification Numbers	PG	Label Codes	Special provisions	Packaging (§173.***)			Quantity limitations		Vessel stowage	
							Exceptions	Non-bulk	Bulk	Passenger aircraft/rail	Cargo aircraft only	Location	Other
(1)	(2)	(3)	(4)	(5)	(6)	(7)	(8A)	(8B)	(8C)	(9A)	(9B)	(10A)	(10B)
				II	4.3	B101, B106	151	212	242	15 kg	50 kg	E	40
				III	4.3	B105, B106	151	213	241	25 kg	100 kg	E	40
G	Water-reactive, solid, oxidizing, n.o.s.	4.3	UN3133	II	4.3, 5.1		None	214	214	Forbidden	Forbidden	E	40
				III	4.3, 5.1		None	214	214	Forbidden	Forbidden	E	40
G	Water-reactive solid, self-heating, n.o.s.	4.3	UN3135	I	4.3, 4.2	B100, N40	None	211	242	Forbidden	15 kg	E	
				II	4.3, 4.2	B101, B106	None	212	242	15 kg	50 kg	E	
				III	4.3, 4.2	B101, B106	None	213	241	25 kg	100 kg	E	
G	Water-reactive solid, toxic, n.o.s	4.3	UN3134	I	4.3, 6.1	A8, B101, B106, N40	None	211	242	Forbidden	15 kg	D	
				II	4.3, 6.1	B105, B106	151	212	242	15 kg	50 kg	E	85
				III	4.3, 6.1	B105, B106	151	213	241	25 kg	100 kg	E	85
	Wheel chair, electric, see Battery powered vehicle or Battery powered equipment.												
	White acid, see Hydrofluoric acid mixtures.												
I	White asbestos (chrysotile, actinolite, anthophyllite, tremolite).	9	UN2590	III	9		155	216	240	200 kg	200 kg	A	34, 40
	Wood preservatives, liquid	3	UN1306	II	3	T7, T30	150	202	242	5 L	60 L	B	40
				III	3	B1, T7, T30	150	203	242	60 L	220 L	A	40
	Xanthates	4.2	UN3342	II	4.2		None	212	241	15kg	50kg	D	40
				III	4.2		None	213	241	25kg	100kg	D	40
	Xenon, compressed	2.2	UN2036		2.2		306	302	None	75 kg	150 kg	A	
	Xenon, refrigerated liquid (cryogenic liquids).	2.2	UN2591		2.2		320	None	None	50 kg	500 kg	B	
	Xylenes	3	UN1307	II	3	T1	150	202	242	5 L	60 L	B	
				III	3	B1, T1	150	203	242	60 L	220 L	A	
	Xylenols	6.1	UN2261	II	6.1	T8	None	212	242	25 kg	100 kg	A	
	Xylidines, solid	6.1	UN1711	II	6.1	T14	None	212	242	25 kg	100 kg	A	
	Xylidines, solution	6.1	UN1711	II	6.1	T14	None	202	243	5 L	60 L	A	
	Xylyl bromide	6.1	UN1701	II	6.1	A3, A6, A7, N33	None	340	None	Forbidden	60 L	D	40
	p-Xylyl diazide	Forbidden											
	Zinc ammonium nitrite	5.1	UN1512	II	5.1		None	212	242	5 kg	25 kg	E	
	Zinc arsenate or Zinc arsenite or Zinc arsenate and zinc arsenite mixtures.	6.1	UN1712	II	6.1		None	212	242	25 kg	100 kg	A	
	Zinc ashes	4.3	UN1435	III	4.3	A1, A19, B108	151	213	241	25 kg	100 kg	A	
	Zinc bisulfite solution, see Bisulfites, inorganic aqueous solutions, n.o.s.												
	Zinc bromate	5.1	UN2469	III	5.1	A1, A29	152	213	240	25 kg	100kg	A	56, 58, 106
	Zinc chlorate	5.1	UN1513	II	5.1	A9, N34	152	212	242	5 kg	25 kg	A	56, 58, 106
	Zinc chloride, anhydrous	8	UN2331	III	8		None	213	240	25 kg	100 kg	A	
	Zinc chloride, solution	8	UN1840	III	8	T7	154	203	241	5 L	60 L	A	
	Zinc cyanide	6.1	UN1713	I	6.1		None	211	242	5 kg	50 kg	A	26
	Zinc dithionite or Zinc hydrosulfite.	9	UN1931	III	None		155	204	240	100 kg	200 kg	A	49
	Zinc ethyl, see Diethylzinc												
	Zinc fluorosilicate	6.1	UN2855	III	6.1		153	213	240	100 kg	200 kg	A	26
	Zinc hydrosulfite, see Zinc dithionite.												
	Zinc muriate solution, see Zinc chloride, solution.												
	Zinc nitrate	5.1	UN1514	II	5.1		152	212	240	5 kg	25 kg	A	
	Zinc permanganate	5.1	UN1515	II	5.1		152	212	242	5 kg	25 kg	D	56, 58, 69, 106, 107
	Zinc peroxide	5.1	UN1516	II	5.1		152	212	242	5 kg	25 kg	A	13, 75, 106
	Zinc phosphide	4.3	UN1714	I	4.3, 6.1	A19, N40	None	211	None	Forbidden	15 kg	E	40, 85
	Zinc powder or Zinc dust	4.3	UN1436	I	4.3, 4.2	A19, B109, N40	None	211	242	Forbidden	15 kg	A	
				II	4.3, 4.2	A19, B109	None	212	242	15 kg	50 kg	A	
				III	4.3, 4.2	B108	None	213	242	25 kg	100 kg	A	
	Zinc resinate	4.1	UN2714	III	4.1	A1	151	213	240	25 kg	100 kg	A	
	Zinc selenate, see Selenates or Selenites.												
	Zinc selenite, see Selenates or Selenites.												
	Zinc silicofluoride, see Zinc fluorosilicate.												
	Zirconium, dry, coiled wire, finished metal sheets, strip (thinner than 254 microns but not thinner than 18 microns).	4.1	UN2858	III	4.1	A1	151	213	240	25 kg	100 kg	A	
	Zirconium, dry, finished sheets, strip or coiled wire.	4.2	UN2009	III	4.2	A1, A19	None	213	240	25 kg	100 kg	D	
	Zirconium hydride	4.1	UN1437	II	4.1	A19, A20, N34	None	212	240	15 kg	50 kg	E	
	Zirconium nitrate	5.1	UN2728	III	5.1	A1, A29	152	213	240	25 kg	100 kg	A	
	Zirconium picramate, dry or wetted with less than 20 percent water, by mass.	1.3C	UN0236	II	1.3C		None	62	None	Forbidden	Forbidden	B	1E, 5E

Appendix A

§172.101 HAZARDOUS MATERIALS TABLE—Continued

Sym-bols	Hazardous materials descriptions and proper shipping names	Hazard class or Division	Identification Numbers	PG	Label Codes	Special provisions	(8) Packaging (§173.***)			(9) Quantity limitations		(10) Vessel stowage	
							Exceptions	Non-bulk	Bulk	Passenger aircraft/rail	Cargo aircraft only	Location	Other
(1)	(2)	(3)	(4)	(5)	(6)	(7)	(8A)	(8B)	(8C)	(9A)	(9B)	(10A)	(10B)
	Zirconium picramate, wetted with not less than 20 percent water, by mass.	4.1	UN1517	I	4.1	23, N41	None	211	None	1 kg	15 kg	D	28, 36
	Zirconium powder, dry	4.2	UN2008	I	4.2		None	211	242	Forbidden	Forbidden	D	
				II	4.2	A19, A20, N5, N34	None	212	241	15 kg	50 kg	D	
				III	4.2		None	213	241	25 kg	100 kg	D	
	Zirconium powder, wetted with not less than 25 percent water (a visible excess of water must be present) (a) mechanically produced, particle size less than 53 microns; (b) chemically produced, particle size less than 840 microns.	4.1	UN1358	II	4.1	A19, A20, N34	None	212	241	15 kg	50 kg	E	
	Zirconium scrap	4.2	UN1932	III	4.2	N34	None	213	240	Forbidden	Forbidden	D	
D	Zirconium sulfate	8	NA9163	III	8	N34	None	213	240	50 kg	No limit	A	
	Zirconium suspended in a liquid	3	UN1308	I	3		None	201	243	Forbidden	Forbidden	B	
				II	3		None	202	242	5 L	60 L	B	
				III	3	B1	150	203	242	60 L	220 L	B	
	Zirconium tetrachloride	8	UN2503	III	8		154	213	240	25 kg	100 kg	A	

Appendix B

Sample Written Hazard Communication Program

(To adapt this written hazard communication program for your company, substitute your company's name throughout this sample and include any additional company company-specific information where appropriate.)

Hazard Communication Program

I. Introduction

The OSHA Hazard Communication Standard was promulgated to ensure that all chemicals in the workplace would be evaluated and that information regarding the hazards of these chemicals would be communicated to employers and employees. The goal of the standard is to reduce the number of chemically related occupational illnesses and injuries.

To comply with the Hazard Communication Standard, this written program has been established for (insert company name). All departments and work centers are included within this program. Copies of this written program will be available for review by any employee in the following locations:

(List all Locations)

_____   _____
_____   _____
_____   _____

Department standard operating procedures (SOPs) work in conjunction with this basic document in providing the safest possible environment to all employees.

265

II. **Responsibilities**
   A. Department directors will be responsible for implementing and ensuring the compliance of their departmental personnel with this hazard communication program. They will assign appropriate supervisors with the responsibility of ensuring compliance.

   B. The safety director has the following responsibilities under this hazard communication program:
      1. Develop and modify as necessary this hazard communication program.
      2. Annually check and review the effectiveness of the overall program and all work center programs.
      3. Inspect quarterly each work center's hazardous chemical inventory list and corresponding material safety data sheet to ensure that they are current and complete.
      4. Receive and review all incoming or updated editions of MSDS and distribute them to pertinent work centers.
      5. Maintain a current master MSDS file.
      6. Train supervisors in requirements for hazard communication program and assist in training personnel as required.

   C. Assigned supervisors have the following responsibilities under this hazard communication program:
      1. Reporting receipt of all new chemicals to safety division.
      2. Ensuring that no chemical is used at the work center until it is listed on the hazardous chemical inventory list, the corresponding MSDS is inserted into the "right to know" station binder, and each employee has received the appropriate hazcom training.
      3. Maintaining current work center "right to know" work stations, which include ensuring the MSDS and chemical inventory list are current and accurately reflect chemicals used on site.
      4. Ensuring proper labeling practices for all hazardous chemicals in accordance with this program.
      5. Forwarding to the safety division all MSDS received from sources other than the safety division.
      6. Ensuring that every employee is trained on this program and the hazards involved with chemicals used in the work center.
      7. Ensuring that all on-site contractors receive copies of the hazard communication program and all work center MSDSs. In turn, the work center

Appendix B 267

supervisor is responsible for obtaining MSDSs from the contractor about chemicals that he or she may bring to the site that employees may be exposed to.

D. Personnel will be responsible for familiarizing themselves with the hazard communication program, and with complying with instructions contained with the hazard communication program.

III. Definition of Terms

The hazard communication program defines various terms as follows (These terms either appear in this hazard communication program or are definitions appropriate to MSDS):

Chemical	Any element, compound, or mixture of elements and/or compounds.
Chemical Name	The scientific designation of a chemical in accordance with the nomenclature system developed by the International Union of Pure and Applied Chemistry (IUPAC) or the Chemical Abstracts Service (CAS) Rules of Nomenclature, or a name that will clearly identify the chemical for the purpose of conducting a hazard evaluation.
Combustible Liquid	Any liquid having a flashpoint at or above 100°F (37.8°C) but below 200°F (93.3°C).
Common Name	Any designation or identification such as code name, code number, trade name, brand name, or generic name used to identify a chemical other than its chemical name.
Compressed Gas	(1) A gas or mixture of gases in a container having an absolute pressure exceeding 40 psi at 70°F (21.1°C); or
	(2) A gas or mixture of gases in a container having an absolute pressure exceeding 104 psi at 130°F (54.4°C), regardless of the pressure at 70°F (21.1°C); or
	(3) A liquid having a vapor pressure exceeding 10 psi at 100°F (37.8°C), as determined by ASTM D-323-72.
Container	Any bag, barrel, bottle, box, can, cylinder, drum, reaction vessel, storage tank, or the like that contains a hazardous chemical.
Explosive	A chemical that causes a sudden, almost instantaneous release of pressure, gas, and heat when subjected to sudden shock, pressure, or high temperature.

Exposure	The actual or potential subjection of an employee to a hazardous chemical in the course of employment, through any route of entry.
Flammable Aerosol	An aerosol that when tested by the method described in 16 CFR 1500.45 yields a flame projection exceeding 18 inches at full valve opening, or a flashback (flame extending back to the valve) at any degree of valve opening.
Flammable Gas	A gas that at ambient temperature and pressure forms a flammable mixture with air at a concentration of 13 percent by volume or less, or a gas that at ambient temperature and pressure forms a range of flammable mixtures with air wider than 12 percent by volume regardless of the lower limit.
Flammable Liquid	A liquid having a flashpoint of 100°F (37.8°C).
Flammable Solid	A solid (other than a blasting agent or explosive as defined in 29 CFR 1910.109 (a)) that is likely to cause fire through friction, absorption of moisture, spontaneous chemical change, or retained heat from manufacturing or processing, or which when ignited, burns so vigorously and persistently as to create a serious hazard. A chemical shall be considered to be a flammable solid if, when tested by the method described in 16 CFR 1500.44., it ignites and burns with a self-sustained flame at rate greater than one-tenth of an inch per second along its major axis.
Flashpoint	The minimum temperature at which a liquid gives off a vapor in sufficient concentration to ignite.
Hazard Warning	Any words, pictures, symbols, or combination thereof appearing on a label or other appropriate form of warning that convey the hazards of the chemical(s) in the container.
Hazardous Chemical	Any chemical that is a health or physical hazard.
Hazardous Chemical Inventory List	An inventory list of all hazardous chemicals used at the site and the date of each chemical's MSDS insertion.
Health Hazard	A chemical for which there is statistically significant evidence based on at least one study conducted in accordance with established scientific principles that acute or chronic health effects may occur in exposed employees.
Immediate Use	The use of a chemical under the control of and only by the person who transfers the hazardous chemical from a labeled container, and only within the work shift in which it is transferred.

Appendix B

Label	Any written, printed, or graphic material displayed on or affixed to containers of hazardous chemicals.
Material Safety Data Sheet (MSDS)	Written or printed material concerning a hazardous chemical, which is developed in accordance with 29 CFR 1910.
Mixture	Any combination of two or more chemicals if the combination is not, in whole or in part, the result of a chemical reaction.
NFPA Hazardous Chemical Label	A color-code labeling system developed by the National Fire Protection Association (NFPA), which rates the severity of the health hazard, fire hazard, reactivity hazard, and/or special hazard of the chemical.
Organic Peroxide	An organic compound that contains the bivalent 0-0 structure (which may be considered to be a structural derivative) and hydrogen peroxide, where one or both of the hydrogen atoms has been replaced by an organic radical.
Oxidizer	A chemical (other than a blasting agent or explosive as defined in 29 CFR 1910.198 (a)) that initiates or promotes combustion in other materials, thereby causing fire either of itself, or, through the release of oxygen, of other gases.
Physical Hazard	A chemical for which there is scientifically valid evidence that it is a combustible liquid, a compressed gas explosive, flammable, an organic peroxide, an oxidizer, pyrophoric, unstable (reactive), or water reactive.
Portable Container	A storage vessel (such as a drum, side-mounted tank, tank truck or vehicle fuel tank) that is mobile.
Primary Route of Entry	The primary means (such as inhalation, ingestion, skin contact, etc.) whereby an employee is subjected to a hazardous chemical.
"Right To Know" Work Station	A document location, which provides employees with a central information work station where they can have access to site MSDS sheets, hazardous chemical inventory list, and this written hazard communication program.
"Right To Know" Station Binder	A station binder located in the "right to know" work station that contains this hazard communication program, the hazardous chemicals inventory list, corresponding MSDS, and

	the hazard communication program review and signature form.
Pyrophoric	A chemical that will ignite spontaneously in air at a temperature of 130°F (54.4°C) or below.
Stationary Container	A permanently mounted chemical storage tank.
Unstable (Reactive Chemical)	A chemical that in its pure state or as produced or transported will vigorously polymerize, decompose, or condense, or will become self-reactive under conditions of shock, pressure, or temperature.
Water reactive (Chemical)	A chemical that reacts with water to release a gas that is either flammable or presents a health hazard.
Work center	Any convenient or logical grouping of designated unit processes or related maintenance actions.

IV. "Right to Know" Work Stations

Each work center will establish an employee "right to know" work station. This "right to know" work station will be accessible to employees during their work hours. The "right to know" work station will contain a "right to know" station binder. This station binder will contain this hazard communication program, hazardous chemical inventory list and corresponding MSDS, and the hazard communication program review and signature form.

V. Hazardous Chemical Inventory List

A list of all hazardous chemicals or fuels used or produced at each work center will be maintained in each work center's hazardous chemical inventory list. This hazardous chemical inventory list (see Figure A.1) is to be filed in the front of each work center's "right to know" station binder, maintained within its "right to know" work station.

The hazardous chemical inventory list will also show the date of the most recent MSDS insertion for each chemical. Only the hazardous chemicals or fuels actually used within each work center will be listed in that work center's hazardous chemical inventory list. The master hazardous chemical inventory list for all chemicals used within the company will be maintained by the safety office.

Each work center supervisor is to ensure that its hazardous chemical inventory list is accurate, updated, and available for employee use. Work centers receiving new chemicals or chemicals not on their current hazardous chemical inventory list shall follow these procedures:

Appendix B

a. Add the hazardous chemical to the hazardous chemical inventory list.

b. Procure and insert the chemical MSDS into the "right to know" station binder.

c. Train employees on the hazards associated with the chemical.

The hazardous chemical inventory list for each work center will be reviewed by each work center at least quarterly; a verifying signature for this quarterly review will be made on the hazardous chemical list. The hazardous chemical inventory list for each work center will be reviewed quarterly by the safety division to ensure that it is current, and that corresponding MSDS are available.

Hazardous Chemical Inventory List Form

Name of Work Center

Chemical	Date of MSDS insertion
_____	_____
_____	_____
_____	_____
_____	_____
_____	_____
_____	_____

Figure B.1 Chemical inventory list form

VI. Material Safety Data Sheet (MSDS)

The material safety data sheets (MSDS) are a set of individual data sheets providing related safety information for each hazardous chemical used or produced at the work center. Material safety data sheets are filed in each work center's "right to know" station binder located in the "right to know" work stations. Each chemical listed on the hazardous chemical inventory list must have a corresponding MSDS. MSDSs are provided to work centers by the safety division any time manufacturers forward new copies or new editions. Work center supervisors are responsible for insuring that MSDS are current and available for all chemicals listed on their work center's hazardous chemical inventory list, and that chemicals are not used unless this information is available.

The material safety data sheets should contain the following information:

- Identity of hazardous chemical
- Identity of hazardous ingredients in a hazardous chemical mixture
- Chemical and physical characteristics of the hazardous chemical
- Chemical and physical hazards of the hazardous chemical

- Acute and chronic health hazards, including signs and symptoms of exposure and medical conditions that are generally aggravated by exposure to the hazardous chemical
- Primary route of entry
- Personal exposure limits in terms of maximum duration and concentration
- Protective measures and special precautions
- Emergency procedures and first aid procedures
- Date of preparation of the material safety data sheet
- Identification of person or agency responsible for the information contained on the material safety data sheet.

The material safety data sheet shall not contain any blank spaces. Not applicable or unknown information should be indicated as such.

The purchasing department shall specify that MSDSs will be required with all orders. Supervisors using local supply orders (LSOs) to obtain chemicals should ensure that a MSDS is available for the product prior to or at receipt of the product.

Any MSDS received by the work center should be forwarded to the safety office with notations on which work center forwarded the sheet, and an indication as to whether the chemical is in use or not. If the work center is using a chemical and does not have an existing MSDS, the safety division should be notified immediately. The safety division will procure the needed MSDS or will generate a generic form. The safety division will review all incoming or self-generated MSDS for completeness before forwarding copies to pertinent work centers.

VII. Hazard Warnings and Labeling

Hazard warnings are individual warnings on hazardous chemical containers that provide related safety information for each respective hazardous chemical utilized or produced within the company.

    A. Hazard warnings should be displayed on or affixed to **all** hazardous chemical containers, providing the following information:

Note: Personnel should not remove or deface existing hazard warnings or labels on hazardous chemical containers received or used at the work center unless the container is immediately marked with the required information.

        1. Each portable container should be labeled, tagged, or otherwise marked as follows:

            a. Chemical or common name of the hazardous chemical

            b. Hazard warnings

Appendix B 273

      c. Name and address or the chemical manufacturer

Note: Supervisors should verify that all containers received for use are appropriately labeled.

   2. Each stationary container shall be labeled as follows:

      a. Chemical or common name of the hazardous chemical stenciled with six (6) inch block letters

      b. NFPA hazardous chemical label

   3. Where applicable, all chemical piping should be labeled in accordance with the piping identification code, or with standard industry color codes. (Hazcom does not require the labeling of pipes, but when they contain hazardous chemicals, labeling is recommended.)

B. NFPA hazardous chemical labels should be affixed to hazardous chemical containers wherever appropriate and/or informative.

C. NFPA "0-4" number rating system for each chemical can be obtained from the safety division.

VIII. Training

The hazard communication program requires periodic training of personnel to ensure that the program requirement's safety precautions are properly conducted. Supervisors should consider personnel training as a primary responsibility.

A. The hazard communication program training duties and responsibilities for personnel are established as follows:

   1. The safety division shall be responsible for training supervisors in the program requirements, and will be available by appointment to present to work center personnel the required hazard communication program brief.

   2. Supervisors shall be responsible for training personnel in the program requirements.

   3. Personnel shall be responsible for familiarizing themselves with the program requirements.

B. The hazard communication program training should be planned so that:

   1. Training is given both in the program requirements and in related safety information, including protective measures, special precautions, and emergency procedures.

   2. Training is given both in the classroom and on the job.

   3. Training is given that uses group participation during both discussion and question and answer periods.

4. Training is ongoing, to preserve the continuity and integrity of program and work center safety.

C. The hazard communication program training should be provided as follows:

1. Training is conducted prior to assignment of an new employee to his work duties.

2. Training is conducted whenever a new hazardous chemical is used or produced at the work center.

3. Training is conducted prior to starting work on non-routine tasks involving hazardous chemicals.

D. The hazard communication program training is to be conducted by the supervisor, according to the recommended outline below:

1. Locate and identify the hazard communication program, hazardous chemical inventory list, and material safety data sheets contained in the "right to know" work stations.

2. Discuss the objective and content of the hazard communication program.

3. Explain that the hazardous chemical inventory list lists all hazardous chemicals at the work center concerned.

4. Discuss the physical and health effects of each hazardous chemical at the individual work center as contained in the MSDS.

5. Discuss the methods and techniques used to determine the presence or release of hazardous chemicals at the individual work center.

6. Discuss the protective measures and special precautions used to lessen or prevent exposure to hazardous chemicals at the individual work center.

7. Discuss emergency procedures for each hazardous chemical at the individual work center as contained in the MSDS.

8. Discuss the hazardous warning and labeling system.

Note: Hazard communication program training shall be recorded in each employee training record (see Figure B.2). These records will be examined during safety division's quarterly safety inspections.

Employee Training Record
Hazard Communication Training

Date	Employee Name	Work center
_____	_____	_____
_____	_____	_____
_____	_____	_____
_____	_____	_____

Figure B.2 Sample training record

XI. On-site Contractors/Visitors

A. Work center supervisors and the engineering division should notify all outside contractors and vendors, etc. who perform work within company work centers of the hazard communication program as follows:

1. The safety division, when directed by work center supervisor/engineering division, will provide a copy of this hazard communication program to regulatory agencies, consulting engineers, contractors, etc., as appropriate upon their initial visit on site to perform each specific project.

Note: A copy of the hazard communication program should be provided upon the initial visit for each specific project. It is not necessary to provide additional copies for ongoing visits to accomplish the same specific project.

2. All written requests for proposals or quotations, all written specifications, and all written contracts and work orders should include a written notification of the hazard communication program as follows:

(insert company name) is required in accordance with 29 CFR 1910.1200 to inform company and contractor personnel that work centers within the company have hazardous chemicals on site. Company and contract personnel may be exposed to these hazardous chemicals while working at company work centers. A written hazard communication program has been developed to inform personnel of the specific hazardous chemicals at the work center and the related safety information (including protective measures, special precautions, and emergency procedures to be observed). The hazard communication program (including material safety data sheets (MSDS) for each hazardous chemical at the work center) will be made available to contractors. Contractors are responsible for communicating the information contained in the material safety data sheets to their personnel working at the work center.

B. It is the responsibility of work center supervisors to provide a copy of the hazard communication program to other supervisors or personnel outside of the work center. It is the responsibility of other supervisors or personnel outside the work center to familiarize themselves and their personnel with the objective and content of the hazard communication program. Whenever personnel visit another work center to perform maintenance, provide assistance, or perform another activity, they should acquaint themselves with the information contained in the work center's "right to know" work station.

The work center supervisor should provide a copy of the hazard communication program and training to the other departments and personnel as appropriate.

Appendix C

Sample Substance Abuse Policy

I. Purpose and Overview

It is the intent of the company to establish and maintain a workplace environment that is conducive to the safe and efficient performance of job duties as well as promoting the health and well-being of all employees. The company is committed to eliminating the negative effects of substance abuse from work centers to ensure the quality of its services and its reputation. It is company policy that employees shall not:

1. *Report to or work on-the-job while under the influence of illegal drugs, or impaired by alcohol.*

   "Impaired by alcohol" is considered to be a blood alcohol concentration (BAC) equal to or exceeding 0.04. "Under the influence" of illegal drugs (or their metabolites) is considered to be when their presence is detected at a specified level as currently determined by the U.S. Department of Transportation (DOT).

   Illegal drugs are substances that are not legally obtained and include but are not limited to: all forms of narcotics, depressants, stimulants, hallucinogens or their derivatives, or prescribed drugs not legally obtained or not being used as prescribed.

2. Use or possess alcohol while on company premises, while in any company vehicle, or report to work impaired while on paid standby.

   Employees may have alcohol in their personal vehicle, provided the container has not been opened.

3. Manufacture, use, purchase, transfer, sell, possess, distribute, or accept illegal drugs while on the job, while on company premises, or while in a company vehicle.

II. Employee Assistance

The company's employee assistance program (EAP) is available to help employees resolve substance abuse problems. The company encourages employees to seek help once they recognize they might have a problem, prior to any negative affect it may have on job performance or behavior. When employees initiate contact with the EAP, the EAP keeps confidential both the employee's identity and anything said. The only exceptions are legal ones requiring the EAP to notify appropriate parties when someone is a threat to themselves, a threat to others, or has a substance abuse problem and is in designated safety sensitive position, currently defined as a position requiring a commercial drivers license (CDL). The company will not know about anyone who voluntarily contacts the EAP, for any reason, other than these exceptions, unless the employee chooses.

However, any employee advising the company of a substance abuse problem, or the possibility of one, shall be formally referred to the EAP. In these cases, on an unannounced basis, followup drug testing shall be imposed.

III. Medically Prescribed Medications

Employees are responsible for their safe behavior on the job. Employees should discuss the effects of their medication in relation to their company duties with their physician, and request sick leave if they know or believe their prescription or non-prescription drugs may impair their ability to safely and effectively perform their duties.

IV. Pre-Employment Testing

The company shall require pre-employment substance abuse testing of all applicants selected for employment (including temporary agency employees) and for all positions requiring a CDL as follows:

A. During the application process, the company shall inform applicant(s) of the pre-employment testing requirement.

B. After the job offer, the newly selected applicant or employee who is selected for a position requiring a CDL shall report to the medical examination facility for testing as directed.

C. The offer of employment shall be withdrawn in instances where the applicant (or employee) refuses to sign a testing examination consent form or refuses to cooperate with the testing examination process.

D. The company shall pay the medical facility's cost of the test and examination, but not for the applicant's time, travel, parking, or any other expenses.

E. Transportation to and from the testing medical examination facility is the responsibility of the applicant.

F. Any applicant who has an unacceptable test result shall be ineligible for employment by company for three years.

V. Requirements and Substance Abuse Testing for Employees Required to Hold a Commercial Drivers License (CDL)

A. Requirements.

In addition to other requirements outlined in this policy, employees who currently drive or are in an official standby position to drive a company vehicles requiring a CDL must comply with requirements outlined in this section, which are based on requirements established by DOT.

CDL drivers are prohibited from reporting to duty, remaining on duty, or performing any function associated with CDL driving responsibilities in any of the following situations:

1. with an alcohol concentration of 0.02 or greater
2. within four hours of using alcohol
3. when using controlled substances, except when the use is pursuant to the instructions of a physician who has specifically advised the employee that the substance does not adversely affect the employee's ability to safely operate a commercial motor vehicle
4. when refusing to cooperate with required testing procedures

CDL drivers who are taking prescription medicine may not drive without a release from their doctor stating that the substance does not affect their ability to drive. Failure to follow this requirement is cause for discipline.

CDL employees may not use alcohol until post-accident or reasonable suspicion testing is done, or for a period of eight hours, whichever comes first.

Supervisors who know that any specific driver is in violation of these CDL requirements cannot permit the driver to work.

B. Substance Abuse Testing, Unannounced

DOT regulations require unannounced random substance abuse testing of CDL drivers. The number of employees to be tested is established by federal regulations. The names of all employees who must have a CDL for their position are placed in a pool. Each month, a certain number will be randomly selected for substance abuse testing from the pool of all eligible names. Since all CDL drivers' names are in the pool each time, selection is random; any employee may be subject to testing more than once per year. Each year, the personnel division will notify CDL drivers of their eligibility for random substance abuse examination and the number of CDL drivers to be randomly tested during the upcoming year.

VI. Post Accident and Reasonable Suspicion Substance Abuse Testing

The company will require substance abuse testing where trained supervisor(s) have cause to test in accordance with DOT and company guidelines, or where trained supervisor(s) or medical authorities have reasonable suspicion that an employee's physical or mental condition in the performance of his or her duties is affected by alcohol and/or drugs. Determination for testing will be based on one or more of the types of conditions outlined in company's *Post Accident and Reasonable Suspicion Report* (see Figure C.1).

VII. Substance Abuse Testing Procedures

Reasonable suspicion and post accident substance abuse tests for illegal drugs should occur as soon as possible, but within 32 hours of a determination that testing is required. Tests for alcohol should occur within two hours; if the test has not been administered within eight hours, attempts at testing will cease. A record will be maintained explaining the reasons that the testing was not conducted within the two hour time frame.

Any time a breath analyzer screen or substance abuse test is required, a "Reasonable Suspicion Report" (see Figure C.1) must be completed and signed by the supervisor(s) authorizing the screen/test, and a copy given to the employee prior to taking the test. The original copy of the report must be submitted to the human resource manager no later than the following workday. This report will document the conditions that formed the basis for the test of the employee and indicate any witnesses to the employee's behavior. Whenever possible, DOT prefers that the supervisor obtain a witness to the employee's behavior. During office working hours, the human resources manager should coordinate substance tests with the testing facility.

Tests for drug or alcohol levels shall be conducted by a company-approved collection facility that follows the collection and testing procedures contained in current DOT guidelines.

    A. Alcohol and drug testing shall normally incorporate the following steps:

        1. Alcohol levels will be determined at an approved facility by a certified breath alcohol technician (BAT), using a testing device approved by the National Highway Safety Administration.

        2. Illegal drug levels shall be determined using the following two steps:

Appendix C

POST INCIDENT AND REASONABLE SUSPICION REPORT
(CONFIDENTIAL)

EMPLOYEE: _____ WORK CENTER: _____ DATE/TIME OF INCIDENT: _____

A. NATURE OF INCIDENT/CAUSE OF SUSPICION (Check all that apply – use Section B for further explanation)

- ☐ Observed possession or use of a prohibited substance or drug paraphernalia just prior to or during work hours.
- ☐ Observed in the company of those possessing or using a prohibited substance or drug paraphernalia just prior to or during work hours.
- ☐ Odor of marijuana.
- ☐ Odor of alcoholic beverage. Breath analyzer screen results _____ (Employees option)
- ☐ An On-The-Job-Injury (OJI) or accident that results in:
  - ☐ Any death ☐ A citation and an injury requiring immediate medical attention for any involved party ☐ A citation and a vehicle being towed
- ☐ Recommendation of a medical authority or following arrest on illegal drug charges.
- ☐ A single unusual or suspicious accident or OJI, or a trend over time of accidents or OJI in conjunction with the recommendation of two of the following: employee's supervisor, Human Resources Manager, Automotive Superintendent or Safety Manager.
- ☐ Three (3) or more OJIs requiring medical attention in a 12 month period.
- ☐ Physical signs, such as nystagmus or needle tracks.
- ☐ Specific, clearly stated, observations concerning unusual, erratic or extremely uncharacteristic behavior of an individual. Check all that apply.

CONSIDER THE EMPLOYEE'S BEHAVIOR AND APPEARANCE OR WHETHER THE BEHAVIOR IS APPROPRIATE FOR THE SITUATION.

1.	SPEECH:	☐ Normal ☐ Whispering	☐ Incoherent ☐ Unusually talkative or silent	☐ Confused	☐ Slurred
2.	BALANCE:	☐ Normal	☐ Swaying	☐ Staggering	
3.	WALKING:	☐ Normal ☐ Reaching for support	☐ Stumbling	☐ Swaying	☐ Arms raised
4.	STANDING:	☐ Normal	☐ Feet wide apart	☐ Sagging at the knees	☐ Rigid
5.	EYES:	☐ Normal ☐ Blood shot ☐ Droopy	☐ Glassy ☐ Closed	☐ Dilated pupils ☐ Watery	☐ Constricted pupils ☐ Lack of convergence
6.	FACE:	☐ Normal	☐ Flushed	☐ Pale	☐ Sweaty (unseasonable or unrelated to work)
7.	DEMEANOR:	☐ Normal ☐ Sarcastic ☐ Hyperactive ☐ Shakiness or twitching	☐ Crying ☐ Calm or irritable ☐ Tremors	☐ Mood swings ☐ Violent ☐ Excited or euphoric	☐ Argumentative or withdrawn/detached ☐ Angry or belligerent
8.	AWARENESS:	☐ Normal ☐ Drowsy or sluggish	☐ Confused/forgetful ☐ Lack of coordination	☐ Paranoid ☐ Partial or full break with reality	☐ Dazed, blank stare ☐ Lethargic/slow reactions
9.	PHYSICAL SYMPTOMS:	☐ Normal ☐ Faster breathing	☐ Vomiting ☐ Poor muscle control	☐ Bleeding or runny nose	☐ Nauseous

B. **WRITTEN SUMMARY**

Summarize the facts and circumstances of the incident including what lead to the discovery of evidence or the noted behavior, the employee's response, supervisor's actions taken, and any other pertinent information not previously noted. Note the date, times and location of reasonable cause testing or note if the employee refused the test. Insure the employee has been informed that failure to consent to a test will be considered the same as a positive test result. Attach additional sheets as needed.

_____

_____

_____

_____

_____

_____

Printed name of Supervisor #1	Signature of Supervisor #1	Date/Time
Printed name of Supervisor #2	Signature of Supervisor #2	Date/Time
Signature of Employee	Copy received	

Figure C.1. Post incident and reasonable suspicion report

Step 1: Screening Test: an initial determination using an immunoassay technique on a urine sample provided by an applicant or employee.

Step 2: Confirmation Test: a confirmation of any positive screening result. The confirmation shall use a portion of the same sample as initially provided by the applicant or employee and shall be analyzed with a GC/MS (gas chromatography/mass spectrometer).

3. Illegal drugs shall be tested using the cut-off levels approved by the Department of Health and Human Services (DHHS). These levels are subject to change by the DHHS as advances in technology or other considerations warrant. Information on the specific cut-off levels in effect at any time are available from the human resource division. Illegal drugs other than the five mentioned in DOT guidelines may be tested for if recommended by a medical authority conducting a fit-for-duty examination.

B. The following procedures shall be used to administer the substance abuse testing process:

1. The supervisor shall inform the employee of the reasons for testing, and that failure to take or cooperate with the test procedure shall be considered a positive result.

2. The supervisor shall ensure transportation for the employee to the test site and then to home after the test.

3. Prior to administration of the test, the employee shall sign a substance abuse test consent form authorizing the medical testing facility to obtain a specimen and to release the test results to the company human resource manager.

4. An employee's time involved in the test shall be considered time worked.

5. The designated testing facility's medical review officer (MRO) shall:

   a. Review and interpret drug test results

   b. Consider an individual's medical history and relevant medical factors to determine if a confirmed positive test resulted from legally prescribed medication. If necessary, the MRO may require a re-analysis of the original sample to determine the accuracy of the reported results.

   c. Provide an opportunity for the employee to discuss medical reasons for a positive result, prior to company notification. If the MRO determines there is a legitimate documented medical reason for the positive result, the test may be reported to company. If the MRO is unable to make contact with the employee in a reasonable amount of time, the human resource manager shall be contacted. The employee shall then be instructed to contact the MRO. If the employee fails to contact the MRO within the agreed upon timeframe, the results shall be provided to company human resource manager.

6. The employee may request an independent confirmatory test be conducted by another Department of Health and Human Services (DHHS) certified laboratory or by the same laboratory from the original test sample within 72 hours after the employee has been notified of a positive result by the MRO. To ensure that payment for these tests is not delayed, the company will pay for the test. If a positive test result is obtained, the cost of the test will be payroll deducted.

7. Test results will be communicated to the employee's supervisor through the human resource manager.

8. The human resource manager will maintain confidential records of all test results. These records will be secured and kept separate from personnel files. Disclosures without an employee's or applicant's consent may only occur when:

    a. The information is compelled by law or by judicial process.

    b. The information has been placed at issue in a formal dispute between the employer and applicant.

    c. The information is needed by medical personnel for the diagnosis or treatment of the individual who is unable to authorize disclosure.

9. Other details of the testing procedure are available from the human resource manager.

VIII. Return to Work

A. After a random substance abuse test, CDL drivers may return to work immediately if released by the testing facility.

B. After a reasonable suspicion or post accident substance abuse test, the employee shall be placed on administrative leave until released for duty by the testing facility or the company's authorized medical authority. If the test result is positive, the time away from work will be charged to the employee's annual leave balance and/or counted as part of the leave without pay suspension in conjunction with the appropriate disciplinary action.

C. CDL drivers with a BAC level between 0.02 but less than 0.04 shall not be allowed to return to work for a 24-hour period.

D. Employees with a BAC of 0.04 or greater, or employees that test positive for illegal drugs will be:

   1. unable to return to work until retested and released by a substance abuse professional (SAP)

   2. subject to a minimum of six random substance abuse tests for a period of one year upon return to work.

3. required to undergo and complete an SAP and company approved drug/alcohol treatment program.

IX. Training

Educational materials and training on alcohol and drug abuse and the company's substance abuse policy and related procedures will be provided to all supervisors and CDL employees. Minimum training times for supervisors shall conform to DOT requirements. Questions on the policy can be directed to the human resource manager. Upon completion of training on company's substance abuse policy, each company employee shall be issued a copy of the policy and will be required to sign the acknowledgment form (see Figure C.2). The signed form will be made a permanent part of each employee's personnel record.

X. Disciplinary Action

The company's disciplinary action policy will apply in cases of substance abuse. An employee will not be disciplined until an unacceptable substance abuse test result is communicated to the company. If the employee is engaged in an activity that would normally result in disciplinary action, the appropriate disciplinary action will be taken regardless of whether the employee was impaired or under the influence of an illegal substance at the time.

Acknowledgement Form

Receipt of company and U.S. Department of Transportation Employee Training and Substance Abuse Policy

I, _____, acknowledge receipt of company's Substance Abuse Policy and associated training materials. I verify that I have been briefed on both the Policy and the training materials, and that I have had the opportunity to ask questions. I further verify I understand the briefing, and I know where to go for assistance if needed.

_____
Signature

_____
Date

_____
Work Center

Figure C.2. Sample substance abuse policy acknowledgment form

Appendix D
Hazmat Definitions

Trying to learn a new subject is difficult unless we first learn the jargon. In short, as Voltaire said, *"If you wish to converse with me, please define your terms."*

OSHA defines many of the terms it uses in its Hazard Communication Standard, at least to a degree. But OSHA doesn't define all of them, and that presents a problem. How many of your workers might not be familiar with any of the following terms?

agent	*ALARA*	*allergen*	*ANSI*
asphyxia	*ceiling limit*	*CERCLA*	*flashpoint*
LFL	*etiological agent*	*miscible*	*tachycardia*
vapor density	*vapor pressure*	*Z list*	

Even though OSHA does not define these terms in the Hazcom Standard, we really do need to know all of them. They are important to Hazcom, especially in reading and interpreting MSDS jargon.

You and your workers are not required to memorize the terms listed above, or the more than one hundred terms that can be used to described various chemicals and chemical mixtures or compounds. However, workers who handle hazardous materials do need to be familiar with such terms. Such terms, their definitions, and how they apply to worker health and safety should be made part of the training presentation for all affected employees. Remember, one of the main purposes of Hazcom (and the MSDS in particular) is to tell workers and/or other users critical information about the material's physical properties or fast-acting health effects that make it dangerous to handle.

More importantly, you and your workers must know where to find information when it is required—for example, when working with or around hazardous materials. Your training program must be designed to provide employees with the information they need to be able to read and decipher the jargon used on MSDSs. This is best accomplished by having someone (a good candidate for the job would be the person responsible for Hazcom training) review the facility's MSDSs and make a list of all of the technical terms on them. For the MSDS to fulfill its important function, the person who uses the MSDS must understand the terms used.

Most of the terms you will run into are presented and defined below. For those terms not listed and defined here, some research may be necessary. Table D.1 lists several resources to refer to in this process—ones that also provide other important information about chemicals and chemical compounds.

Table D.1. Chemical Data/Information Sources

Sources

1. Company files
2. Chemical suppliers
3. *Condensed Chemical Dictionary*, Van Nostrand Reinhold Co, New York
4. *The Merck Index: An Encyclopedia of Chemicals and Drugs*, Merck and Company, Inc., Rahway, NJ
5. *IARC Monographs on the Evaluation of the Carcinogenic Risk of Chemicals to Man*, World Health Organization
6. *Industrial Hygiene and Toxicology*, by F. A. Patty, John Wiley and Sons, Inc., New York
7. *Clinical Toxicology of Commercial Products*, Gleason, Gosselin, and Hodge
8. *Casarett and Doull's Toxicology; The Basic Science of Poisons*, Doull, Klassen, and Amdur, Macmillan Publishing Co., New York
9. *Industrial Toxicology*, by Alice Hamilton and Harriet L. Hardy, Publishing Sciences Group, Inc., Acton, Ma.
10. *Toxicology of the Eye*, by W. Morton Grant, Charles C. Thomas, Springfield, Illinois
11. *Recognition of Health Hazards in Industry*, William A. Burgess, John Wiley and Sons, New York
12. *Chemical Hazards of the Workplace*, Proctor, N. H., and Hughes, J. P., J. P. Lippincott Company, New York
13. *Handbook of Chemistry and Physics*, Chemical Rubber Company, Cleveland, OH
14. *Threshold Limit Values for Chemical Substances and Physical Agents in the Work Environment and Biological Exposure Indices with Intended Changes*, ACGIH, Cincinnati, OH
15. National Fire Protection Association for information on physical hazards of chemicals, Boston, MA
16. The following documents may be obtained from the Superintendent of Documents, U.S. Government Printing Office, Washington, DC
    Occupational Health Guidelines, NIOSH/OSHA Pub No. 81-123
    NIOSH Pocket Guide to Chemical Hazards, NIOSH Pub. No. 90-117
    Registry of Toxic Effects of Chemical Substances
17. Miscellaneous Documents published by the National Institute for Occupational Safety and Health:
    Criteria documents
    Special Hazard Reviews
    Occupational Hazard Assessments
    Current Intelligence Bulletins
18. National Technical Information Service, Springfield, VA

Appendix D

Hazcom Definitions

absolute — A chemical substance relatively free of impurities, e.g., absolute alcohol.

absolute pressure — Gauge pressure plus atmospheric pressure.

absorb — To soak up or drink in, as when a hazardous material is taken into a plant, textile, or soil.

ACGIH — American Conference of Governmental Industrial Hygienists. ACGIH develops and publishes recommended occupational exposure limits for chemical substances and physical agents (see TLV).

acid — A hydrogen-containing compound that reacts with water to produce hydrogen ions; a proton donor; a liquid compound with a pH less than or equal to 2.

acidosis — A condition of decreased alkalinity of the blood and tissues. Symptoms may include sickly sweet breath, headache, nausea, vomiting, and visual disturbances; usually the result of excessive acid production.

acrid — Irritating and bitter (smell).

action level — The exposure level at which OSHA regulations to protect workers take effect; for example, programs including workplace air analysis, employee training, medical monitoring, and recordkeeping. Exposure at or above the action level is termed occupational exposure.

active ingredients — A term used in the development of pesticides. The chemical that has toxic potential. Active ingredients are listed on a pesticide label in order as a percentage of weight or pounds per gallon of concentrate.

acute health effects — Physical reactions that occur or develop rapidly after exposure to a substance.

acute toxicity — Adverse effects resulting from a single dose of or exposure to a material. Ordinarily used to denote effects observed in experimental animals.

adsorb — To collect, as a gas or liquid, in condensed form on the surface of another substance.

aerosol — A system in which liquid or solid particles are distributed in a finely divided state through a gas, usually air. Particles within aerosols are usually less than 1μm (0.001 mm) in diameter, and are more uniformly distributed than in a spray.

agent — Any substance, force, radiation, organism, or influence that affects the body.

ALARA — Radiation acronym for "as low as reasonably achievable."

alkali — Any compound that forms the hydroxyl ion in its water solution.

allergen — A substance that can cause an allergic reaction.

ambient — Surrounding conditions, primarily used in reference to climatic conditions; ambient temperature or sound

anosmia — Loss of the sense of smell.

ANSI — American National Standards Institute: A privately funded organization that identifies industrial public national consensus standards and coordinates their development.

aqueous — Indicates water is present in the solution.

anhydrous — Without water.

article — A manufactured item other than a fluid or particle that is formed to a specific shape or design during manufacture, that has end use function(s) in whole or in part upon its shape or design during end use, and that under normal conditions of use does not release more than very small quantities (e.g., minute or trace amounts) of a hazardous chemical and does not pose a physical hazard or health risk to employees.

aseptic — Free of disease-causing organisms.

asphyxia — Condition arising when the blood is deprived of adequate supply of oxygen; loss of consciousness or death may follow.

asphyxiant — Gas or vapor that when inhaled may lead to asphyxia.

atm (atmosphere) — A unit of pressure equal to the average pressure the air exerts at sea level.

atrophy — Reduction in size or function of tissue, organs, or the entire body, caused by lack of use.

autoignition temperature — Minimum temperature to which a substance must be heated to initiate self-sustained combustion, independent of any open flame.

base	Substance containing group-forming hydroxide ions in water solution.
biodegradable	An organic material's capacity for decomposition as a result of attack by microorganisms; biodegradable materials do not persist in nature.
biological agent	Microorganisms (primarily bacteria) added to the water column or soil to increase the rate of biodegradation of spilled oil.
BLEVE	Boiling Liquid Expanding Vapor Explosion. Normally associated with fires that involve compressed gases in cylinders; rapid rupture of a vessel caused by over-pressure accompanied by rapid burning of the tank contents.
boiling point	Temperature when a substance changes from a liquid to a gas.
BTU	British Thermal Unit. A measuring unit of heat.
buffer	A substance that reduces the change in hydrogen ion concentration (pH), otherwise produced by adding acids or bases to a solution. A pH stabilizer.
bung	A cap or screw-type device to close the small opening in the top of a metal drum or barrel.
°C	Centigrade degrees; degrees Celsius.
calibration	A procedure to ensure the accuracy of instrument measurements.
carcinogen	A substance or agent capable of producing cancer.
CAS Registration Number	An assigned number used to identify a chemical. CAS stands for Chemical Abstracts Services, an organization that indexes information published by the American Chemical Society, and provides index guides by which information about particular substances may be located in the abstracts.
catalyst	A substance that when present even in small quantities affects the rate of a chemical reaction but remains unchanged by that reaction.
caution	A signal word used on pesticide/chemical labels to denote slightly toxic pesticides.
ceiling limit	The concentration not to be exceeded at any time.

CERCLA	Comprehensive Environmental Response, Compensation, and Liability Act or Superfund law.
CFC	Chlorofluorocarbon. Any of a group of compounds that contain carbon, chlorine, fluorine, and sometimes hydrogen that are often used as refrigerants or aerosol propellants and are associated with damage to the earth's ozone layer.
CFR	Code of Federal Regulations. A collection of the federal regulations established by law.
chelating agent	A chemical compound capable of forming multiple chemical bonds to a metal ion. Used to treat metal poisoning.
chemical	Any element, chemical compound, or mixture of elements and/or compounds.
chemical family	A group of single elements or compounds of a common general type. For example, acetone, methyl ethyl ketone (MEK), and methyl isobutyl ketone (MIBK) are of the ketone family.
chemical formula	The number and kind of atoms comprising a mole of a material.
chemical manufacturer	An employer with a workplace where chemical(s) are produced for use or distribution.
chemical name	A chemical's scientific name.
classification of fires	Class A - Fires in combustible solids, such as wood, paper, or cardboard. Class B - Fires in combustibles and flammable gases or liquids. Class C - Fires involving energized electrical equipment. Class D - Fire involving combustible metals such as magnesium, etc.
chronic toxicity	A material's property that produces chronic health effects, usually resulting from repeated doses of or exposures to the material over a relatively prolonged time.
combustible liquid	Any liquid having a flashpoint at or about 100°F but below 200°F. According to the NFPA, any liquids with a flashpoint above 140°F.
common name	Any designation or identification such as a code name, code number, trade name, brand name, or generic name used to identify a chemical other than by its chemical name.

compressed gas	A gas or mixture of gases having, in a container, an absolute pressure exceeding 40 psi at 70°F.
	A gas or mixture of gases having, in a container, an absolute pressure exceeding 104 psi at 130° regardless of the pressure at 70°.
	A liquid having a vapor pressure exceeding 40 psi at 100°F.
container	Any bag, barrel, bottle, storage tank, or the like that contains a hazardous chemical.
corrosive	A chemical that causes visible destruction of or irreversible alterations in living tissue by chemical action at the site of contact, or which causes a severe corrosion rate in steel or aluminum.
critical pressure	The pressure at which a gas becomes a liquid.
critical temperature	The temperature at which a gas becomes a liquid.
cryogenic	Relating to extremely low temperature, as for refrigerated gases.
density	Ratio of weight (mass) to volume of a material.
dermal	Through or by the skin; of or pertaining to the skin.
dermatitis	Inflammation of the skin.
designated representative	Any individual or organization to whom an employee gives written authorization to exercise the employee's rights.
distributor	A business (other than a chemical manufacturer or importer) that supplies hazardous chemicals to other distributors or employers.
DOT identification number	Four-digit number [preceded by UN (United Nations) or NA (North America)] used to identify particular materials for regulation of their transportation.
dyspnea	A sense of difficulty in breathing; shortness of breath.
EC50	Effective concentration. The concentration of a material in water, a single dose of which is expected to cause a biological effect on 50 percent of a group of test animals.
employee	A worker who may be exposed to hazardous chemicals under normal operating conditions or in foreseeable emergencies.
employer	A person engaged in a business where chemicals are either

	used, distributed, or produced for use or distribution, including a contractor or subcontractor.
emulsifier	A chemical that helps a liquid form tiny droplets and thus remain suspended in another liquid.
endothermic	A chemical reaction that absorbs heat.
etiology	All of the factors that contribute to the cause of a disease or an abnormal condition.
evaporation rate	Rate at which a material vaporizes (volatilizes, evaporates) from the liquid or solid state when compared to a known material's vaporization rate.
explosive	A chemical that causes a sudden, almost instantaneous release of pressure, gas, and heat when subjected to sudden shock, pressure, or high temperature.
exposure	Amount of a chemical that is a physical or health hazard to which an employee is subjected in the course of employment. It includes potential (e.g., accidental or possible) exposure. "Subjected" in terms of health hazards includes any route of entry (e.g., inhalation, ingestion, skin contact, or absorption).
exothermic	A chemical reaction that gives off heat. Exothermic materials produce matter with less total energy than the reacting substances.
fire diamond (NFPA 704 hazard rating)	Visual system that provides a general scheme of the inherent hazards and severity of materials relating to **fire** prevention, exposure, and control. Preferred reading order: health (blue), flammability (red), reactivity (yellow), special (white).

blue diamond: (health hazard). Indicates degree of hazard.
0 = ordinary combustible hazards in a fire
1 = slightly hazardous
2 = hazardous
3 = extreme danger
4 = deadly

red diamond: (fire hazard, flammability). Indicates susceptibility to burning.
0 = will not burn
1 = Will Ignite if Preheated
2 = Will Ignite if Moderately Heated
3 = Will Ignite at Most Ambient Conditions
4 = Burns Readily at Ambient Conditions

yellow diamond (reactivity, instability). Indicates energy released if burned, decomposed, or mixed.
0 = stable and not reactive with water
1 = unstable if heated
2 = violent chemical change
3 = shock and heat may detonate
4 = may detonate

white diamond (special hazard). Indicates a specific hazard.
oxy = oxidizer
acid = acid
alkali = alkali
cor = corrosive
w = water reactive, use no water!
radiation symbol = radiation hazard

flammable A chemical that falls into one or more of the following categories:

aerosol, flammable. An aerosol that when tested by the method described in 16 CFR 1500.45 yields a flame projection exceeding 18 inches at full valve opening, or a flashback (a flame extending back to the valve) at any degree of valve opening.

gas, flammable. A gas that, at ambient temperature and pressure, forms a flammable mixture with air at a concentration of 13 percent by volume or less, or a gas that, at ambient temperature and pressure, forms a range of flammable mixtures with air wider than 12 percent by volume, regardless of the lower limit.

liquid, flammable. Any liquid having a flashpoint below 100°F except any mixture having components with flashpoints of 100°F or higher, the total of which make up 99 percent or more of the total volume of the mixture.

solid, flammable. A solid (other than a blasting agent or explosive) that is liable to cause fire through friction, absorption of moisture, spontaneous chemical change, or retained heat from manufacturing or processing, or which can be ignited readily and when ignited burns so vigorously and persistently as to create a serious hazard. A chemical shall be considered a flammable solid if, when tested by the method described in 16 CFR 1500.44, it ignites and burns with a self-sustained flame at a rate greater than one-tenth of an inch

	per second along its major axis.
flashpoint	Lowest temperature at which a flammable liquid gives off sufficient vapor to form an ignitable mixture with air near its surface or within a vessel. Combustion does not continue. Flashpoint is determined by laboratory tests in special cups.
foreseeable emergency	Any potential occurrence including (but not limited to) equipment failure, rupture of containers, or failure of control equipment that could result in an uncontrolled release of a hazardous chemical into the workplace.
hazardous chemical	An explosive, flammable, poisonous, corrosive, reactive, or radioactive chemical requiring special care in handling because of hazards it poses to public health and the environment.
hazard class	A group of materials designated by the DOT that shares a common major hazardous property, i.e., radioactivity, flammability (see hazardous materials categories).
hazardous material	A substance in a quantity or form posing an unreasonable risk to health, safety, and/or property when transported in commerce.
hazardous materials categories (DOT categories)	**explosive:** Any chemical compound, mixture, or device, the primary or common purpose of which is to function by explosion, with substantially instantaneous release of gas and heat. **flammable liquid:** Any liquid having a flashpoint below 100°F as determined by tests listed in CFR 49, Section 173.115(d). **combustible liquid:** Any liquid having a flashpoint above 100°F and below 200°F as determined by tests listed in CFR 49, Section 173.115. **flammable gas:** Any gas that in a mixture of 13 percent or less by volume with air is flammable at atmospheric pressure, or its flammable range with air at atmospheric pressure is wider than 12 percent (by volume) regardless of a lower flammability limit. **nonflammable gas:** Any compressed gas other than a flammable gas.

Appendix D

flammable solid: Any solid material (other than an explosive) that is liable to cause fires through friction or retained heat from manufacturing or processing, or that can be ignited readily and when ignited burns so vigorously and persistently as to create a serious transportation hazard.

oxidizer: A substance that yields oxygen readily to stimulate the combustion of other material.

organic peroxide: An organic compound that may be considered a derivative of hydrogen peroxide where one or more of the hydrogen atoms have been replaced by organic radicals, and that readily releases oxygen to stimulate the combustion of other materials.

poison A (gas): A poison gas, extremely dangerous gases, or liquids of such nature that a very small amount of the gas or vapor of the liquid mixed with air is dangerous or lethal to life.

poison B (liquids or solids): Liquids or solids (including gases, semi-solids, and powders other than irritating materials) known to be so toxic to man as to afford a hazard to health.

irritating materials: Liquid or solid substances that upon contact with fire or when exposed to air give off dangerous or intensely irritating fumes, but not including any Class A poisonous material.

radioactive material (or radiological material): Any material or combination of materials that spontaneously emit ionizing radiation, and have a specific gravity greater than 0.002 micro-curies per gram.

corrosive material: Any liquid or solid, including powders, that causes visible destruction of human skin tissue, or a liquid that has a severe corrosion rate on steel or aluminum.

etiological agent: A viable microorganism or its toxin that causes or may cause human disease.

consumer commodity (drugs and medicines): A material packaged or distributed in a form intended and suitable for sale through retail sales agencies for use or consumption by individuals for purposes of personal care or household use.

hazard warning Any words, pictures, symbols, or combination thereof appearing on a label or other appropriate form of warning that

	convey the specific physical and health hazard(s), including target organ effects, of the chemicals(s) in the container(s).
health hazard	A chemical for which exists statistically significant evidence based on at least one study conducted in accordance with established scientific principles that acute or chronic health effects may occur in exposed employees. The term "health hazard" includes chemicals that are carcinogens, toxic or highly toxic agents, reproductive toxins, irritants, corrosives, sensitizers, hepatoxins, nephrotoxins, neurotoxins, agents that act on the hematopoietic system, and agents that damage the lungs, skin, eyes, or mucous membranes.
highly toxic	A chemical falling within any of the following categories:
	A chemical that has a median lethal dose (LD50) of 50 milligrams or less per kilogram of body weight when administered orally to albino rats weighing between 200 and 300 grams each.
	A chemical that has a median lethal dose (LD50) of 200 milligrams or less per kilogram of body weight when administered by continuous contact for 24 hours (or less if death occurs within 24 hours) with the bare skin of albino rabbits weighing between two and three kilograms each.
	A chemical that has a median lethal concentration (LC50) in air of 200 parts per million by volume or less of gas or vapor, or 2 milligrams per liter or less of mist, fume, or dust, when administered by continuous inhalation for one hour (or less if death occurs within one hour) to albino rats weighing between 200 and 300 grams each.
identity	Any chemical or common name that is indicated on the material safety data sheet (MSDS) for the chemical. The identity used must permit cross-references to be made among the required list of hazardous chemicals, the label, and the MSDS.
IDLH	Immediately dangerous to life and health. The maximum concentration from which one could escape within 30 minutes without any escape-impairing symptoms or irreversible health effects. Used to determine respiration selection.
ignition temperature	The temperature to which a substance must be heated to initiate self-sustained combustion (burning).

immediate use	The hazardous chemical will be under the control of and used only by the person who transfers it from a labeled container, and only within the work shift in which it is transferred.
immiscible	Not capable of mixing or being mixed (as oil and water).
importer	The first business with employees within the customs territory of United States that receives hazardous chemicals produced in other countries for supplying them to distributors or employers within the United States.
irritant	A noncorrosive material that causes a reversible inflammable effect on living tissue by chemical action at the site of contact as a function of concentration or duration of exposure.
label	Any written, printed, or graphic sign or symbol displayed on or affixed to containers of hazardous chemicals. A label should identify the hazardous material, appropriate hazard warnings, and name and address of the chemical manufacturer, importer, or other responsible party.
LC50	Lethal Concentration 50. Concentration of an active ingredient in the air that when inhaled kills half of the test animals exposed to it; expression of a compound's toxicity when present in the air as a gas, vapor, dust, or mist; generally expressed in ppm when a gas or vapor, and in micrograms per liter when a dust or mist; often used as the measure of active inhalation toxicity. The lower the LC50 value, the more poisonous the pesticide.
LD50	Lethal Dose 50. Dosage or amount of an active ingredient that when taken by mouth or absorbed by the skin kills half of the test animals exposed; an expression used to measure acute oral or acute dermal toxicity.
LD100	Lethal Dose 100. The dose of an active ingredient taken by mouth or absorbed by the skin that is expected to cause death in 100 percent of the test animals so exposed.
lower explosive limit, lower flammable limit (LEL, LFL)	The lowest concentration of gas or vapor (percent by volume in air) that burns or explodes if an ignition source is present at ambient temperatures.
material safety data sheet (MSDS)	Written or printed material concerning a hazardous chemical, prepared in accordance with paragraph (g) of 29 CFR 1910.1200.

melting point	The temperature above which a solid changes to a liquid.
miscible	When two liquids or two gases are completely soluble in each other in all proportions. While gases mix with one another in all proportions, the miscibility of liquids depends on their chemical natures.
mixture	A heterogeneous association of materials that cannot be represented by a chemical formula and that does not undergo chemical change because of interaction among the mixed materials. The constituent materials may or may not be uniformly dispersed, and can usually be separated by mechanical means (as opposed to a chemical reaction). Uniform liquid mixtures are called solutions.
molecular weight	The sum of atomic weights of the atoms in a molecule. For example, water (H_2O) has a molecular weight of 18.015, the atomic weights being hydrogen = 2(1.008) + oxygen = 15.999.
organic peroxide	An organic derivative of the inorganic compound hydrogen peroxide.
oxidizer	Substance that yields oxygen readily to stimulate the combustion of organic matter and inorganic matter.
PEL	Permissible exposure limit. Established by OSHA, this may be expressed as a time-weighted average (TWA) limit or as a ceiling exposure limit. A ceiling limit must never be exceeded instantaneously, even if the TWA exposure limit is not violated. Note that OSHA's PELs have the force of law, while on the other hand, ACGIH's threshold limit values TLVs are recommended exposure limits and do not have the force of law.
percent volatile	Percent volatile by volume. The percentage by volume of a liquid or solid that evaporates at an ambient temperature of 70°F, unless some other temperature is stated. For example, gasoline is 100 percent volatile; its evaporation rate varies, but over a period of time it evaporates completely. This physical characteristic indicates or reflects the potential for releasing harmful vapor into the air.
physical hazard	A chemical for which there is scientifically valid evidence that it is a combustible liquid, a compressed gas, explosive, flammable, organic peroxide, oxidizer, pyrophoric, unstable (reactive), or water-reactive.

produce	To manufacture, process, formulate, blend, extract, generate, emit, or repackage.
pyrophoric	A chemical that will ignite spontaneously in air at a temperature of 130°F or less.
responsible party	Someone who can provide additional information on the hazardous chemical and appropriate emergency procedures, if necessary.
sensitizer	A chemical that causes a substantial proportion of exposed people or animals to develop an allergic reaction in normal tissue after repeated exposure to the chemical.
specific chemical identity	The chemical name, Chemical Abstract Service (CAS) registry number, or other information that reveals the precise chemical designation of the substance.
solubility in water	A term expressing the percentage of a material (by weight) that will dissolve in water at ambient temperature. Solubility information is useful in determining cleanup methods for spills and fire-extinguishing methods for a material.
specific gravity	The ratio of the mass of a substance to the same volume of a reference substance, at a specified temperature. Insoluble materials with a specific gravity of greater than one will float on water; insoluble materials with a specific gravity of more than one will sink in water.
tachycardia	Excessively rapid heartbeat, with a pulse rate above 100.
target organ effects	Chemically caused effects from exposure to a material on specific listed organs and systems such as liver, kidneys, nervous system, lungs, skin, and eyes.
threshold limit values (TLVs)	Levels of concentrations of airborne substances. They represent conditions to which it is believed that nearly all workers may be repeatedly exposed day after day without adverse health effects.
Toxic	Describes a material's ability to injure biological tissue.
Trade Secret	Any confidential formula, pattern, process, device, information, or compilation of information that is used in an employer's business and that gives the employer an opportunity to obtain an advantage over competitors who do not know or use it.

upper explosive limit, upper flammability limit (UEL, UFL)	The highest concentration of a material in air that produces an explosion in fire or that ignites when it contacts an ignition source (high heat, electric arc, spark, or flame). A higher concentration of the material in a smaller percentage of concentration in air is too rich to be ignited.
unstable (reactive)	A chemical that in the pure state or as produced or transported will vigorously polymerize, decompose, condense, or become self-reactive under conditions of shock, pressure, or temperature.
use	To package, handle, react, emit, extract, generate as a byproduct, or transfer.
vapor density	The ratio of the mass of a vapor or gas to the mass of an equal volume of air. Materials lighter than air have vapor densities of less than 1.0; materials heavier than air have vapor densities greater than 1.0.
vapor pressure	The pressure a saturated vapor exerts above its own liquid in a closed container. Vapor pressures reported on MSDSs are usually stated in millimeters of mercury (mm Hg) at 68°F. The lower a substance's boiling point, the higher its vapor pressure; the higher the vapor pressure, the greater the material's tendency to evaporate into the atmosphere.
water-reactive	A chemical that reacts with water to release a gas that is either flammable or presents a health hazard.
work area	A room or defined space in a workplace where hazardous chemicals are produced or used, and where employees are present.
workplace	An establishment, job site, or project at one geographical location containing one or more work areas.
Z list	OSHA's Toxic and Hazardous Substances Tables Z-1-A, Z-2, and Z-3 of air contaminants, (29 CFR 1910.1000). These tables record PELs, TVAs, and ceiling concentrations for the materials listed. Any material on these tables is considered hazardous.

Index

Active ingredients, 49, 287

Acute health effects, 287

Acute toxicity, 287

Aerosol, 26, 268, 287, 290, 293

Agent, 28, 32, 84, 124, 135, 145, 268-269, 285, 288-290, 293, 295

Agricultural product, 26

Air blank, 145

Aircraft unit load, 29, 85, 88, 105, 107

Airline Deregulation Act, 8

ALARA, 285, 288

Alcohol, 22, 34, 38, 143-172, 174-183, 186-188, 277, 279-280, 284, 287

Alcohol abuse, 143

Alcohol concentration, 145-146, 148, 151, 153, 163-165, 168-169, 174, 176, 178-181, 277, 279

Alcohol content, 153

Alcohol testing, 143, 145-148, 153, 156-159, 161-164, 172, 183, 186-188

Aliquot, 145

Alkali, 288, 293

ANSI, 285, 288

Asphyxiant, 26, 33, 60-62, 64-65, 116, 288

Atmosphere, 26, 61, 63-64, 288, 300

Autoignition temperature, 61, 288

Bags, 8, 118

Barrels, 27, 118

Base, 97-98, 289

Bill of lading, 120

Blasting agents, 32, 131

BLEVE, 289

Blind sample or blind performance test specimen, 145

Boiling point, 27, 33, 61, 289, 300

Bottle, 26, 148, 267, 291

Boxes, 118

Break-bulk, 27, 85

Breath alcohol technician, 145, 166, 183, 186, 280

Bulk packaging, 27-28, 34, 40-41, 75, 83, 98, 100, 102, 106-107, 121

Bureau of Motor Carriers, 12-13

Bureau of Public Roads, 3, 5, 12

Bureau of Transportation Statistics, 2, 9

Canceled or invalid test, 145

CANUTEC, 110

Carcinogen, 50, 62, 289

Cargo, 27, 41-42, 82-83, 85, 88, 94-95, 101, 106, 124, 126-129, 134, 140-141, 157

Cargo aircraft only, 82, 88

Cargo heaters, 127

Cargo tank motor vehicle, 27, 127

Carrier, 11-16, 18, 27, 83, 85, 103-104, 107, 110, 125-127, 134-135, 139, 148-149, 156, 163, 172

CAS registration number, 289

Caution signs, 70

CFC, 290

Chain of custody, 145, 166, 185

Chemical, 25, 28-30, 32, 38, 40-41, 46-52, 55-57, 60-62, 65, 72-73, 89, 104, 109, 115, 146, 266-275, 285-294, 296-300

Chemical inventory list, 50, 52, 266, 268-271, 274

Chemical manufacturer, 46, 48-49, 72, 273, 290-291, 297

Chemical name, 56, 60, 72, 267, 290, 299

Chemical Transportation Emergency Center, 109

CHEMTREC, 109-110

Chief Financial Officers Act, 10

Civil Aeronautics Board Sunset Act, 8

Class, 26-29, 31-39, 54, 64-65, 75-76, 82-90, 93-99, 101-102, 104, 108, 110-111, 117, 120, 122, 125-128, 131, 140, 290, 294-295

Class entry, 122

Classification, 25-26, 30-31, 113, 123, 137, 141, 290

Clean Air Act, 17

Clean Water Act, 17

Coast Guard, 1, 3, 5-6, 11, 149

Cocaine, 143-144, 146, 183-184

Collection container, 145

Collection site, 145-146, 148, 167, 184-185

Combination packaging, 27, 116

Combustible, 27, 32, 34-35, 63, 78, 84, 88, 90, 95, 101-102, 267, 269, 290, 292, 294, 298

Commercial drivers license, 141, 150, 161, 278-280, 283-284

Commercial motor vehicles, 148-154, 163, 183

Commercial Space Launch Act, 8

Compatibility group, 27, 34, 75-76, 91-92, 101-102

Compatibility group letter, 27, 75-76, 91-92, 102

Composite Packagings, 118

Comprehensive Environmental Response, Compensation, and Liability Act (CERCLA), 17, 285, 290

Compressed Gas, 27, 32-33, 63, 85, 267, 269, 291, 294, 298

Confirmation test for alcohol testing, 146

Consignee, 84, 105, 126

Consolidated packaging, 86

Consortium, 146, 148, 160, 168, 170

Container, 27, 29, 46, 50, 54-55, 61, 64, 72-73, 75, 83-86, 88, 91, 98, 100-107, 110, 126, 129, 140, 145, 148, 267-270, 272-273, 277, 291, 296-297, 300

Contract motor carrier, 85, 149

Controlled substances, 143, 146-150, 152-175, 177-184, 279

Corrosive, 25, 27, 30, 32, 36, 45, 50, 57, 63, 82, 84-85, 89, 98, 103, 108, 116, 125-126, 131, 291, 293-295

Cradle-to-grave, 26, 42, 67

Cryogenic, 17, 27, 33, 291

Cylinder, 27, 61, 63-64, 85, 87, 125, 267

Danger Signs, 70

Dangerous, 13-14, 16-17, 25-27, 29-30, 32, 35, 39, 50, 55, 79, 83-84, 88, 91, 96, 98, 103, 110, 115-117, 122, 126, 131, 133, 285, 295-296

Dangerous when wet, 27, 32, 35, 79, 84, 88, 96, 103, 122

Density, 61, 119, 285, 291, 300

Department of Defense (DOD), 85, 171

Department of Transportation Act, 5, 15

Department of Transportation Enabling Act, 4

Designated representative, 291

Disabling damage, 146, 155

Divisions, 32-35, 39

Domestic employers, 151

Index

Domestic transportation, 27, 37, 39, 102, 117

DOT emergency response, 108

DOT identification number, 291

Drug abuse, 143, 145, 148, 179, 284

Drug testing, 143, 145-147, 150, 183-184, 278, 280

Drums, 28-29, 44, 118

Duplicate labeling, 87

Evidential breath testing device, 145, 147, 187

Elevated temperature material, 28-29, 36, 122

Emergency equipment, 134-135

Emergency Planning and Community Right to Know Act, 17

Emergency response, 16-17, 23, 39, 44-45, 55, 69, 107-112, 121, 123-124

Employee, 28, 43-47, 49, 51-56, 66-67, 69, 73, 108-109, 123-124, 128, 145, 147-148, 152, 161, 171, 177-179, 181-182, 185-186, 265-266, 268-270, 274-275, 278-280, 282-284, 287, 291-292

Employer, 28-29, 43-44, 46-48, 50, 52, 66-67, 72-73, 123, 144-147, 149-158, 160-184, 186, 283, 290-291, 299

Empty label, 83

Endothermic, 292

Etiologic agent, 28, 32, 84

Evaporation rate, 292, 298

Exceptions, 13, 29, 40, 83, 85, 87, 89, 100, 106, 149, 278

Exothermic, 34, 292

Explosive, 16, 27-28, 32-34, 57, 73-76, 101-102, 104, 125-127, 131, 267-269, 292-295, 297-298, 300

Exposure, 16, 25-26, 47, 52, 57, 60-63, 88, 107, 268, 272, 274, 286-287, 292, 297-299

Extremely insensitive detonating substances, 32-33

Familiarization training, 123-124

Federal aviation, 1, 3, 5-6, 8-9, 11-12

Federal Highway Administration (FHWA), 1, 5-7, 9, 12, 15, 22, 150, 153, 156, 158-160, 163, 167-168

Federal Managers' Financial Integrity Act, 10

Federal motor vehicle safety standards, 14

Federal Railroad Administration, 1, 5, 8

Federal Register, 158-159

Federal Transit Administration, 5, 9, 149

Fire diamond, 74-75, 292

Flammable, 17, 28, 30, 32-35, 38, 45, 57, 61, 77-78, 84, 86, 94-95, 101, 104, 111, 116, 125-127, 131, 140, 268-270, 290, 293-295, 297-298, 300

Flammable aerosol, 268

Flammable and combustible liquid, 32

Flammable gas, 28, 32-33, 77, 84, 94, 101, 104, 125, 127, 140, 268, 294

Flammable liquid, 28, 34, 38, 78, 84, 86, 95, 111, 125-127, 268, 294

Flammable solid, 28, 32, 78, 95, 125-126, 268, 293, 295

Flashpoint, 34, 61, 122, 267-268, 285, 290, 293-294

Flask, 85

Follow-up testing, 164, 180-182

Forbidden explosives, 32

Forbidden materials, 32

Foreign employers, 151

Foreseeable emergency, 294

Freight container, 27, 29, 75, 83-85, 88, 91, 98, 100-105, 107, 129

General awareness training, 42-43, 123, 131

General Duty Clause, 73

Government Performance and Results Act, 10

Hazard class, 27-29, 31, 34, 36-38, 54, 65, 75, 84, 86-88, 90, 108, 110-111, 117, 120, 122, 131, 294

Hazard communication training, 51, 53, 275

Hazard determination, 48-50

Hazard warnings, 72, 272, 297

Hazardous chemical, 47-48, 50-52, 55, 60, 72, 266-275, 288, 291, 294, 297, 299

Hazardous chemical inventory list, 52, 266, 268-271, 274

Hazardous material, 15-16, 18, 25-26, 28, 30-32, 36-41, 44, 49-50, 52, 54, 56, 65, 69, 74-75, 83-87, 90, 98, 101-108, 112-115, 117-118, 120-121, 124-127, 133-134, 141, 287, 294, 297

Hazardous material description, 120

Hazardous material information system, 65, 74

Hazardous material regulation, 118

Hazardous Materials Table, 31, 36, 38-39, 42, 54, 83-84, 86, 115, 120-121, 123-124, 191

Hazardous Materials Transportation Control Act, 15

Hazardous Materials Transportation Uniform Safety Act, 17, 43

Hazardous substance, 28, 31, 36-37, 121

Hazardous wastes, 54

Hazcom, 44-48, 50-53, 55, 57, 66-67, 73, 89-90, 266, 273, 285-287

Hazmat employee, 28, 43-45, 54, 69, 108-109, 123-124

Hazmat employer, 28-29, 43-44, 46, 123

Health hazard, 50-51, 60, 62, 268-270, 292, 296, 300

Highway Safety Act, 5-6, 14-15

Household Goods Regulatory Reform Act, 7

Hydrostatic pressure, 117

Identification numbers, 29, 36, 39, 102, 160

Identity, 56-57, 67, 72, 173, 177, 185, 271, 278, 296, 299

Ignition temperature, 296

Immiscible, 297

Importer, 48, 72, 291, 297

Infectious substance, 32, 80, 84

Etiologic agent, 28, 32, 84

Initial isolation, 111-112

Initial response guides, 110

Inspectors General Act, 7

Intermodal Surface Transportation Efficiency Act, 9, 11

International Airlines Reform Act, 7

International Maritime Organization, 37, 83

Interstate Commerce Commission, 4, 11-15, 18, 22

Irritant, 297

Irritating material, 29

Jerricans, 27, 118

Knowledgeable employees, 49

Label design, 84

Label modifications, 86

Label placement, 87

Labeling, 14, 23, 36, 46-48, 52-55, 69, 72, 74-75, 83-87, 89-90, 112, 116, 123-124, 266, 269, 272-274

Labeling systems, 74

Leakproofness, 117

Licensed medical practitioner, 147, 152

Limited quantity, 122

Loading, 14, 18, 43, 98, 124-127, 131, 134, 140-141, 148, 157

Manifests, 54, 124

Marine pollutant, 29, 31, 36, 64, 121

Maritime Administration, 2, 7, 9

Marking, 23, 29, 54, 69, 72, 83, 87, 99, 107, 118-119, 123-124, 127

Material safety data sheet (MSDS), 48-53, 55, 55-60, 65-67, 109, 266-272, 274-275, 285-286, 296-297,

Materials Transportation Bureau, 15

Medical review office, 145, 147, 153, 166, 168, 172, 282-283

Melting point, 298, 29, 61

Minimum annual percentage rate, 158-161

Miscellaneous hazardous material, 32, 36

Miscible, 285, 298

Mixed packaging, 86

Mixture, 27-30, 33, 50, 89, 114, 121, 131, 145, 267-269, 271, 290-291, 293-294, 298

Molecular weight, 61, 145, 298

Index

Motor carrier, 11-16, 18, 85, 103, 126-127, 134-135, 148-149, 156, 163, 172

Motor Carrier Safety Act, 11

Motor Carrier Safety Inspectors, 13-14

Motor Carrier Safety Regulations, 11-13, 15-16, 18, 134-135

Moving America, 9

National Fire Protection Association, 27, 65, 74, 112, 269, 273, 286, 290, 292

NFPA 704 Hazard Rating, 292

National Highway Safety Act, 14-15

National Highway Traffic Safety Administration, 1, 5, 11, 22, 147

National Response Center, 110

National Traffic and Motor Vehicle Safety Act, 14-15

National transportation policy, 8-9, 12

National Transportation Safety Board, 5-6, 171

NFPA hazardous chemical label, 269, 273

Non-bulk package, 83, 88

Non-flammable compressed gas, 32

Non-flammable gas, 62, 77, 87, 93, 101-102

Non-specific chemical labels, 73

North American Emergency Response Guidebook, 55, 109-112

Not otherwise specified, 29, 38, 121

Numerical codes, 40

Occupational Safety & Health Act, 17

Occupational Safety and Health Administration (OSHA), 16, 17, 22, 28, 44-46, 48-51, 53-54, 56, 60, 62, 65-67, 69-70, 72-74, 188, 265, 285-287, 298, 300

Office of Motor Carrier Information Management and Analysis, 16

Office of Motor Carrier Safety Field Operations, 16

Office of Motor Carrier Standards, 16

Office of Motor Carriers, 16

Office of Program Management Support, 16

Oil Pollution Act, 17

OMB control number, 167

On-duty use, 152

Orange Book, 117

Organic Peroxide, 32, 35, 79, 84, 96, 269, 295, 298

Other regulated material (ORM-D), 29, 31-32, 36, 39, 90, 105

Overpack, 29, 39, 83, 85-87, 116

Oxidizer, 29, 32, 35, 39, 50, 79, 84, 87, 96, 101, 269, 293, 295, 298

Oxidizing gas, 29, 33

Oxygen, 26, 29-30, 34-35, 62-65, 87, 93, 101-102, 269, 288, 295, 298

Packaging, 23, 26-29, 33-34, 36, 40-41, 44, 54, 75, 83-84, 86, 89, 98, 100-102, 106-107, 112-119, 121-125

Packaging authorizations, 40

Packaging codes, 118

Packaging compatibility, 114

Pallet, 29

Penalties, 14, 45, 176

Performance standard code, 119

Performance-oriented packaging, 117-118

Physical hazard, 268-269, 288, 298

Placarding, 23, 54, 69, 90, 98-106, 112, 123-124

Poisonous, 32-33, 35, 85, 103-104, 116, 122, 125-126, 140, 294-295, 297

Poisonous gas, 32

Poisonous materials, 32, 35, 126, 140

POP design tests, 117

POP marking codes, 118

Portable container, 269, 272

Portable tank, 27, 83, 85, 88, 94-95, 101, 103, 106, 126, 128-129, 134, 141

Pre-duty use, 152

Primary hazard, 29, 39, 86-87

Primary route of entry, 269, 272

Product information, 60

Proper shipping names, 36-39

Protective action distances, 111

Purchasing agents, 124

Pyrophoric, 29, 35, 269-270, 298-299

Qualified person, 126, 128, 140

Quantity limits, 36

Radioactive, 31-32, 36, 81, 84-85, 87-89, 93, 97, 99, 103-105, 108, 117, 122, 125, 128, 140, 294-295

Railroad crossings, 134, 138

Railroad Regulatory Act, 7

Railroad Safety Appliance Act, 11

Random testing, 149, 158-161, 181

Reasonable suspicion testing, 161-162, 279

Reportable quantity (RQ), 29, 65, 120-121

Research and Special Programs Administration, 2, 7, 17, 23, 110

Residue, 84, 101, 104, 106, 121

Resource Conservation and Recovery Act (RCRA), 17

Responsible party, 72, 110, 297, 299

Return-to-duty testing, 164

Right-to-know, 67

Routing, 134, 138-139

Safe Drinking Water Act, 17

Safety-sensitive function, 147, 152-153, 160-161, 164, 175, 179, 181

Saint Lawrence Seaway Development Corporation, 2-3, 5

Screening test, 146, 148, 165, 167-169, 187, 282

Segregation table, 129-131

Self-heating material, 35

Sensitizer, 299

SETIQ, 110

Shipping container, 148

Shipping papers, 23, 38, 49, 54-55, 90, 109-110, 120-122, 124

Single packaging, 116

Solid, 17, 27-30, 32, 35, 78, 87-90, 95, 107-108, 118-119, 125-126, 268, 287, 292-293, 295, 298

Special provisions, 23, 36, 40, 115

Specific chemical identity, 299

Specific gravity, 295, 299, 61, 119, 184

Specification packaging, 115

Specimen bottle, 148

Spontaneously ignitable materials, 30

Square background placard, 93

Stacking, 117

Staggers Rail Act, 7

Standard packaging, 115-116

Stationary container, 270, 273

Storage, 18, 30, 45, 48, 63, 124-125, 128, 131, 145, 185-186, 267, 269-270, 291

Stowage, 27, 36, 41-42, 124

Subsidiary hazard, 29, 38-39, 54, 85-87, 99, 103, 122

Substance abuse guidelines, 134

Substance abuse professional, 148, 164, 166, 169-170, 179-183, 283-284

Substance abuse testing, 144, 278-280, 282

Surface Transportation Administration, 7

Surface Transportation Board, 2

Table of placards, 110-111

Tank, 27, 83, 85, 88, 94-95, 101, 103-104, 106, 121, 125-129, 134, 140-141, 267, 269-270, 289, 291

Temperature, 28-29, 35-36, 61, 122, 126-127, 140, 267-268, 270, 288-289, 291-294, 296, 298-300

Title 49, 17, 21-23, 31, 176

Toxic Substances Control Act, 17

Toxins, 296

Trade secrets, 60

Training, 23, 42-48, 51-56, 66-67, 69, 108, 112, 123-124, 131, 133-134, 137, 141, 147, 149-150, 165-167, 169-170, 177-179, 185, 188, 266, 273-276, 284-287

Transport index number, 128

Transport vehicle, 27, 29, 75, 84-85, 91, 98, 100-103, 107, 128-129, 131

Transportation Administrative Services Center, 2, 11, 19

Transportation of Explosives Act, 11, 14

Transportation policy, 3, 6, 8-9, 12

Transportation Systems Center, 6-7

Trip lease driver, 148

Truck Regulatory Reform Act, 7

Unit load devices, 90, 100-101, 103, 105

UN POP, 117

Unloading, 14, 18, 43, 124-128, 131, 134, 140-141, 148, 157

Unstable, 34, 116, 269-270, 293, 298, 300

Reactive chemical, 270

Urban Mass Transportation Administration, 5, 7, 9

United States Environmental Protection Agency (U.S. EPA), 17, 22, 28, 54, 89

Valves, 104, 124, 126-127, 140

Very insensitive explosives; blasting agents, 32

Vessel stowage, 36, 41-42

Vibration standard, 118

Violation rate, 148, 158-159

Warning signs, 70

Water-reactive, 298, 300

Workplace, 17, 25, 45-52, 55-56, 65-67, 69, 72-73, 183, 188, 265, 277, 286-287, 290, 294, 300

Written hazard communication program, 47, 51-52, 66-67, 265, 269, 275

Z List, 285, 300

Government Institutes Mini-Catalog

PC #	ENVIRONMENTAL TITLES	Pub Date	Price
629	ABCs of Environmental Regulation: Understanding the Fed Regs	1998	$65
627	ABCs of Environmental Science	1998	$49
672	Book of Lists for Regulated Hazardous Substances, 9th Edition	1999	$95
579	Brownfields Redevelopment	1998	$95
4100	CFR Chemical Lists on CD ROM, 1999-2000 Edition	1999	$150
4089	Chemical Data for Workplace Sampling & Analysis, Single User Disk	1997	$159
512	Clean Water Handbook, 2nd Edition	1996	$115
581	EH&S Auditing Made Easy	1997	$95
673	E H & S CFR Training Requirements, 4th Edition	1999	$99
4082	EMMI-Envl Monitoring Methods Index for Windows-Network	1997	$537
4082	EMMI-Envl Monitoring Methods Index for Windows-Single User	1997	$179
525	Environmental Audits, 7th Edition	1996	$95
548	Environmental Engineering and Science: An Introduction	1997	$95
643	Environmental Guide to the Internet, 4th Edition	1998	$75
650	Environmental Law Handbook, 15th Edition	1999	$89
353	EH&S Dictionary: Official Regulatory Terms, 7th Edition	2000	$95
652	Environmental Statutes, 2000 Edition	2000	$105
4097	OSHA CFRs Made Easy (29 CFRs)/CD ROM	1998	$159
4102	1999 Title 21 Food & Drug CFRs on CD ROM-Single User	1999	$325
4099	Environmental Statutes on CD ROM for Windows-Single User	1999	$169
570	ESAs Made Easy	1996	$75
689	Fundamentals of Site Remediation	2000	$85
515	Industrial Environmental Management: A Practical Approach	1996	$95
588	International Environmental Auditing	1998	$179
510	ISO 14000: Understanding Environmental Standards	1996	$85
551	ISO 14001: An Executive Report	1996	$75
518	Lead Regulation Handbook	1996	$95
554	Property Rights: Understanding Government Takings	1997	$95
582	Recycling & Waste Mgmt Guide to the Internet	1997	$65
615	Risk Management Planning Handbook	1998	$105
603	Superfund Manual, 6th Edition	1997	$129
566	TSCA Handbook, 3rd Edition	1997	$115
534	Wetland Mitigation: Mitigation Banking and Other Strategies	1997	$95

PC #	SAFETY and HEALTH TITLES	Pub Date	Price
547	Construction Safety Handbook	1996	$95
553	Cumulative Trauma Disorders	1997	$75
663	Forklift Safety, 2nd Edition	1999	$85
539	Fundamentals of Occupational Safety & Health	1996	$65
612	HAZWOPER Incident Command	1998	$75
535	Making Sense of OSHA Compliance	1997	$75
589	Managing Fatigue in Transportation, ATA Conference	1997	$75
558	PPE Made Easy	1998	$95
598	Project Mgmt for E H & S Professionals	1997	$85
552	Safety & Health in Agriculture, Forestry and Fisheries	1997	$155
669	Safety & Health on the Internet, 3rd Edition	1999	$75
597	Safety Is A People Business	1997	$65
668	Safety Made Easy, 2nd	1999	$75
590	Your Company Safety and Health Manual	1997	$95

Government Institutes
4 Research Place, Suite 200 • Rockville, MD 20850-3226
Tel. (301) 921-2323 • FAX (301) 921-0264
E-mail: giinfo@govinst.com • www.govinst.com

Please call our Customer Service Department at (301) 921-2323 for a free publications catalog.

CFRs are now available online. Call the Publishing Department at (301) 921-2355 for information.

Government Institutes Order Form

4 Research Place, Suite 200 • Rockville, MD 20850-3226
Tel (301) 921-2323 • Fax (301) 921-0264
www.govinst.com • E-mail: giinfo@govinst.com

4 EASY WAYS TO ORDER

1. **Phone:** (301) 921-2323
   Have your credit card ready when you call.
2. **Fax:** (301) 921-0264
   Fax this completed order form with your company purchase order or credit card information.
3. **Mail:** **Government Institutes Division**
   ABS Group Inc.
   P.O. Box 846304
   Dallas, TX 75284-6304 USA
   Mail this completed order form with a check, company purchase order, or credit card information.
4. **Online:** www.govinst.com

PAYMENT OPTIONS

- ❑ **Check** *(payable in US dollars to* **ABS Group Inc. Government Institutes Division***)*
- ❑ **Purchase Order** *(This order form must be attached to your company P.O.* Note: *All International orders must be prepaid.)*
- ❑ **Credit Card**   ❑ VISA   ❑ MasterCard   ❑ American Express

Exp. ___ /___
Credit Card No. _____
Signature _____

(Government Institutes' Federal I.D.# is 13-2695912)

CUSTOMER INFORMATION

Ship To: (Please attach your purchase order)
Name _____
GI Account # *(7 digits on mailing label)* _____
Company/Institution _____
Address _____
(Please supply street address for UPS shipping)
_____
City _____ State/Province _____
Zip/Postal Code _____ Country _____
Tel ( ) _____
Fax ( ) _____
E-mail Address _____

Bill To: (If different from ship-to address)
Name _____
Title/Position _____
Company/Institution _____
Address _____
(Please supply street address for UPS shipping)
_____
City _____ State/Province _____
Zip/Postal Code _____ Country _____
Tel ( ) _____
Fax ( ) _____
E-mail Address _____

Qty.	Product Code	Title	Price

15 DAY MONEY-BACK GUARANTEE
If you're not completely satisfied with any product, return it undamaged within 15 days for a full and immediate refund on the price of the product.

Subtotal _____
MD Residents add 5% Sales Tax _____
Shipping and Handling (see box below) _____
Total Payment Enclosed _____

All prices and publication dates are subject to change. Please call for current prices and availability.

SOURCE CODE: BP 02

Shipping and Handling
Within U.S:
1-4 products: $6/product
5 or more: $4/product
Outside U.S:
Add $15 for each item (Global)

Sales Tax
Maryland 5%
Texas 8.25%
Virginia 4.5%